2R

D0151676

sw.co 2014 64.95

ew

ENCYCLOPEDIA OF
FORENSIC SCIENCE

A COMPENDIUM
OF DETECTIVE
FACT AND FICTION

BARBARA GARDNER CONKLIN
ROBERT GARDNER
DENNIS SHORTELLE

ORYX PRESS
Westport, Connecticut • London

The rare Arabian Oryx is believed to have inspired the myth of the unicorn.
This desert antelope became virtually extinct in the early 1960s. At that time,
several groups of international conservationists arranged to have nine
animals sent to the Phoenix Zoo to be the nucleus of a captive breeding
herd. Today, the Oryx population is over 1,000, and over 500 have
been returned to the Middle East.

Library of Congress Cataloging-in-Publication Data

Conklin, Barbara Gardner.
  Encyclopedia of forensic science : a compendium of detective fact and fiction / Barbara
Gardner Conklin, Robert Gardner, and Dennis Shortelle.
    p. cm.
  Includes bibliographical references and index.
  ISBN 1-57356-170-3 (alk. paper)
    1. Forensic sciences—Encyclopedias.   2. Criminal investigation—Encyclopedias.   I.
Gardner, Robert.   II. Shortelle, Dennis.   III. Title.
  HV8073.C595   2002
  363.25'03—dc21   2001036638

British Library Cataloguing in Publication Data is available.

Copyright © 2002 by Barbara Gardner Conklin, Robert Gardner, and Dennis Shortelle

All rights reserved. No portion of this book may be
reproduced, by any process or technique, without the
express written consent of the publisher.

Library of Congress Catalog Card Number: 2001036638
ISBN: 1-57356-170-3

First published in 2002

Oryx Press, 88 Post Road West, Westport, CT 06881
An imprint of Greenwood Publishing Group, Inc.
www.oryxpress.com

Printed in the United States of America

The paper used in this book complies with the
Permanent Paper Standard issued by the National
Information Standards Organization (Z39.48–1984).

10 9 8 7 6 5 4 3 2 1

Contents

Preface

The *Encyclopedia of Forensic Science: A Compendium of Detective Fact and Fiction* provides readers with information about various aspects of forensic science. The entries pertain primarily to events that occurred and people who lived during the nineteenth and twentieth centuries in the United States, although there are topics from other countries, primarily Great Britain, the source of our legal system.

The Encyclopedia covers the major scientific techniques and devices that forensic scientists use in analyzing evidence collected at a crime scene. Readers will find information about a variety of instruments and techniques used in forensic science, including the polygraph, the Breathalyzer, the mass spectrograph, fingerprint analysis, and analysis of skeletal evidence, as well as forensic science's latest breakthrough—the use of DNA evidence in convicting the guilty and in freeing others who have been unjustly imprisoned. Included, too, are many of the investigators, scientists, officials, and others who helped in the development, growth, and use of forensic science.

On the other side of the criminal ledger, readers will find accounts of some infamous criminals, such as Jack the Ripper, "Son of Sam" (David Berkowitz), and Richard Crafts, the "woodchipper" murderer, many of whom were caught or convicted as a result

of forensic evidence. Included, too, are famous criminal cases—the Mormon forgery murders, the St. Valentine's Day Massacre, the Cleveland torso murders, and many more. Accounts are given of certain types of crimes and criminals, such as arson, forgery, rape, murder, and serial killers, and there are entries that discuss the type of evidence that criminals leave behind and how it is used in forensic analysis. In addition to fingerprints, blood, and DNA evidence, readers will find explanations of how forensic scientists use trace evidence, paint analysis, hair and fiber evidence, and the marks left by tools, tires, shoes, palms, and lips.

The preponderance of news pertaining to crimes, criminals, police investigations, and court proceedings that has permeated radio and television programs as well as newspapers and magazines offers ample evidence of public interest in crime and matters related to it. Each decade seems to have at least one famous case that draws wide public attention. In the 1930s, for example, families gathered around their radios to hear the latest information about the Lindbergh kidnapping and later the trial of Bruno Hauptmann, who was convicted of kidnapping and murdering the famous aviator's baby. In the 1950s, the case of Dr. Sam Sheppard, who was accused and convicted of killing his wife Marilyn, was in the limelight. In the 1990s, public interest in crime-related matters was evident from the millions of people who watched the televised court proceedings and reports pertaining to the trial of football great O.J. Simpson, who was accused and, after a lengthy and controversial trial, found not guilty of murdering his former wife and her friend.

As Sir Arthur Conan Doyle, author of the Sherlock Holmes stories, discovered in the late nineteenth century, the public's fascination with crime, law enforcement, courts, lawyers, and detectives is not limited to real-world crime. Murder mysteries have always filled library shelves and have been widely read. Television dramas related to crime solving, crime investigators, and courtrooms that used to include *Columbo, Quincy*, and *Perry Mason* and currently comprise such programs as *CSI: Crime Scene Investigation, NYPD Blue, Law and Order*, and *The Practice*, among others, attract large audiences. In response to that aspect of forensic science, this Encyclopedia includes entries that pertain to fiction with a forensic-science bent as well as some of the authors who created the fiction.

In order to provide a broad base for understanding forensic science, we have included entries that touch on the disciplines of criminal justice, police investigations, criminal and civil law, literature, and history. A comprehensive list of all the topics related to forensic science would require a multivolume work. Consequently, we have tried to select entries that are interesting and that reflect the diverse ways that forensic science impacts our society. In an effort to constrain both the book's size and price, we devoted a greater number of words to people we believe played more important roles in the development of forensic science and fewer words to those whose contributions were, in our opinion, less significant.

Of course, crime investigations involve many kinds of evidence other than DNA or fingerprints. These types of evidence include ballistics, dead-body, skeletal, insect, glass, dental, document, and various types of trace evidence, as well as evidence derived from the analysis of hair, handwriting, inks, tool marks, voiceprints, and other materials. All of these and others are included here, together with many of the diverse instruments used to examine evidence such as comparison and electron microscopes, several types of chromatography, spectrometry, and photometry, along with breathalyzers, polygraphs, and many other devices. The reader will find an abundance of information about forensic science, whether real or fictional. Of course, criminals, criminalists, and cases, both real and fictional, are all too numerous. We have tried to choose those that best illustrate the use of forensic science in solving crimes.

At the end of many entries are "See also" listings that refer readers to other related entries. References that follow most entries and an extensive bibliography offer further information for readers who wish to pursue certain topics related to forensic science in greater depth and detail. The appendix provides a list of Web sites that contain valuable information about forensic science.

Introduction

Forensic science is the application of scientific knowledge and techniques to criminal investigations in order to obtain the evidence needed to solve crimes. The evidence collected can then be used in court proceedings to help prove a suspect either guilty or innocent. The forensic scientists who collect and analyze evidence may also serve as expert witnesses during court proceedings.

As the entries in this encyclopedia reveal, forensic science draws upon the knowledge found in many scientific disciplines—psychology, pathology, toxicology, odontology, and physical anthropology as well as physics, chemistry, biology, and geology, to name a few. As a result, scientists from diverse fields of research and technology contribute to forensic science and the solution of crimes.

The forensic scientists most frequently involved in crime-scene analysis include pathologists, toxicologists, anthropologists, odontologists, psychiatrists, chemists, and biologists. Forensic pathologists examine unexpected deaths and perform autopsies to determine both the cause and time of death. Forensic toxicologists provide pathologists with information about possible drugs or poisons that may be present in the tissues of a corpse, an injured person, or a suspect. Forensic dentists (odontologists) use dental evidence and records to identify people or to determine who inflicted bite marks. Forensic anthropologists who identify bones and skeletal remains are able to determine if bones are human as well as the gender, age, size, and build of their source if they are human. Forensic psychiatrists and psychologists analyze behavior and personality issues pertaining to criminal acts in an effort to determine a suspect's state of mind at the time of the crime and whether or not the person indicted is competent to stand trial. Forensic chemists and biologists, as well as general criminalists, usually work in crime laboratories where they examine and analyze such evidence as hair, fibers, soil, blood, paint chips, firearms, bullets, documents, and fingerprints.

The use of science in solving crimes was first popularized by Sir Arthur Conan Doyle, whose fictional hero, Sherlock Holmes, used science, keen observation, and superb logic in conducting his investigations. Holmes was often depicted in a chemistry laboratory analyzing blood, soil, poisons, or other evidence found at the scene of a crime. The first Sherlock Holmes story, *A Study in Scarlet*, was published in *Beeton's Christmas Annual* in 1887, and it has often been said that forensic science was initiated by Doyle's stories. History, however, suggests a different account, although it is true that forensic science did not play a significant role in crime detection until shortly after Doyle's stories appeared.

INTRODUCTION

As early as the 1820s, the Czech physiologist Johann Purkinje noticed the unique nature of fingerprints, but did not use them to solve crimes. Even earlier, in 1813, Mathieu Joseph Bonaventure Orfila published his *Treatise on Poison or General Toxicology*, the first scientific treatise on the detection and effects of poisons, which was instrumental in making forensic toxicology a legitimate discipline. In 1830, James Marsh developed a test that allowed chemists to identify the presence of arsenic in the tissues of victims of poisoning. The test was used in 1840 to show that Marie Lafarge was guilty of poisoning her husband (*see* Arsenic).

In 1848, police in Birmingham, England, began using the newly developed technique of photography to keep photographic records of criminals. In the two decades that followed, most police departments in Europe and the United States developed "rogues' galleries" that they used to help witnesses identify suspects. During the 1850s, William Herschel, working in India, applied inked imprints of the fingers of illiterate natives to legal documents in order to prevent fraud. In 1865, at his last cabinet meeting, Abraham Lincoln established the U.S. Secret Service in an effort to combat the growth of counterfeiting. In the same year, Alfred Swaine Taylor published *The Principles and Practice of Medical Jurisprudence*, which was the first comprehensive document on forensic medicine in the English language.

In 1880, well before Doyle published *A Study in Scarlet*, Henry Fauld secured the release of a man suspected of thievery by showing police that fingerprints found at the crime scene did not match those of the man they were holding. Twelve years later, Sir Francis Galton published *Finger Prints* and demonstrated that fingerprints, in addition to being unique, do not change over time. His system of classifying these prints was soon improved by Edward R. Henry, and in 1901 Scotland Yard hired Henry to establish a fingerprint division. Following Scotland Yard's lead, police departments throughout the world began to use fingerprints as a means of identification. However, Argentina, in 1896, became the first nation to adopt fingerprint identification as the official means of identifying criminals after Juan Vucetich, a Buenos Aires detective, used bloody fingerprints to solve an 1892 murder. It was 1901 before the New York City Civil Service Commission became the first organization in the United States to officially use fingerprints as a means of identification. Three years later, the New York City police department required everyone it arrested to be fingerprinted—a practice that soon spread to other police departments throughout the country.

In 1905, the first criminals in England (the Stratton brothers) were convicted on the basis of fingerprint evidence. In the United States, Joseph A. Faurot, in 1906, was the first detective to obtain a confession based on fingerprints, and in 1911, a year after fingerprint testimony was first admitted into a U.S. court, Faurot was the first to obtain a conviction based on such evidence.

Prior to Galton's investigation, professional police forces, such as Scotland Yard, had used the Bertillon system to identify criminals. This system, devised by Alphonse Bertillon, consisted of a photograph of the criminal as well as body measurements that included the dimensions of the head, the height, and the lengths of arms, legs, and feet. Using mathematics, Bertillon demonstrated that the chance of two people being identical for all measurements was 1 in 286,435,456. The measurements were tedious to make, however, and when it was discovered that identical twins could be differentiated by their fingerprints but not by Bertillonage, fingerprints quickly replaced Bertillon's system as the key to identifying people.

In 1889, Alexandre Lacassagne successfully identified a corpse using bones and dental comparisons. In the same year, he was the first to recognize that the grooves left on the surface of a bullet might be used to identify the gun that fired it. A little more than three decades later, Philip O. Gravelle invented the comparison microscope, which allowed Calvin Goddard to develop ballistic techniques that allowed experts to identify firearms by examining the markings left on bullets. In 1925, Sir Sydney Smith published *Textbook of Forensic Medicine*, which established Goddard's work on ballistics as a new branch of criminalistics.

During the first year of the twentieth century, Karl Landsteiner discovered that humans have two kinds of blood antigens on their red blood cells, which were designated A and B. His discovery led to the realization that there are four blood types (A, B, AB, and O) that are inherited and can be used to establish innocence, as well as possible guilt in paternity and other court cases. Later, as other blood factors were discovered, blood evidence could be used to establish probability of guilt. In 1938, New York medical examiner Thomas Gonzales established this country's first serological laboratory, under Alexander Weiner, to study blood and bloodstains collected as evidence at crime scenes. Today, DNA analysis of very small quantities of blood can be used to positively identify people.

In 1910, Edmond Locard established the world's first crime laboratory in Lyon, France. Locard's work made it evident that a scientific approach to collecting and analyzing evidence found at the scene of a crime was a fruitful means of solving crimes. As a result, forensic laboratories were established throughout Europe during the third decade of the twentieth century. In America, the first U.S. crime laboratory was founded in Berkeley, California, in 1923. A second crime laboratory was organized at Northwestern University six years later. In 1924, the Federal Bureau of Investigation was established, but it was not until November 24, 1932, that J. Edgar Hoover, the director of the bureau, opened the Federal Crime Laboratory. Today, that laboratory carries out thousands of analyses each week for FBI agents and law-enforcement agencies across the nation. In an effort to enhance the use of science in solving crimes, Hoover founded the National Police Academy in 1935 to train law-enforcement personnel who would have the knowledge needed to make good use of the technology and scientists available at the Federal Crime Laboratory. A year earlier, shortly before his death, Charles Norris, New York City's first medical examiner, succeeded in creating the first Department of Forensic Medicine in the United States at the New York University College of Medicine.

Today there are about 400 crime laboratories and nearly 40,000 forensic scientists in the United States. The use of modern techniques and sophisticated machinery enables forensic scientists to identify minute quantities of chemical evidence and determine the source of fibers, glass particles, paint chips, and other physical evidence collected at a crime scene. These scientists can identify bloodstains and determine their source. Lasers and ultraviolet light can be used to reveal fingerprints that would never have been detected a decade ago. Computers allow the storage of vast amounts of data, including fingerprints, that make it possible to share and exchange information among the nation's many law-enforcement agencies.

After FBI headquarters were moved close to the Smithsonian Institution in the 1930s, anthropologists at the institution and then elsewhere began to serve as forensic scientists. Physical anthropologists are the scientists who are most often called upon to interpret the bone evidence found at crime scenes. During the 1990s, Clyde Snow and other anthropologists, acting as forensic scientists, were sent to investigate mass graves in the former Yugoslavia by a war-crimes tribunal established by the United Nations Security Council.

During the 1980s, building on the model of DNA first devised by James Watson and Francis Crick three decades earlier, scientists found ways to extract and examine an individual's unique DNA from small quantities of human cells found in blood, semen, hair, and skin. The first conviction (Tommie Lee Andrews) in the United States resulting from DNA evidence occurred in 1987. Since then, many people have been convicted or shown to be innocent on the basis of DNA testing.

DNA testing is but one of many recent developments in forensic science. For example, computerized imaging technology has produced the Integrated Ballistic Identification System (IBIS), which makes it possible to store bullet and cartridge surface markings in a manner similar to the automated fingerprints identification system (AFIS) developed by the FBI. Despite the sophisticated equipment and techniques available to forensic scientists, solving crimes still requires ingenuity, clear thinking, careful questioning, and thorough, careful collection of the evidence that is later examined in a crime lab.

Guide to Selected Topics

This guide is designed to help readers find topics related to a particular facet of forensic science. All of the entries in the encyclopedia are listed here by broad topic. Please also consult the index at the back of the book for more narrow topics. We have grouped together and listed here the entries in this encyclopedia that pertain to (1) various types of forensic evidence; (2) much of the science and technology used by forensic scientists; (3) various poisons and drugs commonly involved in forensic science; (4) some of the people who have made significant contributions to the field of forensic science; (5) some of the types of crimes and criminals encountered by forensic scientists; (6) some famous criminal cases; (7) a number of famous criminals; (8) some serial killers; (9) authors of fiction about forensic science; and (10) some fictional characters.

Types of Forensic Evidence

bite marks

blood

bruises

cause of death

contrecoup

crime scene

disputed documents

DNA evidence

drowning

expert testimony

eye prints

eyewitness account

fiber evidence

fingerprints

footprints

forensic anthropology

forensic engineering

forensic entomology

forensic photography

forensic psychiatry

forensic psychology

forensic-science errors

forensic-science issues

forensic taphonomy

glass evidence

gunshot wounds

hair evidence

insanity defense

lineup

lip prints

lung flotation

morgue

paint evidence

palmprints

rogues' gallery

GUIDE TO SELECTED TOPICS

DeSalvo, Albert
Mudgett, Herman Webster
Palmer, Dr. William
serial killers
Williams, Wayne

Fiction Writers of Forensic Science

Abbot, Anthony
Blochman, Lawrence
Buchanan, Edna
Cornwell, Patricia
Deaver, Jeffery
Doyle, Sir Arthur Conan
Freeman, R. Austin
Gardner, Erle Stanley
Reeve, Arthur Benjamin

Reichs, Kathy
Sayers, Dorothy L.

Fictional Characters in and Television Programs about Forensic Science

Coffee, Dr. Daniel Webster
Colt, Thatcher
CSI: Crime Scene Investigation
Holmes, Sherlock
Kennedy, Craig
Mason, Perry
Pudd'nhead Wilson
Quincy
Scarpetta, Dr. Kay
Thorndyke, Dr. John Evelyn
Wimsey, Peter Death Bredon

A

Abbot, Anthony (1893–1952)

Anthony Abbot was the pen name of C. Fulton Oursler, an American editor, journalist, mystery writer, novelist, and playwright who was very interested in crime and forensic science. On three occasions he was a graduation speaker at the FBI's National Police Academy. From the 1930s to the mid-1940s, using the name Anthony Abbot, Oursler wrote a series of detective novels that featured New York Police Commissioner Thatcher Colt. His novel, *About the Murder of Geraldine Foster*, established the distinctive titles Oursler used. Most of his titles began with the word *about*. Cynics remarked that this was Oursler's way of assuring a prominent place in any alphabetical listing of titles, but he denied it. The novels were generally well received by critics, and *About the Murder of the Clergyman's Mistress* and *About the Murder of the Nightclub Lady* were best-sellers in the mystery genre.

In 1922, Oursler became editor of the MacFadden group of magazines that included *True Detective, True Story*, and *Physical Culture*, among others. In 1931, MacFadden bought the debt-ridden five cent weekly *Liberty*, and the magazine became Oursler's main focus from 1931 to 1942. In trying to revive the fortunes of *Liberty*, he made sure that the magazine always pre-sented a number of fictional and true crime articles. Today, Oursler is best known for his 1949 retelling of the New Testament, *The Greatest Story Ever Told. See also* Colt, Thatcher.

References

"Fulton Oursler, Author, Dies at 59." *New York Times*, May 25, 1952.
"Oursler, Fulton." *Current Biography*. New York: H.W. Wilson, 1942.

Alkaloids. *See* Opium.

Anthropology and Crime. *See* Forensic Anthropology.

Appel, Charles. *See* FBI Crime Laboratory.

Arsenic Poisoning

Arsenic poisoning is the best-known, but a seldom-used, method of poisoning people. In fact, arsenic, the twentieth most abundant element in the earth's crust, is probably best known as a poison. Although it was once the main ingredient of rat poison, arsenic gained its notoriety as a substance used to kill humans in real life and in works of literature, such as *Strong Poison*, a book by Dorothy

1

Sayers. Arsenic was also the ingredient used to poison elderberry wine in Joseph Kesselring's enduring play *Arsenic and Old Lace*. In the comedy, two elderly women operate a boarding house that caters to old men who have no family or friends. The sisters carry out what they believe to be acts of kindness by poisoning the men with elderberry wine. The play was also made into a 1944 movie starring Cary Grant and directed by Frank Capra.

In real life, arsenic has been used to kill, or attempt to kill, a number of people. It was Marie Lafarge's poison of choice when she murdered her husband in 1840, and there is evidence that it was used to poison Napoleon Bonaparte. Scottish researcher J.M.A. Lenihan and his associates tested samples of Napoleon's hair using neutron-activation analysis and found abnormally high concentrations of arsenic. In recent times, its use has diminished because forensic scientists can easily detect it in body tissues.

Arsenic is a brittle, steel-gray metal. It has no practical uses as a metal but is used in alloys, particularly with lead and copper. Compounds of arsenic are used as pharmaceuticals, insecticides, weed killers, and wood preservatives and in cattle and sheep dips. The substance most commonly referred to as arsenic is actually arsenic trioxide (As_2O_3 or As_4O_6), which is also known as white arsenic. Interestingly, according to an Associated Press story released in November 1998, a pilot study at New York's Memorial Sloan-Kettering Cancer Center found that this common poison (arsenic trioxide) was effective in treating patients with acute promyelocytic leukemia (APL). Eight of 12 people believed to be in a terminal stage of the disease appeared to be cancer free 10 months after treatment according to Dr. Robert E. Gallagher of New York's Montefiore Medical Center.

Toxicologists think that an excess of arsenic inhibits the action of the body's enzymes, causing severe gastric effects leading to convulsions, coma, and death. A large dose can cause death within 24 hours. It affects the intestinal walls, causing nausea, vomiting, diarrhea, cold sweats, a burning sensation in the throat, and finally a comatose condition. An effective immediate antidote for acute arsenic poisoning is freshly prepared ferric hydroxide, which is made by mixing ferric chloride with milk of magnesia. The mixture reacts with the poison to form an insoluble nontoxic compound. At a later stage, treatment involves flushing the stomach and intestine with water coupled with standard treatment for dehydration and shock. Kidney dialysis may also be helpful.

Neither arsenic nor white arsenic is soluble in cold water. White arsenic is more soluble in hot water; however, it falls out of solution as the solvent cools. Consequently, fictional stories in which a character drinks tea poisoned with arsenic are truly fiction. There are reports that some people who lived in the Austrian province of Styria used to eat arsenic to improve their complexion and make their hair lustrous. Supposedly, they developed an immunity to the poison by gradually increasing the dosage.

Prior to his death on May 5, 1821, Napoleon Bonaparte, emperor of France, who had been exiled to the island of St. Helena off the west coast of Africa, was convinced that he was being slowly murdered by an English assassin. Most historians believed that he died of cancer, but shortly after the middle of the twentieth century the diaries of Louis Marchant were published. Marchant, who had been Napoleon's valet, described his master's illness in detail. The symptoms—loss of hair, obesity, muscular weakness coupled with periods of paralysis, and drowsiness, together with alternating insomnia and chills—suggested that Napoleon's death was the result of chronic arsenic poisoning.

When locks of Napoleon's hair were found, they were tested for arsenic. The analysis provided clear proof that he had indeed received doses of arsenic over an extended period. But who murdered him? Some believe that it was Charles Montholon, who had been excused from an embezzlement charge by the Bourbons and had then joined Napoleon as an attendant. The Bourbons, who had been deposed by Napoleon early in

1815, returned to power following his defeat at Waterloo later that year. Marchant's diary reveals that Montholon, shortly before Napoleon's death, persuaded doctors to administer drugs that would remove any arsenic from his gastrointestinal tract during an autopsy.

Although white arsenic has been used to poison people, often by placing it in cakes or other food, it is seldom used as a murderous poison today. It accumulates in body tissue, particularly in nails and hair, and is easily detected by forensic toxicologists. The first method of testing for arsenic poisoning was developed by Johann Metzger of Königsberg (Germany) in 1787. He found that tissue containing arsenic would, when heated, produce a dark deposit on a strip of copper held over the tissue.

Another test, known as the Marsh test, was developed by James Marsh in England in 1830. He found that when tissue containing arsenic was added to acid and zinc, it produced arsine gas (AsH_3). The gas could be forced through a nozzle and heated. When arsine is heated, it is decomposed into arsenic and hydrogen. Marsh found that a cold porcelain dish held above the hot gas would become coated with a black deposit of arsenic metal.

The Marsh test was first used to provide evidence in court during the famous case of Marie Lafarge. Police claimed that while her husband, Charles, was on a business trip to Paris, Marie Lafarge sent him a cake laced with arsenic. When he returned sick, but still alive, she added the same substance to his milk. Evidence for this action was provided by the family maid, who told investigators that she had seen white flakes floating on the milk's surface (a result of arsenic's low solubility). Later, she reported, she had seen Marie Lafarge adding pinches of the white powder to her husband's food. Following the death of Charles Lafarge on January 16, 1840, the suspicious maid delivered some of the white powder to the police, and Marie Lafarge was soon under arrest.

At the trial, a local chemist testified that he had found traces of arsenic on the glass that held the milk Charles Lafarge had drunk and on the box that held the white powder the maid had seen Marie Lafarge adding to food, as well as in Charles's stomach. However, the defense called to the stand toxicologist Mathieu Joseph Bonaventure Orfila. Orfila disputed the chemist's tests and claimed that they were inconclusive with regard to arsenic. Through Orfila, the prosecution learned about the Marsh test and requested him to apply the test to the evidence available. When Orfila applied the Marsh test using the remains of Charles Lafarge, he found arsenic in the stomach, liver, heart, brain, and intestines. On the basis of this new evidence, Marie Lafarge was found guilty and was sentenced to hard labor for life.

When the Marsh test was refined, it revealed that all human tissue has trace amounts of arsenic. The average person is about one ten-millionth arsenic by mass, and people who eat lots of seafood or who work in vineyards may have arsenic levels three times as large. Modern toxicologists can readily and conclusively identify arsenic in tissue by using atomic absorption spectrophotometry, emission spectroscopy, or X-ray diffraction.

The René Castellani Case

On July 11, 1965, Canadian Esther Castellani from Vancouver died. She had been undergoing serious bouts of abdominal pain, dizziness, nausea, and vomiting, as well as numbness and pain in her hands and feet, for more than a year. Her family doctor suspected gastritis, but he could not find a medicine that would relieve her attacks. She sought a second opinion from Dr. Bernard Moscovich. He realized that she was seriously ill, but was unable to diagnose the cause as she began to suffer from heart failure shortly before her demise. An autopsy indicated that she died of heart failure. But why? wondered Dr. Moscovich. After a careful review of all the symptoms his patient had exhibited, he was able to rule out everything except arsenic poisoning.

In an effort to confirm his posthumous diagnosis, Moscovich had a toxicologist analyze samples of her tissues that had been preserved. He found that they contained arsenic at a concentration of 24 ppm, 1,000 times more than normal. This evidence was enough to convince authorities to exhume her body. A pathologist found that her peripheral nerves had been severely damaged, indicating small repeated doses of arsenic. A large lethal dose of the poison will damage nerve cells in the brain and spinal cord but not the peripheral nerves. Arsenic enters a person's hair almost immediately after being ingested. Because pathologists found arsenic in the roots of the hair, they concluded that she had received arsenic while she was in the hospital. The presence of arsenic near the tips of the hair examined revealed that she had been receiving small doses of arsenic for at least seven months prior to death.

While this testing was going on, police learned that René Castellani was having an affair with a young woman who worked in the office of his construction business. He willingly allowed police to search his house, where they discovered a bottle of weed killer that contained a high concentration of sodium arsenate. The can had been opened and some was missing. Castellani said that he had thought that the can, which was near some shoe polish and household cleaners, contained charcoal starter fluid. Could the samples of hair be used to show that the concentration of arsenic in the victim's body at different times corresponded to the times when she was most ill? In an effort to find out, forensic chemist Norm Erickson sliced hair samples from the victim into lengths of 1.0 cm, 0.5 cm, and, finally, 0.25 cm. These are the lengths that hair normally grows in a month, two weeks, and one week, respectively. Erickson then used neutron-activation analysis to analyze the samples. He placed the short lengths of hair in a nuclear reactor to make them radioactive. He then analyzed the gamma rays released by the samples to determine when the peaks due to arsenic were greatest. His results showed beyond reasonable doubt that Esther Castellani's periods of illness occurred at the same times the concentrations of arsenic in her hair were peaking.

Meanwhile detectives had learned from hospital personnel that René Castellani had frequently taken milkshakes to his wife during her stays there. Furthermore, the victim's sister-in-law had observed him trying to force his wife to eat dinner and had asked her several times to ask his wife's mother to leave so that he could feed his wife without interference from her mother. But wouldn't weed killer mixed with food make the food unpalatable? Forensic scientists found that mixing a little weed killer with meat or a milkshake had no effect on the food's taste or odor.

All the evidence indicated that Castellani had poisoned his wife, slowly but effectively. He was convicted and sentenced to death in 1966. He obtained a second trial in 1969 and was again convicted; however, by this time Canada had done away with the death penalty. After serving 13 years, Castellani was released in 1979. He later admitted to killing his wife but claimed that he did so because he knew that she had cancer.

A Recent Case of Arsenic Poisoning

Although poisoning by arsenic is very uncommon today, a couple from Wakayama, Japan, Masumi and Kenji Hayashi, were arrested in October 1998, and charged with poisoning a friend. The friend became very sick and was hospitalized for four months but finally recovered. After it was determined that the man was suffering from arsenic poisoning, police conducted an investigation and found that the Hayashis, with whom he had dined before becoming sick, had, unbeknownst to him, taken out insurance policies on his life that were worth more than $1 million. They also learned that Masumi Hayashi had been one of the cooks who prepared a beef curry at a festival following which 4 people died of arsenic poisoning and 63 became very ill. *See also* Neutron-Activation Analysis; Orfila, Mathieu Joseph Bonaventura; Poisons; Sayers, Dorothy; Toxicologists; Toxicology.

References

American Jurisprudence Proof of Facts Annotated. Vol. 15. San Francisco: Bancroft-Whitney, 1964, 115–148.

Barrett, Sylvia. *The Arsenic Milkshake and Other Mysteries Solved by Forensic Science.* Toronto: Doubleday Canada, 1994.

Cyriax, Oliver. *Crime: An Encyclopedia.* North Pomfret, VT: Trafalgar Square, 1996.

Evans, Colin. *The Casebook of Forensic Detection: How Science Solved 100 of the World's Most Baffling Crimes.* New York: Wiley, 1996.

Weiner, Eric, and Renee Montagne. "Japanese Poison Suspects Arrested," *Morning Edition* (National Public Radio), Oct. 5, 1998.

Arson

Arson is the illegal burning of property. In 1994, it accounted for 14 percent of all fires involving buildings and 21 percent ($1.4 billion) of all property loss from such fires. However, the number of fires attributed to arson varies from year to year because arson is a recession-related crime. A struggling business can often recover its losses if a fire "accidentally" destroys the company's insured building. Arson is difficult to establish in court. Consequently, insurance companies have difficulty proving criminal action. Ships with large mortgages are also frequently destroyed by arson during an economic recession, particularly if the owners have been able to insure the vessel for more than it is worth.

Arson is a difficult crime to solve because it is usually well planned and is committed at a time determined by the criminal, who is able to leave the scene long before police arrive. It is also a crime that by its very nature destroys evidence and one that could be caused in a variety of ways that involve no crime, such as faulty wiring, a carelessly placed cigarette, an overheated electric motor, or a defective heating system.

A key factor in establishing arson is timing. The investigation must be started as soon as possible because an accelerant that may have been used to start the fire will evaporate quickly. Furthermore, in *Michigan v. Tyler* (1978), the U.S. Supreme Court held that "entry to a fire requires no warrant, and once in the building, officials may remain there for a reasonable time to investigate the cause of a blaze. Thereafter, additional entries to investigate the cause of a fire must be made pursuant to the warrant procedures. . . . Consequently, if police are called to investigate a fire after others have left the scene, they must obtain a search warrant or risk having a court deny the admission of any evidence uncovered during the investigation."

The first thing investigators look for is the fire's origin. It will usually be near the lowest burned region because fires move upward with the convection currents created by the heat. If arson is involved, an investigator may find containers that held the accelerant, the igniter that was used, evidence of breaking and entering, or theft. Someone nearby may have seen someone or heard sounds. The most common accelerant used by arsonists is gasoline or kerosene, and although these substances burn or evaporate quickly, trace amounts are often trapped in porous materials such as carpets or upholstery, cracks in concrete, walls, ceilings, and the like. Of course, a match is generally used to ignite the fuel, and sometimes it is found. In one case, the base of the match discarded at the scene of a fire, which was torn from a matchbook, was found to match the base of a match that had been removed from the matchbook found on the suspected arsonist.

Accelerants can be located with a vapor detector known as a "sniffer." The device acts like a vacuum cleaner. It pulls in air from around a suspected sample and carries it over a hot filament. Any combustible vapor in the air will be oxidized, releasing heat and raising the temperature of the filament, which will be indicated on a meter.

Some fire and police departments have replaced the mechanical sniffer with a live one. Because of their keen sense of smell, dogs can be trained to detect small amounts of chemicals left at the scene of a fire. In fact, dogs can find small quantities of accelerants that produce no response from machines such as the "sniffer." In training these animals, it has been found that many of them can detect a single drop of an accelerant such as gasoline, kerosene, turpentine, or alcohol even when it is buried in material that has been burned.

Once an accelerant has been detected in a sample, the porous material is placed in a metal or glass container (vapors can diffuse through plastic bags), sealed to prevent further evaporation, and taken to a laboratory. There the container is warmed to increase the concentration of the vapor before a syringe is used to remove a volume of the vapor. Sometimes, to capture a larger volume of vapor, the container is heated and its contents are flushed out with a stream of nitrogen gas that forces the vapors through an absorbent such as charcoal. The vapor can then be extracted with a solvent such as carbon disulfide, or the charcoal can be heated to drive out the vapor. Whichever method is used, the vapor or liquid is then injected into a gas chromatograph, where the components are separated and tentatively identified.

If gasoline is detected through gas chromatography, the chromatogram produced by the sample can be compared with samples found on a suspect's clothing or in a can believed to hold the accelerant used. If testing is done quickly, the chromatogram can be compared with samples taken from local gasoline stations. Sometimes the station that sold the fuel can be identified and an employee will remember who bought the fuel or be able to identify a suspect.

In addition to those who set fires for financial reasons or those who are hired to perform that service, there are pyromaniacs who do it for the pleasure it brings them. After his arrest in 1931, Peter Kürten, who was called the monster of Düsseldorf (Germany) for his arson, reported that he always had sexual pleasure when he watched big fires. Others convicted of arson report similar behavior and the thrill of watching the fires they set.

Many psychiatrists characterize pyromaniacs as unhappy people who suffer from feelings of inferiority and rejection. To them, flames are a symbol of purity and a means of "burning away" their guilt and offering a new beginning after their feelings of oppression have been removed. Psychiatrists can point to historical evidence that suggests that people believe that total destruction wrought by fire can solve problems. For example, Nazis burned their victims' remains in ovens, and earlier societies burned witches and heretics.

Arsonists resort to ingenious schemes in seeking monetary gains. In 1955, George Fisher of Chicago insured his huge home for $75,000, installed a complicated electrical device in his telephone, and left for Florida, where he was careful to register at a hotel and converse extensively with personnel to establish an alibi. When he called his home at a previously calculated time, the device he had installed set off a fire that burned his home. Fisher's reputation led his insurance company to be suspicious. Their forensic investigators discovered the remains of the gadget's wiring in Fisher's phone and refused to pay the claim.

In 1998, during the appeal trial of Peter G. Saunders, who had been convicted of arson, Saunders offered as a defense the fact that he had a net worth of $3 million. Why would anyone who was a millionaire risk arson for the small gain involved in recovering insurance for an ordinary dwelling? Saunders's attorney called on forensic accountant Stanley Dennis to testify as to Saunders's fiscal soundness. However, during cross examination, the prosecution made it known that Saunders had not paid $109,000 due the Internal Revenue Service (IRS) in 1988–1989. Furthermore, the IRS had a tax lien on his property. Despite supporting testimony by the forensic accountant, the judgment against Saunders was affirmed. His failure to pay taxes, the lien against his property, and reports indicating that he had offered the tenant money to set the fire were his undoing. *See also* Chromatography.

References

"Arson Detection Dogs Sniff Out Valuable Evidence." *American City & County*, Dec. 1996, pp. 12–13.

Commonwealth v. Peter G. Saunders. Appeals Court of Massachusetts, Plymouth, 1998.

Crimes and Punishment: The Illustrated Crime Encyclopedia. Westport, CT: H.S. Stuttman, 1994.

Cyriax, Oliver. *Crime: An Encyclopedia.* North Pomfret, VT: Trafalgar Square, 1996.

DeHaan, John D. *Kirk's Fire Investigation.* 4th ed. Upper Saddle River, NJ: Brady, 1997.

Arsonists. *See* Arson.

Atropine

Atropine ($C_{17}H_{23}O_3N$), a colorless, solid alkaloid derived from the deadly nightshade plant (*Atropa belladonna*), has played a role in a number of murder cases. One of the most famous was the Buchanan-Sutherland case (1892), in which Dr. Robert Buchanan, a New York City physician, used atropine to dilate his wife's pupils before she died from an overdose of morphine that he had administered. The medical examiner, upon seeing that the victim's pupils did not have the pinpoint size characteristic of morphine poisoning, dismissed any idea of murder and reported that the cause of death was a stroke.

Early in the twentieth century, a Cambridge, Massachusetts, nurse, Jane Toppan, used atropine in a similar manner while murdering at least 31 people with a mixture of morphine and atropine. She used the atropine to counteract the symptoms of morphine poisoning.

Atropine blocks the effect of the parasympathetic nervous system on glandular cells, smooth muscle, and heart muscle. It opposes the action of acetylcholine, a substance released at the end of a nerve cell following the transmission of a nerve impulse. Anything that prevents acetylcholine from doing its job, such as atropine, is an anticholinergic agent.

Atropine paralyzes visual accommodation (the ability of the eye to adjust the curvature of the lens for viewing objects at different distances) and dilates the pupils of the eyes, which is why Robert Buchanan used it to counteract the pupil-contracting effect of morphine poisoning. Atropine also decreases the secretion of saliva and other digestive activities while increasing both heart and breathing rates. Atropine is often used medicinally to increase heart rate and is a component of a number of medications used to treat the eyes, skin, and gastrointestinal tract.

In excess, it is a poison and has been used as a murder "weapon." Tolerance to atropine varies, but a fatal dose for children is about 20 milligrams (mg). About five times as much is fatal for adults. Symptoms of atropine poisoning include a dry, burning sensation in the mouth, thirst, difficulty swallowing, rapid pulse, fever, hot, dry skin, confusion, delirium, and behavior suggesting insanity. Because of the symptoms that accompany atropine poisoning, medical examiners can often identify the cause of death, which serves as an important clue in determining whether the victim died from suicide or murder.

Hyoscyamine ($C_{17}H_{23}O_3N$), an isomer of atropine, is obtained from the poisonous henbane or jimsonweed plants. Its effects on the body are similar to those of atropine. *See also* Buchanan-Sutherland Murder; Opium.

References

The American Medical Association Encyclopedia of Medicine. Ed. Charles B. Clayman. Heidi Hough, editor-in-chief. New York: Random House, 1989.

Magill's Medical Guide: Health and Illness. Pasadena, CA: Salem Press, 1995.

The Merck Manual of Medical Information. Home ed. Ed. Robert Berkow, MD. Whitehouse Station, NJ: Merck Research Laboratories, 1997.

The Oxford Companion to Medicine. Ed. John Walton, Paul B. Beeson, Ronald Bodley Scott. Oxford and New York: Oxford University Press, 1986.

Turkington, Carol. *The Home Health Guide to Poisons and Antidotes.* New York: Facts on File, 1994.

Automobile Accidents

Automobile accidents are often minor fender benders that are settled by insurance companies, but others involve fatalities. Police routinely photograph vehicles and scenes involving serious accidents. They measure skid marks, record weather and road conditions, and interview drivers and witnesses. Forensic scientists with a background in physics or engineering can often determine a vehicle's speed prior to collision from skid marks and distance traveled. Damage to the car is also indicative of speed. Since the kinetic (motion) energy of any object is proportional to the square of its velocity, the adage that speed kills is a truism. A car traveling 60 mph has four times the energy of a car traveling 30 mph and can, therefore, do four times as much damage.

Safety belts and air bags have reduced injuries associated with car crashes, which is why drivers and passengers are required by law to wear seat belts. A person traveling without a seat belt will obey Newton's first law of motion, a physical law known to every first-year physics student. He or she will continue to move with a constant velocity when a car breaks down or crashes. As a result, that person may travel through the windshield and suffer serious injury or death. A safety belt prevents that from happening, and an air bag provides a cushion between the victim and the vehicle.

One of the most common causes of automobile accidents is alcohol. Forty percent of all deaths due to motor vehicles are attributed to alcohol. A driver under the influence of alcohol has slower reaction times and poorer judgment than a sober driver. Statistics reveal that a person with a blood-alcohol level of 0.08 percent (weight per volume) is four times as likely to have an accident as a sober driver. When that level nearly doubles to 0.15 percent, the likelihood of an accident increases by a factor of 25. Most states define a blood-alcohol level of 0.08 as the point at which a driver may be considered under the influence of alcohol and unfit to drive.

The invention of the Breathalyzer by Robert F. Borkenstein of the Indiana State Police in 1954 provided police with an instrument that allows them to quickly determine the body's blood-alcohol level. According to state laws, a person suspected of driving under the influence of alcohol who refuses to submit to a Breathalyzer or blood test will lose his or her driver's license for a specified period of time. In the case of *Schmerber v. California* (1966) the U.S. Supreme Court upheld the right of a state to do so. Schmerber had argued that when police obtained a sample of his blood in the hospital to which he was taken after an accident, they violated his Fifth Amendment right under the U.S. Constitution to avoid self-incrimination. The Court ruled that the Fifth Amendment applies to testimony and not to physical evidence such as fingerprints, blood, or photographs.

Field Tests for Sobriety

Before requiring a driver to take a Breathalyzer test, police will first ask a driver to perform some simple acts to test for sobriety. To test for nystagmus (involuntary jerky movements of the eyeball), police hold a penlight at eye level and ask a subject to use his or her eyes to follow the light as it is moved slowly back and forth. A person who is intoxicated will have the odor of alcohol on his or her breath and will exhibit nystagmus at a much smaller angle from the frontal line of sight than one who is sober. The subject may be asked to walk heel-to-toe along a straight line, turn 180 degrees, and repeat the process. He or she may also be asked to stand on one leg for 30 seconds while counting backwards from 100. Finally, some police departments have pocket-size breath testers that will indicate simply whether or not a Breathalyzer test should be administered.

Hit and Run

Hit-and-run drivers, frequently under the influence of alcohol, often leave tire prints or trace evidence in the form of glass from a broken headlight or paint chips from a fender. The car, in turn, may carry away small drops of the victim's blood or fibers from his clothes. Such evidence is often enough to allow police to find and convict an offender.

James Wallace Case

James Alan Wallace, twice convicted of driving while impaired (DWI), was staggering when he left a bar in Gibsons, Canada, and got behind the wheel of his 1982 Chrysler in the afternoon of December 26, 1995. On his way home, he struck a group of women out for a walk and kept going. One of the women, Jennifer Stewart, was killed, another suffered a fractured pelvis and vertebra along with severe cuts and bruises, a third had cuts, bruises, and a broken clavicle, and a fourth received a scalp wound.

The next day a neighbor observed Wallace wiping the front fender of his car, which had struck the victims. That observation, given as evidence in court, was ruled hearsay evidence and, therefore, inadmissible. However, police, acting on a tip they received, visited Wallace's home and found that paint left at the accident matched that on Wallace's fender. Faced with the forensic evidence, Wallace pleaded guilty to four counts of DWI, one of which caused the death of Jennifer Stewart. He was sentenced to six years in prison and forbidden to drive in British Columbia for 20 years or 10 years in other parts of Canada. Wallace was not charged as a hit-and-run driver because no one could prove that in his inebriated condition he was aware that his vehicle had struck anyone.

It is well known that a majority of DWI offenders are alcoholics, which suggests that they might be rehabilitated. Evidence that this is possible has been provided by the Addiction Research Foundation (ARF). A long-term ARF study of 650 DWI offenders showed that rehabilitation programs significantly reduced a reoccurrence of a DWI offense. Other methods of reducing the occurrence of a second DWI incident by offenders include impoundment of cars and the installation of alcohol-ignition-interlock systems as a requirement for reinstatement of a driving license. These systems have a breath analyzer connected to the ignition that will prevent a start up if the driver's alcohol is more than 0.04 percent by weight. *See also* Crime Scene Fiber Evidence; Glass Evidence; Paint Evidence; Toxicologists; Toxicology; Trace Evidence.

References

Brook, Paula. "The Ones Who Get Away: Drunks at the Wheel." *Chatelaine*, Dec. 1, 1996, p. 72.

Feldman, Bernard J. "Elementary Physics in a Real Automobile Accident." *Physics Teacher*, Sept. 1997, pp. 335–337.

Lane, Brian. *The Encyclopedia of Forensic Science.* London: Headline, 1992.

Saferstein, Richard. *Criminalistics: An Introduction to Forensic Science.* Upper Saddle River, NJ: Prentice Hall, 1998.

Autopsy. *See* Coroner, Medical Examiner.

B

Ballistics

Ballistics is a branch of physics concerned with the kinematics (motion) of projectiles; however, in forensic science, ballistics has come to mean the study of bullets and firearms. When a bullet is fired from a gun, it travels along a spiral groove in the gun barrel. The purpose of the groove is to make the bullet spin around its long axis. A spinning bullet, like a spiral pass thrown by a quarterback, acquires angular momentum that it will retain after it leaves the barrel, causing it to travel straight and true to its target. The barrels of old muskets, such as those used in the American Revolution, were not spirally grooved. As a result, the paper-packed lead projectiles, which were powered by igniting gunpowder, traveled much like a football thrown by a child who has not yet learned how to make the ball spiral, and muskets were not very accurate beyond a few feet.

During the manufacture of a gun, the barrel is drilled by a cutting tool that leaves striations as well as the curved grooves on the inside surface of the barrel. Because of slight changes in the drilling tool as it works, no two barrels have identical striations. As a result, every rifled (spiraled) barrel will leave characteristic striations on any bullet that travels along it. The shell casing that remains in the barrel will have identifying scratches made by the firing pin, the breech face, the extractor, or the ejector post. Guns of the same make will leave similar but not identical striations and scratches. A ballistics expert using a comparison microscope can often determine whether or not a bullet found at a crime scene or in a victim came from a suspect's gun. A bullet found at the scene of a crime can be compared with a test bullet fired from a suspect's gun. To avoid damaging any markings made by the gun barrel, the test bullet is generally fired into a recovery tank filled with cotton or water.

Gunpowder Residues and Distances

Possible suicides by shooting or claims of firing in self-defense can often be assessed by determining the distance between the bullet hole and the gun that fired the bullet. When a gun is fired, the gunpowder is ignited by the firing pin and burns rapidly, creating an expansion of gases that propels the bullet along the barrel. The gunpowder, however, is never completely burned. As a result, unburned and partially burned particles of the powder emerge from the barrel. The distribution of these particles around the hole made by the bullet can be used to determine the distance between the bullet hole and the point where the bullet left the gun barrel. Because different bullets and guns leave different residue patterns on a target, precise determinations of distances can be made only by comparing the patterns of particles left at different distances when similar bullets are fired from the suspect's gun.

However, there are general clues that can be used to make reasonable estimates of distances between gun and bullet hole. For example, if a gun was fired in contact or very close to the bullet hole, there will be a high concentration of vaporous lead around the bullet hole. If fibers surround the hole, they are likely to show scorch marks and evidence of having been melted by the flame that emerged from the gun barrel. Firings from distances greater than a yard generally leave no particles of powder on the target. Microscopic examination of garments and other articles recovered from a crime scene may reveal a residue of gunpowder particles. Even if such particles are obscured by dried blood, they may become evident through chemical tests or when viewed under infrared light.

Often, residues of gunpowder or initiating explosives (primers) can be detected on the hands of a suspect by the use of chemical tests that detect the barium and antimony in such compounds. A suspect's hands are swabbed with dilute nitric acid, and the swabs are then tested for barium and antimony using absorption spectrophotometry. If a scanning electron microscope is available, it can be used to detect such particles after they have been removed from a suspect's hands by adhesive tape. Paraffin testing, which was used for many years to detect nitrates on a suspect's hands, is no longer used because many other substances in the environment, such as fertilizers, explosives, tobacco, urine, and cosmetics, contain nitrates and nitrites. Consequently, the paraffin test lacks specificity (*see* Paraffin Testing).

The Integrated Ballistic Identification System (IBIS) and DRUGFIRE

Computerized imaging technology makes it possible to store bullet and cartridge surface markings in a manner similar to the automated fingerprints identification system (AFIS) developed by the FBI. The automated system for storing forensic ballistic information developed by the FBI is called DRUGFIRE. A microscopic view of the bullet or cartridge provides a magnified image that is captured by a video camera, digitized, and stored in a database. Networking allows other forensic laboratories to access the FBI database.

The Bureau of Alcohol, Tobacco, and Firearms has developed a similar system called the Integrated Ballistic Identification System (IBIS). It has two software systems— Bulletproof, to analyze bullets, and Brass-catcher, to analyze cartridges. Both IBIS and DRUGFIRE are screening systems that allow police to eliminate many possible matches. The final detailed comparison of bullets and/or cartridges found at a crime scene with those from a suspected gun is still conducted by a firearms examiner.

In 1997, using DRUGFIRE, Baltimore police were able to link three separate killings previously thought to be unrelated to the same 9-mm semiautomatic pistol discovered in a suspect's home. The gun's identity was then confirmed by a firearms examiner using the traditional comparison microscope. "Prior to IBIS, police rarely compared bullet fragments from different cases unless other evidence offered a compelling reason to analyze them. . . . [S]hots fired randomly were ignored outright, the spent shells from them virtually useless to officers.

"Now, science has shortened to seconds a matching process that, if done manually, would have ground investigations to a halt in the past" (Latour).

Other Cases

The earliest case in the United States in which forensic ballistics played a role was probably the 1784 murder of Edward Culshaw. A constable who examined Culshaw's wound found a piece of newspaper that had been used as wadding to pack the powder in the murderer's pistol. Upon searching John Toms, the primary suspect, the constable found a piece of newspaper in his pocket that matched the torn bit of wadding like two pieces from a jigsaw puzzle. The ballistic evidence convinced the jury, and Toms was found guilty of murder.

Nearly a century and a half later, in 1920, Nicola Sacco and Bartolomeo Vanzetti were arrested for the murder of two men shot during a payroll robbery in Braintree, Massachusetts. On the basis of the fact that both men owned guns and that a bullet recovered from one of the victims matched those found in Sacco's pocket, they were convicted. Since both men were immigrants and were suspected of being Communists, left-wing organizations around the world claimed that a fair trial was impossible. Felix Frankfurter, later to become a Supreme Court justice, wrote a book about the case and in 1927 persuaded the governor of Massachusetts to appoint a committee to investigate the trial. By this time the comparison microscope had been invented, and the committee asked Calvin Goddard, America's leading firearms expert at New York's Bureau of Forensic Ballistics, to examine the evidence. In the presence of defense ballistics expert Augustus Gill, Goddard fired a bullet from Sacco's gun into a tank filled with cotton. Through a comparison microscope, he then compared the bullet collected as evidence with the test bullet he had fired. There was no doubt. Gill agreed with Goddard. Both bullets had been fired from the same gun. Despite continuing protests and the distinct possibility that Vanzetti was innocent as he claimed, both men were executed on August 23, 1927.

Forty years later, on December 4, 1969, Chicago police reported a fierce shootout with a gang of Black Panthers who refused to surrender and showered the police with a rain of bullets. When the shooting ended, the building had been pierced by 80 bullets, two Black Panthers were dead, and four were wounded. There were no police casualties. The seven living members of the gang were charged with attempted murder. When the case came to trial, Herbert Leon McDonnell, the ballistics expert for the defense, reconstructed the gunfight on the basis of the paths followed by each of the bullets recovered. He prepared scaled diagrams to show that the first shot had been fired by police

through a partially opened door into the apartment. The second shot had been fired by a Black Panther after the door had been opened. All the other bullets were fired by police. In 1982, the court agreed that the civil rights of the Black Panthers had been violated, and they were awarded nearly $2 million in damages.

Again, it was ballistic evidence that led to the conviction of racist serial killer Joseph Christopher, who murdered black males in the Buffalo, New York, area, beginning on September 22, 1980. Christopher, who was captured while trying to stab a black soldier at Fort Benning, Georgia, on January 6, 1981, had been seen by witnesses carrying his firearm in a paper bag, a scheme that not only concealed the weapon but prevented his leaving shell casings as evidence. However, during several hasty escapes he dropped a total of five shell casings, which police recovered. Following his arrest, he attempted to emasculate himself and was committed to a hospital, where he bragged to a nurse about the many blacks he had killed in New York City as well as in Buffalo. A search of his Buffalo home where he had lived with his mother led to the discovery of a misfired bullet in the basement. The firing-pin imprint on the case was identical to those found on the other shell casings he had left at several murder scenes. At the family's hunting lodge, police found a barrel that had been sawed off a .22 rifle. The remaining part of the weapon could have been concealed in a paper bag. Although the actual murder weapon was never discovered, the other evidence was sufficient to convict him.

Loose Guns

Boston police, using an IBIS database over a five-year period beginning in 1995, discovered that relatively few guns are fueling a large number of violent crimes. In 2000, they discovered that the same two guns—a .38-caliber handgun and a 9-mm handgun— were used in multiple shootings in different parts of the city. In fact, the 9-mm gun has been linked to a dozen shootings over an 18-month period. Police believe that young shooters have learned that "if you have a gun that's already got a body on it, it's already linked to something, that becomes the throwaway gun you pass around to other kids, and that gun becomes very cheap . . . no one is ever going to have possession of it. It will always be in a basement or something for whoever needs it. But it won't be the gun you carry" (Latour).

Groundbreaking New Law

Effective October 1, 2000, for every new handgun sold in Maryland the gun's manufacturer must provide a spent shell casing from the gun to state police. The markings on the shell will be entered into an IBIS database so that police can quickly identify the gun from which a shell is found at a crime scene. A similar law took effect in New York on March 1, 2001. Critics of the law point out that there are no penalties for noncompliance, and gun dealers are free to sell handguns even if the manufacturers do not include shell casings with the guns. However, two gun-making companies, Smith and Wesson and Beretta, have indicated that they plan to comply with the law. *See also* Comparison Microscope; Crime Scene; Electron Microscope; Goddard, Calvin Hooker.

References

Biasotti, A. "A Statistical Study of the Individual Characteristics of Fired Bullets," *Journal of Forensic Sciences* 4(1), 133–140, 1959.

Evans, Colin. *The Casebook of Forensic Detection: How Science Solved 100 of the World's Most Baffling Crimes.* New York: Wiley, 1996.

Latour, Francie. "This Is the Gun," *Boston Globe*, Sept. 20, 2000, p. 1.

Within, George and Katie Helter. "High Tech Crime Solving." *U.S. News and World Report*, July 11, 1994, p. 30.

Benson, Steven (1959–)

Steven Benson, an American pipe-bomb murderer, was convicted in 1985 of killing his mother and an adopted brother, largely through the testimony of federal forensic scientists who were bomb experts. Mrs. Margaret Benson, a 63-year-old widow and heiress to a $10-million tobacco fortune, had family problems. One adopted son, Scott, a 21-year-old would-be tennis pro, had a heavy drug habit and was subject to violent, uncontrollable bouts of anger that often led to the abuse of his mother, sister, or anyone else in the vicinity. Steven, her 33-year-old older son, who apparently was the peacemaker and voice of stability in the family, had two failed marriages and a string of unsuccessful business ventures. In fact, his lavish lifestyle and a contract for an expensive home had raised questions of financial irregularities in his most recent business enterprise. Margaret Benson, as the chief financial investor in her son's enterprise, had recently ordered an audit of the company, a fact that was not lost on the struggling businessman. The two sons lived with their mother in the upscale Quail Creek development of Naples, Florida.

In the early morning of July 9, 1985, Margaret Benson, her two sons, and her daughter, Carol Lynn Kendall, piled into the family's Chevy Suburban to go see the land acquired for Steven Benson's new home. At the last moment, Steven, realizing that he had forgotten a tape measure, flipped the car keys to Scott and returned to the house. At 7:45 A.M., two explosions devastated the Suburban. Scott and Margaret Benson were killed instantly, but Carol Kendall was thrown from the vehicle and was aided by golfers who ran to the home from the adjacent course. Later, during the trial, Kendall recalled that her brother did nothing to help as she lay by the shattered vehicle.

By 11:30 A.M., Bureau of Tobacco, Alcohol, Tax and Firearms (ATF) specialists were at the home. ATF explosives expert Albert Gleason headed the investigation and began a systematic search of the charred Suburban by establishing a 25-foot cordoned-off area around the blast site. Government experts theorized that the explosion had been caused by two one-foot-long pipe bombs placed in the rear console area of the vehicle. Investigators recovered two four-inch pipe end caps stamped with the manufacturer's hallmark, which raised hope of tracing the buyer or seller. As the lawmen turned to the question of the triggering device, they found the remains of four D-cell batteries. Gleason believed that they were the remnants of an electronically controlled trigger, but there was little else at the site to reinforce his conclusion.

While explosives experts combed the rubble at the Benson home, investigators pursued the end-cap lead. They canvassed hardware stores, junkyards, plumbers, and construction sites to find the source of the caps. One Naples store had sold materials similar to the bomb days prior to the attack, and the clerk's description of the buyer was strikingly similar to Steven Benson.

Agents were temporarily stymied by the lack of fingerprints provided by evidence gathered at the site of the wreck. Investigators suggested checking the sales slip from the store. A close analysis revealed that Steven Benson's palmprint and fingerprints were on the paper. When police investigated Benson's business background, they discovered that he had been skimming from the family for a number of years. Consequently, the impending audit posed a real threat for him. The agents theorized that the plot to wipe out the whole family was designed to ensure that Benson would inherit the entire fortune. Benson was arrested on August 21 and charged with murder.

At the trial, Benson's defense team attempted to raise a reasonable doubt about his guilt by casting a drug-crazed Scott Benson as the culprit. The prosecution countered by presenting a nearly insurmountable amount of circumstantial evidence. Agent Gleason, citing the battery remnants and the later discovery of segments from a circuit board, testified that it was his belief that the ignition had not sparked the blast. He insisted that the bomb had been detonated electronically from outside the vehicle. Witnesses testified about Benson's electrical skill, his precarious financial state, and the fear of disinheritance if his illegal skimming from family accounts was uncovered. The fingerprints and palmprint on the sales receipt together with damning documentation from the prosecution's star witness, Benson's sister, only

added to the weight of evidence. After an 11-hour deliberation, the jury returned a guilty verdict recommending against the death penalty. Benson was sentenced to two consecutive life terms with no opportunity for parole before 50 years. He received an additional 37 years for attempted murder and arson.

References

Evans, Colin. *The Casebook of Forensic Detection: How Science Solved 100 of the World's Baffling Crimes.* New York: Wiley, 1996.

Greenya, John. *Blood Relations.* New York: Harcourt Brace Jovanovich, 1987.

Berkowitz, David (1953–)

David Berkowitz, an American arsonist turned serial killer, is an example of someone apprehended not by careful forensic analysis but through his own carelessness. He was convicted, however, by the use of forensic evidence. During a 13-month reign of terror in New York City beginning in 1976, Berkowitz killed six people and seriously injured seven others whom he shot at point-blank range while they sat in a car or on a porch. Berkowitz was dubbed "Son of Sam" by the media because of a letter he left at one crime scene. He wrote, "I am a monster. I am the Son of Sam love to hunt." Later, he wrote a similar note to newspaper columnist Jimmy Breslin in which he claimed to receive messages to kill from a barking dog. It was subsequently learned that the dog was owned by Sam Carr, a neighbor of Berkowitz. In an attempt to prove that he was insane, Berkowitz claimed that the dog was possessed by a demon who required humans to be sacrificed, and he (Berkowitz) was merely the demon's agent. He often killed his victim by shooting her through the window of the car where she was parked with her boyfriend. Sometimes he killed the boyfriend as well. Like many disorganized serial killers, he was a failure, a loner with a menial job who fantasized about sex and felt that he was on a mission. He told police, "The women I killed were filth. . . . I was just cleaning the place up a bit" (Cyriax, p. 201). In court, before being convicted, he saw the parents of one of his victims, Stacy Moscowitz, and shouted, "Stacy is a whore, Stacy is a whore. I'll shoot them all" (Cyriax, p. 202).

Moments after a double murder on June 26, 1977, a witness reported seeing a man running to a car that had been ticketed for illegal parking. When New York police checked their records, they found that the car belonged to Berkowitz. After finding the car, police discovered a .44-caliber pistol and a note on the front seat. The handwriting matched that on the note found by police at a murder scene and the letter sent to Breslin, and ballistics testing showed that the gun was the one used to kill Son of Sam's six victims. A serial killer had been captured because of a traffic violation, but it was the ballistics and document evidence that led to a conviction and a sentence of 365 years. *See also* Serial Killers.

References

Cyriax, Oliver. *Crime: An Encyclopedia.* North Pomfret, VT: Trafalgar Square, 1996.

Zonderman, Jon. *Beyond the Crime Lab: The New Science of Investigation.* New York: Wiley, 1990.

Bertillon, Alphonse (1853–1914)

Alphonse Bertillon, a French criminologist, is considered the father of criminal identification. He developed the first scientific system of personal-identification anthropometry. Anthropometry was based on the theory that the human skeletal system was fixed after age 20 and that no two individuals could have the exact same measurements. Consequently, Bertillon concluded that body measurements could be used as a means of identifying criminal offenders. There were five essential assessments in the system: the length and width of the skull, the length of the left forearm, and the size of the left foot and the left middle finger. Bertillon favored the left side because it was least likely to be affected by work. For two decades, Bertillonage was considered the most accurate identification

system until it was supplanted by the fingerprint method early in the twentieth century.

Bertillon was born in Paris to a respected and successful family whose members included a grandfather who was a renowned mathematician and naturalist. Physically Bertillon was pale, thin, sickly, and prone to a number of maladies, and though he was bright, he struggled scholastically and was expelled from boarding school for disciplinary reasons and fired as a bank clerk for incompetence. His position with the Paris police in 1878 was secured only by virtue of his family's prestige and the persistent efforts of his father, a physician who also was vice president of the Anthropological Society of Paris.

The 25-year-old patronage appointee was quickly relegated to mind-numbing clerical duties. One of these was the daily recording of officers' descriptions of the hoodlums they arrested. Arresting officers used these records and a photographic file to aid in the recognition of repeat offenders. Bertillon soon realized that the system of identification was useless. The descriptions were very general and more like notes intended to jog detectives' memories rather than provide comprehensive identification. The primitive nature of photography produced poor-quality photos, and the archives were not organized in any useful way. The entire French criminal-identification system relied primarily on the memory of individual detectives to determine who had been arrested previously for what and where. Familiar with the anthropological and anthropometric techniques of his father and brother (who pioneered statistics in the social sciences), Bertillon was convinced that there was a better alternative to the existing situation.

Bertillon particularly recalled the statistical theories of Dr. Adolphe Quetele, who believed that the chances of two comparable people having the same body measurement were 4:1. A second measure, reasoned Bertillon, would increase the odds to 16:1. With many measurements, the odds increased immeasurably. The bodily assessments with an organized photo file could make information readily available. Bertillon filed a report detailing his idea, but it was considered a joke and impractical. Once again his family's intercession and his own persistence got Bertillon's system a three-month trial. Since the concept required the acquisition of large amounts of data in order to make judgments, Bertillon's efforts seemed doomed from the beginning.

In February 1883, he got a break with the arrest of a man called Dupont, the favorite alias of Paris toughs. As he measured the man, Bertillon thought that he recognized him. He checked his files and discovered that the man had been arrested three months earlier using the name Martin. This first step in applying some precision to police work was viewed by most as sheer luck, but it did get Bertillon an extension for his trial.

In 1892, the struggle against anarchism, then the scourge of Europe, led to Bertillonage gaining credibility and secured Bertillon's place in forensic history. Since the 1880s, anarchists had wrought havoc in the world with terrorist acts like the assassination of Russia's Alexander II in 1881 or the Haymarket riot in the United States in 1886. France was not exempt from the anarchist assault. On March 3, 1892, there was an explosion at the home of a French judge who had presided at the trial of a number of anarchists the previous year. Responsibility for the bombing was attributed to an unidentifiable anarchist known as Ravachol. In the ensuing investigation, French police got a tip and arrested a suspect. Bertillon himself measured the man police believed was the culprit. His measurements revealed that the man was not an idealistic anarchist but an escaped criminal wanted for burglary, murder, and grave robbery. Bertillon's high-profile role in the case not only gave his system worldwide acceptance but earned him the Legion of Honor and a new position as head of the Bureau of Identification, an office he held until his death in 1914.

Bertillon's forensic fame from his identification system was relatively short-lived. As the twentieth century dawned, the growing prominence of fingerprinting as a method of identification and the decline of his reputation as a criminologist damaged an increasingly outmoded creation. Bertillon suffered two highly public failures of his supposedly foolproof system. The first mistake appeared in 1901 when twins Albert Ebenezer and Ebenezer Albert Fox could only be differentiated by the fingerprint technique. In 1903, a more difficult situation suggested the limitations of Bertillonage. Two convicts at the U.S. Penitentiary in Leavenworth, Kansas, presented identical measurements, had the same name, Will West, and even looked remarkably alike though there was no familial relationship. Once again, only fingerprints could convincingly distinguish the convicts.

Bertillon's personal failings provided more ammunition for his critics. He served as an expert witness for the prosecution in the infamous Dreyfus affair. In his role as a handwriting expert, Bertillon delivered damaging testimony about documents he determined were written by the defendant. They were later proven to have been written by another principal in the case. Worse, in 1913, the Mona Lisa was stolen from the Louvre. The only clues were fingerprints on the painting's case. Bertillon considered fingerprints as a mere supplement to his superior technique. He kept a file of prints but had not addressed the issue of devising a classification system and was unable to use the evidence from the crime scene. The painting was recovered, and, embarrassingly, the culprit had been arrested several times previously in Paris, most recently two years prior to the heist.

Beset by numerous physical maladies, Bertillon died at age 61, bitter at his perceived lack of appreciation. His classification system limped to oblivion shortly after his death. Though most of his ideas were eventually disproved, the field of forensic anthropology that developed from his concepts still thrives. *See also* Fingerprints; Forensic Anthropology.

References

Cyriax, Oliver. *Crime: An Encyclopedia*. North Pomfret, VT: Trafalgar Square, 1996.

Liston, Robert. *Great Detectives*. New York: Platt and Munk, 1966.

Wilson, Colin. *Colin Wilson's World Famous Murders*. London: Robinson, 1993.

Bertillonage/Bertillon System

From the mid-1880s through the turn of the century, Bertillonage, also known as the Bertillon system, was the standard method of criminal identification in most countries. The system, invented by Alphonse Bertillon, used measurements of body parts that Bertillon believed remained constant throughout adulthood. Such measurements included the length and width of the head and the right ear, the length of the left middle and little fingers, and the length of the left forearm and foot. Further, Bertillon determined that statistically no two humans could have the same dimensions the method assessed.

An integral part of Bertillon's scheme, though he considered it only supplemental, was the portrait parlé or description photo, a combination of two photos, front and profile, designed to record facial marks and characteristics. The profile photo centered on the right ear, the structure of which Bertillon believed could be used as another identifying feature. The pioneering use of photography in identification may be the most significant legacy of Bertillonage. Today's mug shot remains a staple part of any criminal investigation.

The system had its flaws. How could an investigator use the system in the field at a crime scene? It was impossible to remember the measurements of every suspect a policeman encountered, and even if an officer did recall some, he could not very well throw a citizen down and begin measuring for comparison. Second, the method required diligence, accuracy, and training if the comparative analysis was to be useful. Not all clerks had the time or inclination to be so involved in the process. The Will West case (see Fingerprints entry) marked the end of Bertillon's system and convinced authorities to switch to the more reliable fingerprint technique as a method of identification. *See also* Bertillon, Alphonse; Fingerprints; Galton, Sin Francis; Lombroso, Cesare.

References

Cyriax, Oliver. *Crime: An Encyclopedia*. North Pomfret, VT: Trafalgar Square, 1996.
Liston, Robert. *Great Detectives*. New York: Platt and Munk, 1966.
Wilson, Colin. *Colin Wilson's World Famous Murders*. London: Robinson, 1993.

Bite Marks

Bite marks are often caused by criminals who sometimes, in a fit of emotion, bite their victims and leave, not fingerprints, but bite prints. Even after bites heal, the marks can be made visible by shining ultraviolet light onto the area involved. The pigmented cells that form around the wounds during healing absorb ultraviolet light more than normal cells, providing a distinct contrast that can be photographed. Some forensic investigators believe that bite marks are not always recognized as such by medical examiners, since they often resemble bruises.

A number of cases have been solved by examining the impressions left by teeth. In one case, detectives found a wad of chewing gum with teeth marks at the scene of a murder. A forensic odontologist made casts of the impressions in the gum by applying silicone to the hardened gum. It was clear that the teeth that chewed the gum did not match the teeth of the victim, nor did they match those of one suspect; however, there was a close match to the impressions taken from another suspect. As confirming evidence, the forensic lab was able to capture enough saliva from the gum to determine that the chewer had type-AB blood, a type found in only 4 percent of the population, and one that matched the suspect's blood type. With this evidence confronting him, the suspect confessed to the crime.

The Gordon Hay Case

On the morning of August 7, 1967, the body of 15-year-old Linda Peacock was found in a cemetery in Biggar, Scotland. She had been struck with a blunt object and then strangled with a rope. Although she had not been raped, her blouse and bra had been disturbed, and her right breast revealed an oval-shaped bruise thought to be a bite mark. Forensic odontologist Dr. Warren Harvey and pathologist Professor Keith Simpson agreed that the bruises were caused by teeth and that one of the teeth was particularly irregular.

The investigation focused on boys in a nearby detention center because Peacock had been seen talking to one of the young men the evening she was killed. Impressions were made of the teeth of a number of suspects and compared with photographic transparencies of the bite marks. The teeth of 17-year-old Gordon Hay matched the bite marks. He had pitted upper and lower right canines with sharp edges that could have produced the ragged marks left on Linda Peacock's breast. The unusual canines were the result of a disorder known as hypocalcination. The bite marks on the breast were upside-down, which correlated with the mud on the knees of the suspect's pants and other forensic evidence indicating that the victim had been strangled from behind. Apparently, Hay, had while kneeling behind his victim, had bent over her shoulder to bite her breast. The jury found the bite mark evidence convincing, and Hay was convicted of murder. Because of his age, he was sentenced "to be detained during her Majesty's pleasure" (Evans, p. 150).

The Theodore Bundy Case

Prior to dawn on January 15, 1978, a club-carrying masked man entered the Chi Omega house on the Florida State University (FSU) campus in Tallahassee, killed two young women, and battered two others. He left little evidence except for a bite mark on one of the dead victim's buttocks. A month later, a man was arrested for stealing a motor vehicle. Though he was using an assumed name, records revealed that he was Theodore Bundy, a convicted felon suspected of being the serial killer who had committed dozens of sex murders throughout the northwestern United States. Bundy had escaped from prison two weeks before the killings at the sorority house at FSU. He had been convicted of abducting

a woman who escaped and later identified him in a lineup, but police had been unable to tie him to the serial murders. When a detective searched Bundy's apartment for fingerprints, there were none, because Bundy had carefully wiped them all away.

After making impressions of Bundy's dentition, forensic dentist Dr. Richard Souviron prepared solid casts of the suspect's teeth. Before the jury, Souviron demonstrated how a transparency of a photograph of Bundy's front teeth matched a same-size photograph of the bite marks on the victim's buttock. For the first time, a jury had clear evidence that Bundy was a murderer. He was convicted on July 23, 1978, and was executed more than a decade later. During that decade, he hinted at, but never confessed to, killing 40 to 50 women, most of whom were young and attractive, with long hair they parted in the middle. The victims of serial killers are often very similar in appearance. In this case, Bundy's victims were similar in appearance to his wife, who had once rejected him.

The Carmine Calabro Case

On Columbus Day 1979, the naked body of Francine Elveson, a 26-year-old teacher, was found on the roof of her apartment building in New York City. She had been strangled and mutilated, and the killer had left bite marks on the victim's thighs. A single pubic hair, identified as Negroid, led police to suspect that a black man was the culprit. With bite marks as a substantial clue, police asked the building's residents to submit impressions of their teeth. None matched the bite marks on Francine Elveson's body. After a months-long search without any success, local police asked the FBI for a profile of the killer. The profile indicated, among other things, that it was likely that the criminal lived in the building, had a serious mental problem that had required treatment, and was probably white.

If the killer was white, what was the source of the Negroid pubic hair? A careful investigation found that the bag used to carry the victim's body to an autopsy had previously been used to carry a murder victim who had

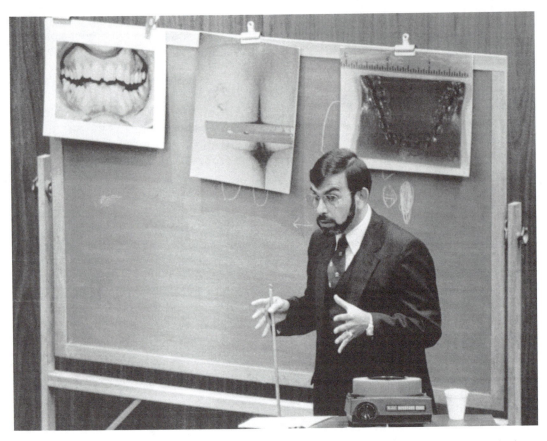

Dr. Lowell J. Levine, a New York forensic ondontologist, testifies that the bite marks found on the buttocks of a Florida State University coed reflect characteristics of Ted Bundy's teeth. Photos of Bundy's teeth and bite markes are displayed on the chalk board behind Dr. Levine during the 1979 trial in Tallahassee, Florida. © Bettmann/CORBIS.

been black. Failure to clean the bag thoroughly must have left some debris from the dead black man.

The police now resumed their search with an emphasis on white, not black, suspects. No one living in the building had been treated for psychoses, but one man had a son, Carmine Calabro, who was being treated at a nearby hospital. Police found that security at the hospital was minimal. Calabro could have left the facility, committed the crime, and returned without anyone noticing his absence. Calabro agreed to a request for an impression of his teeth. A cast made from the impression revealed that Calabro's teeth provided a perfect match for the bite marks left on Francine Elveson's body. He was found guilty and was sentenced to life in prison.

The Irene Kennedy Case

Although bite marks have proved to be convincing evidence, the case of Irene Kennedy reveals the dangers involved in accepting such evidence with absolute confidence. On December 1, 1998, Irene Kennedy, 75, and Thomas Kennedy, 78, began their early-morning walk in Francis William Bird Park in Walpole, Massachusetts. As was their custom, they walked a short distance and then separated to follow different paths before reuniting at the end of their strolls. But Irene Kennedy never rejoined her husband. She was brutally murdered, and her nearly naked body was left covered with bite marks.

Police dogs brought to the crime scene led police to the nearby home of Edmund

Burke, the eccentric brother of the Kennedys' son-in-law. Burke was questioned and was asked to submit samples of blood, saliva, fingerprints, palmprints, and dental impressions. After carefully examining Burke's dental impressions and the bite marks on Irene Kennedy's body, forensic scientist Dr. Lowell Levine reportedly told police that the bite marks on the body were, with reasonable scientific certainty, made by Edmund Burke. On the basis of the available evidence, Burke was arrested on December 10 and charged with murder.

Eight days later, the results of tests comparing the DNA in Burke's saliva with that collected from the bites on Irene Kennedy were released. The DNA samples did not match, but Burke was not released. A month later, a bloody palmprint found on Irene Kennedy's thigh did not match the print of Burke's palm. Burke was released on January 20. DNA and palmprints, two types of evidence more conclusive than bite marks, had shown that Burke was not Irene Kennedy's murderer. The crime remains unsolved. *See also* Bertillon, Alphonse; Crime Scene; Fingerprints; Footprints; Odontology; Palmprints; Psychological Profiling; Serial Killers.

References

Cyriax, Oliver. *Crime: An Encyclopedia*. North Pomfret, VT: Trafalgar Square, 1996.

Evans, Colin. *The Casebook of Forensic Detection: How Science Solved 100 of the World's Most Baffling Crimes*. New York: Wiley, 1996.

Jarriel, Tom, Connie Chung, and Jack Ford. "Mark of a Killer." *ABC 20/20*, Sept. 20, 1999.

Ramsland, Katherine. "Bite Marks as Evidence to Convict." Crime Library. 2000. http://www.crimelibrary.com/forensics/bitemarks/4.htm. Accessed August 21, 2001.

Zonderman, Jon. *Beyond the Crime Lab: The New Science of Investigation*. New York: Wiley, 1990.

Blochman, Lawrence (1900–1975)

Lawrence Blochman was an author of mystery stories whose knowledge of forensic pathology gave his writing a strong scientific basis. He wrote a series of stories about Dr. Daniel Webster Coffee, a pathologist-sleuth who was able to solve difficult cases involving mysterious deaths by using his scientific knowledge.

Blochman graduated from the University of California at Berkeley in 1921 and 30 years later received a certificate in forensic pathology from the Armed Forces Institute of Pathology. He worked abroad in several branches of the U.S. government. He was a member of the Overseas Press Club and the Mystery Writers of America. Being fluent in French, he translated several books from French into English. His work abroad gave him the ability to locate his stories in many different parts of the world. His study leading to a certificate in forensic pathology from the Armed Forces Institute of Pathology provided him with the valuable information that permeated his writing. He used this knowledge to his advantage in the 1960s when Dr. Coffee was featured in a television series, *Diagnosis Unknown*, starring Patrick O'Neal. Blochman also wrote a number of other scripts for radio, television, and films, as well as other mystery fiction that did not include Dr. Coffee.

References

Penzler, Otto, Chris Steinbrunner, and Marvin Lachman, eds. *Detectionary*. Woodstock, NY: Overlook Press, 1977.

Steinbrunner, Chris, and Otto Penzler, eds. *Encyclopedia of Mystery and Detection*. New York: McGraw-Hill, 1976.

Blood

Blood is a key piece of evidence at a crime scene. Such sites are often bloody or marked by blood stains. If such evidence is carefully collected and preserved, it can be analyzed in a forensic laboratory and may provide valuable information. Liquid blood provides the best evidence. Often samples can be drawn from both victim and suspect. Liquid out-of-body blood found at a recent crime scene can be transferred to a test tube by using an eyedropper and can be transported under refrigeration to a lab.

Bloodstains, rather than liquid blood, are

more commonly found at crime scenes. Investigators try not to expose such stains to heat, moisture, bacteria, fungi, or anything that can change the chemicals in the stain. The stains are then photographed and their precise location recorded. Stains are frequently found on the clothes of a victim or suspect. Towels and handkerchiefs may also reveal stains left by a criminal who wiped his hands of blood. Bloodstains may be evident in cracks and crevices in the floor or walls after they were wiped or washed to remove a victim's blood.

If there are spattered bloodstains, an expert will probably be called in to examine the scene. It may be possible to re-create the crime on the basis of the pattern of blood drops. From the appearance of the dried drops, an expert can determine the angle and speed at which the drops struck the surface. From the size, shape, and position of spattered blood, forensic scientists can often determine the angle and force of the blow or bullet that struck a murdered or severely injured victim.

If possible, each stain is preserved by cutting away the surface on which it is located and bringing it to the lab. If the surface cannot be cut, the stain can be scraped onto a moist swab, placed in a manila envelope or paper bag, and taken to a laboratory for analysis. Samples are transported in closed containers because there may be moisture that will enable mold or other organisms to grow on the dried blood and change its chemistry.

Serology

Technically, serology is the study of blood serum, but it has come to mean the laboratory tests used to analyze specific antigen-antibody reactions associated with blood, particularly those used to establish blood types. The first thing to determine about a stain is whether or not it is blood. This can be done by mixing the particles in the stain with phenolphthalein and hydrogen peroxide. A deep pink color indicates that the material is blood. (Some vegetable matter such as potatoes and horseradish will also give a positive reaction, but the likelihood of mis-

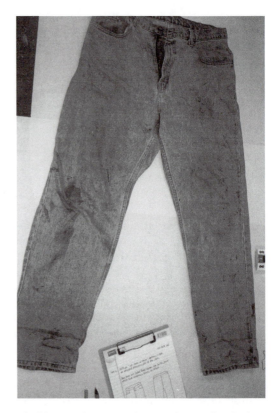

Blood-stained pants and a mandatory sketch that accompanies evidence and helps an investigation. Courtesy of the Division of Forensic Science, Commonwealth of Virginia, Department of Criminal Justice Services.

taking such matter for bloodstains is minimal.) The particles may also be tested with luminol. In the presence of dried blood, the luminol will emit light. To be seen clearly, the luminol test should be conducted in darkness. Often luminol will enable forensic scientists to find stains on clothing and other articles that were not initially visible.

The second step is to establish whether or not the blood is human. The precipitin test is used for this purpose. Animals, usually rabbits, are injected with human blood. The rabbit produces antibodies specific for the antigens found in the blood of a particular species. (Antigens are proteins that cause the body to form antibodies that react with the antigen.) These antibodies can be recovered from the rabbit's blood serum (the light yellow fluid that remains after blood has clot-

Table 1
The Four Blood Types and the Antigens on Their Red Blood Cells and the Antibodies in Their Blood Serum

Blood Type	Antigens on RBC	Antibodies in Serum
A	A	Anti-B
B	B	Anti-A
AB	A and B	Neither antibody
O	Neither antigen	Anti-A and anti-B

ted). For example, a human antiserum can be obtained that will precipitate the antigens in human blood but not the blood of other animals. The test is so sensitive that it can detect human blood in stains as old as 15 years.

Finally, the blood should be typed to see if it can be associated with the blood of a specific suspect. That humans may have two kinds of blood antigens, A and B, on their red blood cells (RBC) was established by Karl Landsteiner in 1901, but it was 1930 before he received the Nobel Prize for his work. These two antigens found on red blood cells cause the formation of antibodies that will react with the antigens.

An individual's red blood cells may contain one, or both, or neither of the two antigens. Blood, therefore, can be one of four types—A, B, AB, or O. As can be seen from Table 1, a person with type-A blood has the A antigen on his or her red blood cells; a person with type-B blood has the B antigen; a person with type-AB blood has both antigens; and a person with type-O blood has neither antigen. Blood serum may contain antibodies that react with the A or B antigens, causing the blood cells to clump together (agglutinate). Agglutination is readily seen through a microscope. Consequently, the blood type can be readily established. The antibody that reacts with the A antigen is called anti-A; the antibody that reacts with the B antigen is called anti-B. As can be seen in Table 1, a person with type-AB blood has neither antibody. If he or she did, his or her antibodies would react with the antigens on his or her own red blood cells, causing agglutination.

To find out what blood type a person has, a small amount of his or her blood is placed on each of two glass slides. A drop of blood serum containing the anti-A antibody is added to one drop; a drop of serum containing the anti-B antibody is added to the other drop. Either anti-A or anti-B will cause type-AB blood to clump. On the other hand, a person with type-O blood carries both antibodies, but neither antigen on his or her red blood cells. Therefore, the blood cells will not clump when blood serum with either antibody is added. Someone with type-A blood has the A antigen and the anti-B antibodies. His or her red blood cells will clump when anti-A serum is added but not when anti-B serum is added. Type-B blood, which contains the B antigen, will clump when anti-B serum is added but not when anti-A serum is added (Table 2).

Blood types can also be identified by adding red blood cells of a known type to blood samples of unknown type in order to determine the type of antibodies present in the samples. Agglutination or lack of it with the two types of red blood cells will reveal the presence or absence of antibodies in the unknown blood. A summary of possible results and what they indicate is shown in Table 3.

Once the type of blood found at the scene of a crime has been determined, that information may or may not be useful. Suppose that the blood type matches that of the victim and that of the blood found on a knife

Table 2
Typing Blood by Adding a Known Antiserum to the Blood in Question

Anti-A Serum + Blood Sample	Anti-B Serum + Blood Sample	Antigen on RBC	Blood Type
+	−	A	A
−	+	B	B
+	+	A and B	AB
−	−	Neither	O

Note: + indicates agglutination; − indicates no agglutination.

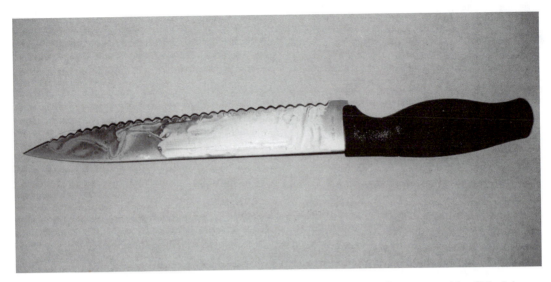

Knife with blood evidence. Courtesy of the Division of Forensic Science, Commonwealth of Virginia, Department of Criminal Justice Services.

in the possession of a suspect. Then the knife could have been the one used in the crime. However, this is not positive proof. On the other hand, if the blood type or types on the knife do not match that of the victim, then the knife was not used on the victim, or if it was, it was cleaned before being used on someone else.

The percentages of the people in the United States who have the four blood types are found in Table 4. Type O is the most common, type AB the rarest, and the frequency of each type is related to race.

Table 3
Typing Blood by Adding a Known Antiserum to the Blood in Question

RBC of Type-A Blood + Blood Sample	RBC of Type-B Blood + Blood Sample	Antibody Present	Blood Type
+	−	Anti-A	B
−	+	Anti-B	A
+	+	Anti-A and anti-B	O
−	−	Neither antibody	AB

Note: + indicates agglutination; − indicates no agglutination.

Finding type-AB blood at the scene of a crime can be very useful in reducing the list of suspects, whereas type-O blood would likely eliminate only half of the suspects.

The blood types A, B, AB, and O are not the only ones used by crime labs. Human blood contains other antigens for which tests have been developed. Crime labs routinely test for Rh (+ or −), M and N, and Le antigens as well as a number of other blood enzymes. By using all of these tests, investigators might clear a high percentage of suspects of any wrongdoing. Unlike fingerprints, blood-type evidence cannot provide proof of guilt. It can only reduce the number of individuals who might be guilty of a crime. However, suppose that a suspect has type-AB blood and is Rh negative. Only 15 percent of the population is Rh negative. The prob-

Table 4
Percentages of People in the United States Who Have Each of the Four Blood Types

Blood Type	Whites (%)	Blacks (%)
O	45	49
A	40	27
B	11	20
AB	4	4

Table 5
Identification of Blood Types by Means of Absorption-Elution

Antibodies and Type of Red Blood Cells Added to the Blood Stain		Antigen That Was on the Stain	Blood Type in the Stain
Anti-A + Type-A RBC	*Anti-B + Type-B RBC*		
+	−	A	A
−	+	B	B
+	+	A and B	AB
−	−	Neither A nor B	O

Note: + indicates agglutination; − indicates no agglutination.

ability of having both type-AB and Rh-negative blood is 0.04 × 0.15 = 0.006, or 6 people in a 1,000. If other blood factors are tested and matches are found between the suspect's blood and the blood found at the scene of the crime, then the probability of guilt may increase to a point where a jury will convict.

Typing Bloodstains

As blood dries, the red blood cells rupture. Consequently, it is impossible to look for agglutination because there are no cells to clump together. Forensic scientists overcome this difficulty by using what is known as the absorption-elution technique. The first step in this technique is to add known antiserum (either anti-A or anti-B antibodies) to a sample of the stain and allow time for the antibodies to react with the antigens that remain in the dry blood. Excess antibodies are then washed away before the antigens and antibodies are separated by elution (heating the material to 56°C), which breaks the bonds between them. Finally, the eluted antibodies are mixed with red blood cells of a known type, and the blood is examined under the microscope to see if agglutination occurs. The identification of the A-B-O type is summarized in Table 5.

Bloodstains can be tested in similar ways for Rh, M and N, Le, and other blood enzymes. Furthermore, about 4 out of 5 people are secretors. This means that their blood antigens are also found in other body fluids such as saliva and semen, so these fluids can be tested for antigens as well. With the recent advances in DNA testing, bloodstain analysis is not used as frequently as it used to be. DNA evidence is usually more definitive in identifying an individual.

Inheritance of Blood Type and Paternity

Blood types are inherited. The genes that cause the production of the A or B antigens on red blood cells are dominant to the gene that produces neither antigen. Since chromosomes are paired, a person carries two genes for blood type. The genes for transmitting blood types can be represented by the letters **A, B,** and **o**. The lower case for **o** is used here to indicate that the gene is recessive. The capitalized **A** and **B** indicate codominant genes. Both are dominant to **o** but not to each other. If both the **A** and **B** genes are present, both A and B antigens will be on that individual's red blood cells. A person's genotype (the genes the individual carries for a particular trait) may be represented by the genes shown in Table 6.

Table 6
Genotypes and Gametes for A-B-O Blood Types

Individual's Blood Type	Possible Genotypes	Possible Genes in Individual's Gametes
A	AA	only A
	Ao	A or o
B	BB	only B
	Bo	B or o
AB	AB	A or B
O	oo	only o

Note: The genes for type-A and type-B blood are dominant to the gene for type-O blood. If both A and B genes are present, both antigens will be present on red blood cells.

Cases

Blood types played an important role in a famous case beginning in 1943 that involved an accusation by a young actress, Joan Barry, who claimed that Charlie Chaplin, the well-known actor-comedian, was the father of her child. The blood tests, which were monitored by a three-doctor panel, showed conclusively that Chaplin was not the child's father. Chaplin, who was type O, and Barry, who was type A, could not have conceived a child who was type B because neither of them carried the gene for type-B blood. Nonetheless, Barry's lawyers pursued their efforts to have Chaplin declared the father of Carol Ann Barry. In 1953, California passed a law that upheld paternity blood tests and would have led to the dismissal of the case against Chaplin, but this was of no help to the accused in the 1940s. The jury ignored the blood tests and found Chaplin responsible. Two appeals failed to change the conviction. It would appear that Chaplin was found guilty of violating conventional morality even though the forensic evidence provided by the blood tests proved that he was not the father of Barry's child.

Early on the morning of October 24, 1983, in Sheffield, England, 18-year-old Nicole Laitner was awakened by a man who had already murdered her parents and brother by repeated stabbings and slashings. After raping her several times, the killer-rapist tied Laitner to her bed and left. She managed to free herself and called the police. The bedrooms of Laitner's parents and brother were covered with the blood of the victims. Laitner had not been cut, but there was blood on her bed and her nightgown, which led forensic scientists to believe that the blood belonged to the killer. Analysis revealed that the combination of antigens in the sample was rare, occurring in only 1 in 50,000 people. It was also the same kind of blood found on barbed wire that had been left by an escaping rapist in nearby Selby a month earlier.

The suspect in the earlier rape was Arthur Hutchinson, who had two previous convictions for sex crimes. Police put out media reports stating that the wire on which the suspect had been cut had been covered with a chemical that could cause gangrene. When Hutchinson checked into an emergency room, he was recognized and captured. At the trial, the evidence included not only the blood statistics but bite marks on cheese at the Laitner home that matched his dentition, a bloody footprint that matched his shoe, a matching palmprint on a bottle, and Laitner's identification of him as the man who had attacked her. He was found guilty of psychopathic murder and rape and sentenced to three consecutive life terms.

In 1990, forensic scientist Herb LeRoy entered the room of a home in Surrey, British Columbia, where a woman had been shot. The shot had caused her head to literally explode, sending blood and bits of flesh outward in all directions to the walls. She was holding a rifle in her hands. It was a bolt-action gun, so she could not have fired twice. However, there were two holes in the door behind her. Despite the two holes and rumors of discord between the woman and her husband, LeRoy was certain that it had been a suicide because there was no shadow pattern on the wall. Had a killer been present, the blood flying outward would have hit him, leaving a "shadow" on the wall much like paint from a spray can aimed at a wall with a hand in front of the surface. But why two bullet holes?

Beyond the door with the two holes, an investigator found a copper-jacketed slug in the wall. A hole in a nearby window indicated that the actual lead-laden bullet had gone outside. LeRoy surmised that a single shot was fired, but when the slug hit the woman's head, it broke into two parts—the copper-jacketed slug and the lead part of the bullet. Both "bullets" then passed through the door and beyond.

In June 2000, jurors in the case of Matthew Mirabal visited the canyon north of Boulder, Colorado, where the body of his wife, Natalie Mirabal, had been found. They also toured the parking lot of a supermarket where the victim's car was found. Matthew Mirabal was charged and later convicted of the strangulation and decapitation murder of his wife. Shortly before her death, Mirabal had taken out a quarter-million-dollar life-insurance policy on her life. Forensic expert Joe Clayton from the Colorado Bureau of Investigation testified that DNA tests indicated that bloodstains on gloves found in the abandoned car matched the blood of both Natalie and Matthew Mirabal. He had found, as well, bloodstains on a pair of Matthew Mirabal's trousers during a search of his apartment. There were also bloodstains on Natalie Mirabal's left shoe and on the handle of Mirabal's car. DNA test results linked the bloodstains from the evidence to Mirabal. *See also* Chaplin, Charlie: Paternity Case; Crime Scene; DNA Evidence; Forensic Photography; Immunoassay; Lattes, Leone.

References

Barrett, Sylvia. *The Arsenic Milkshake and Other Mysteries Solved by Forensic Science.* Toronto: Doubleday Canada, 1994.

Evans, Colin. *The Casebook of Forensic Detection: How Science Solved 100 of the World's Most Baffling Crimes.* New York: Wiley, 1996.

Hansen, Peter. "Queensland's Own Quincy." *Sunday Mail* (Brisbane, Australia), Sept. 3, 2000, p. 40.

James, S.H., and W.G. Eckert. *Interpretation of Bloodstain Evidence at Crime Scenes.* Boca Raton, FL: CRC Press, 1999.

MacDonell, Herbert L. *Bloodstain Pattern Interpretation.* Corning, NY: Laboratory of Forensic Science, 1983.

"Matthew Mirabal Gets Life in Prison for Killing His Wife." *Daily Camera.* http://www.thedailycamera.com/extra/topten00/mirabal.

McCullen, Kevin. "Mirabal Jurors Visit Mountain Crime Scene Wife Was Strangled and Dumped in Canyon." *Denver Rocky Mountain News,* June 21, 2000, p. 26A.

Within, George and Katie Helter. "High Tech Crime Solving." *U.S. News and World Report,* July 11, 1994, p. 30.

Boston Strangler. *See* DeSalvo, Albert.

Breathalyzer

The Breathalyzer, invented by Robert F. Borkenstein in 1954, is an instrument used to measure blood-alcohol levels. It is widely used by police to test people suspected of driving while intoxicated. Drivers can refuse to take the test, but if they do, they automatically lose their license for a designated time.

In *Schmerber v. California,* a case tried in 1966, a blood sample was obtained from Anoncenda Schmerber, by order of the police, while he was being treated in an emergency room for injuries sustained in an accident. On the basis of his blood-alcohol concentration, Schmerber was arrested for drunk driving. Schmerber had objected to having the blood drawn and sued the police, arguing that taking the blood sample violated his right to avoid self-incrimination as stated in the Fifth Amendment of the U.S. Constitution ("No person . . . shall be compelled in any case to be a witness against himself"). On appeal to the U.S. Supreme Court, the Court maintained that the Fifth Amendment only prohibits requiring a suspect to give self-incriminating testimonial evidence. Providing physical evidence, such as fingerprints, photographs, and blood samples, is not prohibited by the amendment.

Ingested ethyl alcohol (C_2H_5OH) is absorbed through the stomach and small intestine into the bloodstream and transported to the cells of the body. In the lungs' alveoli, alcohol diffuses from the blood into the air in the lungs as oxygen diffuses in the opposite direction. The ratio of the concentration of alcohol in blood to alcohol in air is 2,100:1; that is, 1 milliliter (ml) of blood holds as much alcohol as 2,100 ml of air.

When a subject blows into a Breathalyzer, the machine traps 52.5 ml of the air. This volume is $\frac{1}{40}$ of 2,100 ml; consequently, the machine will measure the alcohol in $\frac{1}{40}$ ml of blood. The trapped lung air from the subject is forced into a dilute acidified solution of potassium dichromate, where it is oxidized to acetic acid. The reaction reduces the color intensity of the purplish potassium dichromate. The change in the absorbance of light passing through the colored solution can be measured with a photocell. The change in light intensity provides a measure of the concentration of alcohol in the air that was blown into the machine. The signal generated by the photocell is amplified and used to move a needle on a gauge calibrated to read percentage of blood alcohol. In 18 states, 0.08 percent alcohol by weight is considered the level at which a person can be considered to be under the influence of alcohol. Some European countries, however, regard levels as low as 0.03 to be indicative of a person who is unfit to drive. On October 3, 2000, a House and Senate panel agreed to give all states two years to define 0.08 as the legally drunk level. States that do not conform will be denied federal aid for highway construction. To reach a blood-alcohol level of 0.08 would require a 170-pound man to consume five 12-ounce beers over a period of two hours. A 120-pound woman would reach this level by drinking three 12-ounce beers over the same time period.

Newer devices to measure the alcohol in breath use a wavelength of infrared light that is absorbed by alcohol vapor. The light passes through a sample of the subject's breath, and the change in absorbance is measured with a photocell. Current from the photocell provides a signal to a microprocessor that produces a digital readout of the blood-alcohol percentage.

Breathalyzers may also be connected to a car's ignition system so that a driver whose blood alcohol is over the legal limit will be unable to start his or her car. To activate the ignition, the driver has to breathe into the Breathalyzer. The instrument measures his or her blood-alcohol concentration and blocks the car's ignition system if the alcohol is above the allowed limit.

In the case *Commonwealth v. Charles E. Smith IV* (Appeals Court of Massachusetts, Suffolk, 1993), Charles E. Smith had been convicted of operating a motor vehicle negligently so as to cause the death of two Boston University students and then leaving the scene. In an appeal, the defense argued that the Breathalyzer used to collect information about Smith's blood alcohol had not been tested periodically as required by law. Furthermore, the Breathalyzer test, administered two hours after the accident, showed a level of .06 percent alcohol by weight, which is below the .08 percent regarded as indicating

intoxication. The prosecution had used retrograde extrapolation to determine that Smith's blood alcohol was .09–.10 at the time of the accident. Police officials provided documentation showing that the Breathalyzer had been tested periodically, and forensic experts testified that retrograde extrapolation was not uncommon and that data as to the rate of alcohol metabolism had been tested repeatedly by experimentation. The previous judgment was affirmed. *See also* Automobile Accidents.

References

Brook, Paula. "The Ones Who Get Away: Drunks at the Wheel." *Chatelaine*, Dec. 1, 1996, p. 72.

Commonwealth v. Charles E. Smith IV. Appeals Court of Massachusetts, Suffolk, 1993.

Lane, Brian. *The Encyclopedia of Forensic Science*. London: Headline, 1992.

Saferstein, Richard. *Criminalistics: An Introduction to Forensic Science*. Upper Saddle River, NJ: Prentice Hall, 1998.

Bruises

Bruises, which result from the breaking of small blood vessels beneath the skin, may provide valuable evidence to a coroner or medical examiner investigating a crime or accident. The examiner will look closely for such marks in the neck area, which may have been caused by a strangler's hands, a rope, or another article used to shut off the flow of air to and from the lungs. Bruises in the neck area as well as the groin and other parts of the body are commonly found in rape victims. Elderly people and those taking drugs to reduce blood viscosity will bruise more readily than others.

Initially, a bruise will be reddish or purple in color. With time, the color becomes brownish, then green, and eventually yellow before it disappears. The time for a bruise to heal varies with the individual; however, bruises in various stages might offer evidence that a battered child, spouse, or partner has undergone numerous beatings at the hands of a batterer over time.

It is difficult to tell whether recently inflicted bruises on a corpse occurred before or after death because blood vessels can be broken by battering after death as well as before, and some discoloration may be the result of lividity. Lividity, the discoloration of the skin resulting from gravity's action on blood, begins shortly after death. A person who dies lying on his back will show bruiselike discoloration in the small of the back and on the back of the neck and thighs. A prone corpse will show similarly colored skin on the anterior parts of the body.

During an autopsy, the pathologist may look at the white-blood-cell count in blood at the site of bruises. A high concentration of these cells indicates that the bruising occurred before death and the body responded by sending white blood cells to the area because these cells are involved in the repair of damaged tissue.

In manual strangulations, the victim's hyoid bone is usually broken, and there are bruises about the neck left by thumbs and fingers. There will be bruising, too, in the muscles of the neck and, possibly, in the tongue, lower mouth, epiglottis, and lining of the larynx. Such bruises indicate that the strangulation was the result of homicide, not suicide or accident.

On October 16, 1929, 63-year-old Rosaline Fox was dragged from her hotel room, which was on fire, by a traveling salesman who was in a nearby room. Meanwhile, her son, Sidney, who emerged from an adjoining room, watched the tragedy from a safe distance. Rosaline Fox was pronounced dead as a result of shock and suffocation by a doctor at the scene.

Immediately after his mother's burial, the son was at her insurance company filing his claim on a policy he had purchased. A suspicious agent asked for a more thorough investigation. The case was ultimately referred to Scotland Yard. Rosaline Fox's body was exhumed, and forensic scientist Sir Bernard Spilsbury was called upon to perform a postmortem autopsy. Finding no carbon monoxide in the blood or sooty deposits in the lungs, Spilsbury concluded that Rosaline Fox had died prior to the fire. He detected

bruises at the back of the larynx indicating manual strangulation as the cause of death. Sidney Fox was eventually convicted of murdering his mother.

On the morning of August 7, 1967, Linda Peacock's body was found in a cemetery in Biggar, Scotland. She had been hit with a blunt object and strangled with a rope. She had not been raped, but her right breast revealed an oval-shaped bruise thought to be a bite mark. Forensic odontologist Dr. Warren Harvey and pathologist Professor Keith Simpson agreed that the bruises were caused by teeth, including one that was very irregular. Dental impressions were made of the teeth of several suspects. When photographs of the bite-mark bruises were compared with the impressions, it was found that the teeth of 17-year-old Gordon Hay matched the bruise marks on the victim. The jury found the evidence convincing, and Hay was convicted of murder. Because of his age, he was sentenced "to be detained during her Majesty's pleasure" (Evans, p. 150).

Lack of bruises can lead to an acquittal. In a 1998 New York City trial, a woman claimed that she had been tied up, beaten, and sexually assaulted with a nightstick by a Columbia University graduate student. Doctors who examined the victim said that they could find no bruises to indicate that she had been beaten or abused sexually. What she claimed to be bruises were normal skin coloration, the doctors testified.

Investigators could clearly see the bruises on the body of 8-year-old Joey Torres, found dead in his bed on June 26, 2000, in Hobe Sound, Florida. His mother, Tammy Huff, said that her son was clumsy and accident prone, but the medical examiner concluded that the boy had died of massive internal injuries. Police found marks from four fingers and a thumb on the boy's left arm that matched his mother's hand. Scratches in the boy's genital area were made by adult-size fingernails. Detectives confiscated a belt, a sandal, and tools from the kitchen of a trailer shared by Huff and her boyfriend, Bradley Dial. They also collected powders and creams that were believed to have been used to conceal the boy's terribly bruised groin region. Dial claimed that Huff was responsible for the child's death, but both were arrested and held without bail. Their trial was set for spring 2002. *See also* Crime Scene; Time of Death.

References

Lane, Brian. *The Encyclopedia of Forensic Science.* London: Headline, 1992.

Taylor, Jill, and Pat Moore. "Bruises Led to Charges for Couple." *Palm Beach Post,* July 28, 2000.

Buchanan, Edna (1937–   )

American-born Edna Buchanan is an award-winning writer of crime-related fiction and nonfiction. For 17 years during the 1960s and 1970s, she was the crime reporter for the *Miami Herald.* Her knowledge of how real crimes are solved or not solved is evident in her books. Buchanan uses her reporter's experiences traveling the late shift with the police, visiting the city morgue, or surveying the police crime lab to make her books realistic and accurate in every detail. Many of the situations experienced by her fictional character, Britt Montero, a female crime reporter, are similar to those Buchanan herself has encountered as a reporter. For example, Buchanan had received threatening letters from a rapist she had interviewed. In one of her books, Britt Montero is physically intimidated by a rapist. In her book *Garden of Evil,* Buchanan uses her knowledge gained from interviews with serial killers to write about a female serial killer who kidnaps Montero during an interview.

Buchanan's first two books were nonfiction. The first one, *Carr: Five Years of Rape and Murder,* was published in 1979. It was the story of a rapist who had killed many people throughout the country. Buchanan had interviewed him in prison several times, and her book gives an in-depth look at the criminal. The book has been used as a teaching tool by the FBI. The second book, *The Corpse Had a Familiar Face,* was published in 1987. It contains a series of chapters on different subjects related to her job as a crime reporter. Some chapters are about her personal life at the time, some are about interesting cases she had covered, and some involve personal notes about the people she encountered.

Buchanan says that one of the reasons she enjoys writing is that she gets to solve cases that in real life were never solved. Some cases from her days on the police beat still haunt her because no matter how much evidence was uncovered, some murders were never solved and some missing persons have never been found. Buchanan likes the fact that she can solve all these cases through fictional characters in her books, where justice always prevails. As she told Charles Silet during an interview, "Real life doesn't make sense, but fiction does."

References

Buchanan, Edna. *The Corpse Had a Familiar Face.* New York: Random House, 1987.

Silet, Charles. "An Interview with Edna Buchanan." http://www.mysterypages.com/buchananiv1-9.html.

Buchanan-Sutherland Murder (1892)

The Buchanan-Sutherland murder was one of New York City's most famous crimes, with the sensational Buchanan trial itself being enough to earn Robert W. Buchanan a place in the annals of forensic science. Originally an unlicensed practitioner in Canada, Buchanan left the country and moved to Chicago for a couple of years, only to return to Halifax, Nova Scotia. Back in Canada, he married the daughter of a wealthy manufacturer, who financed the completion of his medical studies in Scotland. In 1887, Buchanan and his wife moved to New York City, where he established a modest practice in Greenwich Village. The practice slowly disintegrated as a result of rumors about Buchanan's roguish lifestyle.

In 1890, Buchanan divorced his wife and sent her back to Halifax. About the same time, the doctor began to pay attention to Anna Sutherland, a woman almost twice his age. Though Sutherland lived in Greenwich Village, she ran a profitable brothel in Newark, New Jersey. Sutherland's most appealing quality to the doctor was her bank account. With her savings and the investment in Newark, she was worth about $50,000, a large sum in 1890. When the doctor proposed marriage, the older woman jumped at the chance for respectability and acceptance.

The marriage was not a happy one, and in 1892 Buchanan announced that he was going to Scotland alone. Four days before he was scheduled to leave, Sutherland became very ill, and Buchanan canceled his trip. Buchanan called a doctor to treat his wife, but it was too late. On April 23, 1892, the 18-month marriage ended with Anna Sutherland Buchanan's death. The attending physician, Dr. B.C. McIntyre, cited the cause of death as apoplexy.

The late Anna Buchanan's friends from New Jersey did not accept McIntyre's decision, especially a friend identified only as Mr. Smith. In May 1892, Smith showed up at the coroner's office with a ringing indictment of Buchanan as a murderer. Smith asserted that the doctor had killed his wife for her money. He insisted that he had independent corroboration of his suspicions. The authorities paid little attention to Smith because he was a shady character attacking a solid citizen. Present at the time of Smith's tirade was Ike White of the *New York World*, who was interested in Smith's allegations. (White's 1943 obituary heralded him as the greatest crime reporter in American history.) Why would a good-looking young man of Buchanan's prospects marry a much older prostitute? White had just achieved fame for his investigative reporting in the celebrated Carlyle Harris murder trial, and he sensed that this case could be just as rewarding.

White began his inquiry with an interview of Dr. McIntyre. White asked if the physician had considered morphine poisoning in his cause-of-death evaluation. McIntyre replied that the major symptom of morphine poison-

ing, pinpoint pupils, was not evident. Further, he maintained that apoplexy produced the same symptoms without the pupil contraction. Finally, McIntyre added that he knew of no drug that would conceal the narrowing of the pupils. White kept digging and turned up some interesting facts that further sparked his curiosity. He discovered that Buchanan had canceled his Scotland trip 10 days prior to his wife's death, not 4 days, as he claimed. A Canadian reporter told an inquiring White that on May 16, 1892, the "grieving" doctor had remarried his first wife and would be returning to New York, where it was rumored that the doctor was coming into some money.

White collected most of the important evidence against Buchanan, and none was more devastating than what he picked up at the doctor's favorite bar. At Mr. Smith's suggestion, White interviewed two men who frequently drank with the physician. Both men told the journalist that the marriage had been characterized by fighting. Most often Sutherland had shouted that she would not give her husband any money unless he respected her as a wife. For his part, Buchanan had complained of his wife's morphine habit, stating that she would certainly die soon if it continued. Sutherland's friends claimed that she had not used morphine, and White found no one who had ever seen her using the drug. Most telling were the men's recollections about conversations concerning the Harris case. Buchanan had told them that he knew the medical student's mistake and that he could get away with murder if he so chose.

White put all this information together and added a memory of his own. In recalling his school days, the reporter remembered a number of children with eye problems going to the doctor and getting drops of belladonna to dilate their pupils before an eye examination. Belladonna or atropine was the key to White's hypothesis. He interviewed Anna Sutherland's nurse, who confirmed that prior to her death Buchanan had put drops in his wife's eyes.

Faced with White's story, the authorities ordered the body exhumed, and a series of toxicological tests were run. Dr. Rudolph Witthaus, of Harris-trial fame, did the postmortem and established two significant facts. Anna Sutherland had not died of a stroke; it was morphine that had claimed her life. Buchanan was arrested and charged with murder.

The trial captured the popular imagination through a dramatic courtroom demonstration. A cat was killed with morphine, and the court noted the distinctive symptoms, narrowing of the pupils. When atropine was dropped in the animal's eyes, the pupils dilated. The toxicological success was followed by Buchanan's inept attempt to testify on his own behalf. He became entangled in a number of contradictions and lies and was simply not credible. It took the jury only 28 hours to return a guilty verdict. The case lasted two more years in the appeals courts, but in the end, the initial verdict was upheld. On July 2, 1895, Robert W. Buchanan was executed at New York's Sing Sing Prison. *See also* Atropine; Harris, Carlyle; Opium.

References

Cyriax, Oliver. *Crime: An Encyclopedia*. North Pomfret, VT: Trafalgar Square, 1996.
Thorwald, Jürgen. *Proof of Poison*. London: Thames & Hudson, 1966.

Bundy, Theodore (1946–1989)

U.S. serial killer Theodore Bundy was sentenced to death on the basis of bite marks he inflicted on a victim who was a member of Florida State University's Chi Omega sorority. Like many organized serial killers, Bundy was an "intelligent" man who carefully planned his murders and tried to avoid leaving any incriminating evidence at the scenes of his crimes. Bundy was a handsome, charming fellow who was well educated. He had been studying law but left law school probably because he was afraid (and probably rightly so) that he was a suspect in a series of murders being committed in the Seattle area.

Ironically, but not uncharacteristic of organized serial killers, he was employed for some time by the Seattle Crime Prevention Advisory Commission and had once chased a mugger to make a citizen's arrest.

As is true of many serial killers, he lived an apparently normal life, usually with a woman to whom he was engaged. His victims were attractive young women who strongly resembled the young woman he had married and then rejected. They were young and attractive, with long, dark hair parted in the middle. His wife believed that his rejection of her stemmed from a desire to punish her for breaking off with him earlier in their relationship (Evans, p. 22).

It was early in 1974, shortly after his wife left, that the rapes and killings began. In July of that year, he is believed to have killed two young women on the same day at Lake Sammamish Park near Seattle. He had a knack for gaining the confidence of women. Often he would have a fake cast on one arm so he could persuade them to help him with a task that required two arms. Once he was alone with his victim, he would often handcuff her and take her to a secluded place where he would rape and kill her.

In August 1975, after several more murders, Bundy was stopped in Colorado for a traffic violation. There he was identified by Carol DaRonch as the man who had tried to abduct her on November 8, 1974, at a shopping mall in Salt Lake City. Strands of hair taken from the vehicle were examined microscopically and found to match those taken from two young women—Melissa Smith, who had been murdered late in 1974, and Caryn Campbell, who had been killed early in 1975. Bundy was convicted only of attempting to abduct DaRonch. Two decades later, DNA from the hair would have matched that from the remains of the victim, and Bundy would probably have been convicted of murder.

Bundy escaped from the Aspen, Colorado, courthouse by simply jumping out the window after being left alone. He was captured a few days later, but shortly after Christmas

1977, he escaped from prison by squeezing through a loose ceiling panel and crawling along the overhead space to a room without bars from which he made his way to freedom.

Following his escape, the killings began again, but in the southeastern rather than the northwestern United States. In the early hours of January 15, 1978, Bundy entered the Chi Omega house on the Florida State University (FSU) campus in Tallahassee, where he killed two young women and battered two others. However, he left evidence in the form of a bite mark on the buttocks of Lisa Levy, one of his victims. A month later, a man using the name Chris Hagen was arrested while driving a stolen motor vehicle. Records revealed his true identity to be Theodore Bundy, the convicted felon suspected of being the serial killer who had committed dozens of sex murders throughout the northwestern United States and now the murders of two Chi Omega sisters at FSU.

Bundy, an organized serial killer, was careful not to leave evidence. When a detective searched his Tallahassee apartment for fingerprints, there were none. Bundy had carefully wiped them all away. In response to the bite-mark evidence, forensic dentist Dr. Richard Souviron prepared solid casts of Bundy's teeth. Before the jury, Souviron demonstrated how a transparency of a photograph of Bundy's front teeth matched a same-size photograph of the bite marks in the victim's buttock. He was convicted of murdering Lisa Levy on July 23, 1978, and was finally electrocuted in January 1989, more than a decade later. During that decade, he hinted at, but never confessed to, murdering 40 to 50 young women. It was reported that when someone asked him if he had killed 36 women, he replied, "Add a one." No one knows whether he meant that

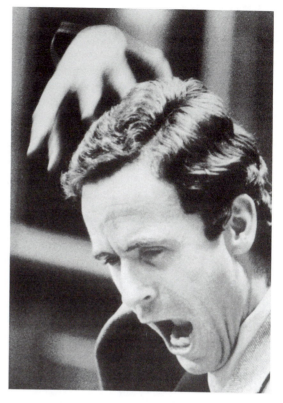

In 1979 convicted murderer Ted Bundy had this reaction after the judge left the courtroom following a jury recommendation that Bundy die in the electric chair. The jury had earlier found Bundy guilty of murdering two Florida State University Chi Omega sorority sisters. © Bettmann/CORBIS.

the number was 37, 136, or 361. *See also* Bite Marks; Odontology; Serial Killers.

References

Cyriax, Oliver. *Crime: An Encyclopedia.* North Pomfret, VT: Trafalgar Square, 1996.

Evans, Colin. *The Casebook of Forensic Detection: How Science Solved 100 of the World's Most Baffling Crimes.* New York: Wiley, 1996.

Wilson, Colin, *The Mammoth Book of True Crime.* New ed. New York: Carroll & Graf, 1998.

C

Camps, Francis (1905–1972)

Francis Camps, a British pathologist involved in many controversial cases in which the forensic evidence was inconclusive, was part of what one newspaper reporter called the "Three Musketeers"—Keith Simpson, Camps, and Donald Teare—all well-known British forensic specialists who did not always agree. In one unusual case in 1953, all three pathologists were involved.

Timothy Evans had been executed in 1950 for the murder of his pregnant wife, Beryl, and his one-year-old daughter, Geraldine. The murders had occurred in November 1949. However, another man, John Christie, confessed to the murder of Beryl Evans when he confessed to six other murders. The public then demanded to know if an innocent man had been executed.

Christie stated that he had gassed Beryl Evans and had then strangled her as he had done in the other murders. Teare had performed the original postmortem examination on Beryls Evans' body in 1949. In the first examination, he had found no indication of carbon monoxide poisoning, which gives a characteristic pink coloration to the body. He had noted a black eye, a bruised lip, and bruises on her thigh, lower leg, and vagina.

Timothy Evans confessed to the murder and said that he had hit his wife before he strangled her. Later, he revoked his confession and said that Christie had murdered her. The bruise on the vagina could have indicated forced intercourse, or it could have been caused by a syringe, as there were allegations that Beryl Evans had tried to abort her fetus. Teare had not taken a vaginal swab in the first examination because the police had told him not to bother because Timothy Evans had confessed to the crime. Christie was using an insanity defense, so he probably thought that seven murders would sound more insane than six. All the other murders involved sexual intercourse with the victims when they were either dead or unconscious from gas, so if semen had been found in Beryl, it could have indicated that Christie did kill her.

Camps wrote *Medical and Scientific Investigations in the Christie Case* in 1953. In the book, he wrote that though there was no indication of carbon monoxide in the tissue at the second autopsy, the teeth showed some pink coloration. He tested them for carbon monoxide and found none; however, he still argued that the pink teeth could indicate carbon monoxide poisoning, suggesting that Christie was the murderer.

Simpson, in his book *Forty Years of Murder*, disagreed. He argued that if Christie had gassed Beryl as he alleged, there would have been some pink coloration of the tissues, and all three pathologists at the second autopsy had agreed there was no pink coloration. The issue of who killed Beryl Evans was never solved, but most individuals involved believed that Timothy Evans, not Christie, was the murderer.

Camps also edited *Recent Advances in Forensic Pathology*, which dealt with the issue of establishing time of death. Isabel Le Bourdais wrote a book about a Canadian murder where the time of death was one of the key pieces of evidence, *The Trial of Steven Truscott*. Camps and his constant rival Keith Simpson both read the book and came to completely different conclusions. In the Truscott case, the conditions of the stomach contents of the murder victim were used to establish the time of death, and Camps disagreed with the conclusion reached by the medical examiner. His opinion was heard in court at the appeal of Truscott's conviction in 1966. In this 1959 case, Lynne Harper, a 12-year-old girl, was raped and murdered. A 14-year-old boy, Steven Truscott, was charged with the murder based on circumstantial evidence. He had been seen with Harper at 7:00 P.M. on the night of the murder, and the medical examiner had estimated the time of death as 7:30 P.M. Truscott had graze marks on his penis that could indicate forcible rape. There were grass stains on his pants even though they had been washed, and Harper's body had been found in a grassy area. The medical examiner, John Penistan, had estimated the time of death as 7:30 P.M. based on the condition of the contents of Harper's stomach. Penistan reasoned that death had occurred within a couple of hours after Harper's last meal at 5:30 P.M. since most of the food was still in her stomach and only partially digested.

Simpson believed that Penistan's estimation of the time of death was correct. Along with Simpson, two other well-known pathologists, Dr. Milton Helpern, New York's Chief medical examiner, and Dr. Samuel Gerber from Ohio, believed that the conclusion reached by Penistan was the right one. Camps, on the other hand, thought that death could have occurred anywhere between 1 and 10 hours after Harper's last meal, which left plenty of time for someone else to have abducted, raped, and murdered Harper, as Truscott's defense had alleged. Camps based his belief on the facts that people have different metabolic rates, and various factors can affect the digestion of food. Fear and injury have been found to hinder digestion. The court, however, did not think that this was sufficient reason to give Truscott a new trial, and the original decision remained in effect. *See also* Helpern, Milton.

References

Evans, Colin. *The Casebook of Forensic Detection: How Science Solved 100 of the World's Most Baffling Crimes.* New York: Wiley, 1996.

Thorwald, Jürgen. *The Century of the Detective.* New York: Harcourt, Brace & World, 1965.

Wilson, Colin. *Clues! A History of Forensic Detection.* New York: Warner Books, 1991.

Capone, Al. *See* St. Valentine's Day Massacre.

Carbon Monoxide

Carbon monoxide is a colorless, odorless gas produced during the incomplete combustion of carbon-containing compounds, such as the gasoline in automobile engines. If there are suspicions of murder or suicide at the scene of a fire, forensic scientists commonly check the carbon monoxide levels in the blood of victims. Murderers often attempt to cover their crime by burning their victim's home or car to make it appear as though fire was the cause of death. Lack of carbon monoxide in the blood is convincing evidence that the victim did not breathe the vapors produced by the combustion and was dead before the fire started.

It was the absence of carbon dioxide in the lungs of Sidney Fox's mother that led investigators to realize that her murder was not accidental. On October 16, 1929, Fox and his mother Rosaline Fox checked into the Metropole Hotel in Margate, England. Later that evening, fire broke out in their adjoining rooms. Naked except for his shirt, Fox burst into the hallway shouting a warning. A heroic traveling salesman, at first repulsed by the dense smoke, managed to drag the unconscious Rosaline Fox from the fire; however, attempts to resuscitate her were futile. A doctor at the scene initially reported the cause of death to be shock and suffocation. This conclusion was reinforced when an inquest determined that the death was accidental.

On the afternoon of his mother's funeral, Fox was at the insurance company filing his claim. A suspicious agent, aroused by Fox's lack of grief, asked the main office to investigate. When Rosaline Fox's body was exhumed, the renowned Sir Bernard Spilsbury performed a postmortem. Finding no carbon monoxide in her blood and no sooty deposits in her lungs, Spilsbury concluded that Rosaline Fox had been killed prior to the fire. He noted bruising at the back of the larynx, which indicated that she had been strangled. This was a very controversial conclusion because the body presented none of the usual physical signs of strangulation, most notably a fracture of the hyoid bone in the neck. Spilsbury was adamant, and, given his forensic reputation, the decision took on nearly infallible proportions. Despite his denials, Fox was charged with murder and convicted.

Carbon monoxide is a major air pollutant in cities where automobile traffic is heavy. Its concentrations in urban areas may reach 40 parts per million, 95 percent of which is from the exhaust of automobile engines. The gas is a relatively unreactive substance; nevertheless, it is a very poisonous gas and one that is frequently encountered in forensic-science laboratories. After being inhaled into a person's lungs along with air, it, like oxygen, combines with the hemoglobin in red blood cells to form carboxyhemoglobin. (Oxygen combines with hemoglobin to form oxyhemoglobin.) Unfortunately, hemoglobin has a significantly greater affinity for carbon monoxide than it does for oxygen. As a result, much of the hemoglobin that would normally be available to carry oxygen to body cells becomes tied up with carbon monoxide. When about half the hemoglobin is bound to carbon monoxide, the body essentially suffocates from lack of oxygen. Death is preceded by drowsiness and headache.

The concentration of carbon monoxide in blood can be determined by spectrophotometry or by using a reagent that releases the gas from a known volume of blood. The carbon monoxide released can then be measured using gas chromatography. Although death is common when 50 percent of the hemoglobin has been converted to carboxyhemoglobin, a level 10 to 15 percent less may be fatal if the victim has a high level of alcohol in his or her blood. People who smoke may normally have 5 to 10 percent of their hemoglobin bound to carbon monoxide.

Breathing automobile exhaust fumes is a common method of suicide. A car's engine may be allowed to run in a closed garage, or a hose may be used to carry fumes from the car's tailpipe to its interior. If someone is found where carbon monoxide poisoning is suspected, such as a garage where a gasoline engine is running, that person should be moved immediately into fresh air.

In the case *Western Alliance Insurance Company v. Jarnail Singh Gill and Others* before the Supreme Judicial Court of Massachusetts in 1997, Western Alliance Insurance Company maintained that its policy did not cover Katya Fels, who had been permanently damaged by carbon monoxide poisoning while dining at the India Gate Restaurant in Cambridge, Massachusetts. The company felt that a pollution exclusion clause exempted it from paying damages. An expert witness testified that the high level of carbon monoxide in the restaurant was the result of operating a kitchen fan while all doors and windows were closed. As a result, a tandoori oven "starved for air" emitted carbon monoxide rather than carbon dioxide. The court held that the restaurant owners could not have anticipated such an event and ruled that the company should pay. *See also* Atropine; Chromatography; Heinrich, Edward Oscar;

Opium; Poisons; Spectrophotometry; Strychnine.

References

Chang, Raymond. *Chemistry*. 2nd ed. New York: Random House, 1984.

Saferstein, Richard. *Criminalistics: An Introduction to Forensic Science*. Upper Saddle River, NJ: Prentice Hall, 1998.

Western Alliance Insurance Company v. Jarnail Singh Gill and Others. Supreme Judicial Court of Massachusetts, Middlesex, 1997.

Cause of Death

Cause of death is one of the first things that must be determined in any death that may be suspicious. Ultimately, death is the result of damage to the brain, heart, lungs, or blood vessels (loss of blood), and in some cases the cause of death is quite evident. The corpse may have a bullet wound to the head, heart, or a major artery, or there may be a knife in the heart or aorta. In many cases, the cause of death is not obvious, but there is suggestive evidence. There is a red line around the neck, or the trachea is broken, indicating that the person was strangled. Perhaps the skull is fractured and the body bruised, indicating that a blunt instrument was used to strike fatal blows.

Surrounding circumstances may indicate cause of death. A body in a car with the motor running in a closed garage suggests carbon monoxide poisoning, which can be determined by analyzing a sample of the person's blood. It may also indicate suicide if the person was free to escape, or murder if the blood concentration of carbon dioxide is low, indicating that the victim died before the motor was started.

A charred body in a building that has burned will be carefully examined by a coroner to see if the blood contains carbon monoxide—the result of smoke inhalation. If there is no evidence of a high level of carbon monoxide, the person was probably killed before the fire. The fire may have been set to cover up a murder.

A body found in water may have drowned. If that is the case, a fine frothy foam will fill the respiratory system and the stomach will hold water swallowed during the drowning process. In addition, the blood on the left side of the heart will be diluted as a result of water that seeped in from the lungs. The heart's blood will not be diluted if the victim died prior to entering the water.

An autopsy of a drowning victim will reveal microscopic plants (diatoms) in the body tissues. These organisms enter the lungs during attempts to breathe, enter the blood, and are carried to tissues throughout the body. If the victim died before entering the water, diatoms will be found in the lungs but not elsewhere.

In still other cases, an apparently young and healthy body may reveal no obvious cause of death. An autopsy may disclose evidence of a stroke or heart failure. A toxicologist will examine organs and tissues for evidence of poison or drugs. In some cases, the cause of death may never be determined; in others, such as the murder of Georgi Markov, where ricin was believed to be the poison used, the cause may be assumed but never proven.

In one of the longest delays to determine cause of death, the remains of Captain Charles Hall were exhumed from a frozen grave in 1968. Hall, captain of the *Polaris*, which had anchored off the northwest coast of Greenland in the autumn of 1871, was on an expedition to find the North Pole. After he returned from a scouting trip in late October, he drank a cup of coffee and soon after became ill. Reports of animosity between Hall and the ship's physician, Dr. Emil Bessels, who served the coffee and treated Hall after he became ill, led some to believe that Hall had been poisoned. Despite the fact that Hall's symptoms, as reported by shipmates, were typical of someone suffering from arsenic poisoning, an inquiry, held after the crew had returned home, concluded that Hall had died of natural causes.

In the summer of 1968, pathologists Franklin Paddock and Professor Chauncy Loomis found Hall's grave and his reasonably well preserved corpse. Paddock removed samples of tissue and sent them to two separate laboratories for evaluation, including neutron-activation analysis. The results showed that the tip of Hall's fingernail held arsenic at a concentration of about 25 parts per million (ppm), which was comparable to the concentration of arsenic in the soil around Hall's grave. At the base of the nail, however, the concentration was more than 75 ppm. The conclusion was that Hall had received massive doses of arsenic during the last two weeks of his life.

A victim's complete body is not always needed to establish cause of death. Following the disappearance of Helle Crafts in 1986, police became convinced that she had been bludgeoned by her husband, Richard Crafts, who then froze her body, cut it into smaller sections, and ran it through a woodchipper on the shore of Connecticut's Housatonic River. They reached this conclusion on the basis of mounting evidence. A stain on the mattress in the master bedroom was found to be blood of the same type as Helle Crafts's. A passerby reported seeing someone using a woodchipper on the shore of the river one November night a few days after Helle Crafts's disappearance. Divers found a discarded chain saw, its numbers filed away, with bits of flesh and cloth fibers between its teeth. Restoration of the numbers revealed that it had been purchased by Richard Crafts. Credit-card receipts for a freezer, new bedding, and a rented woodchipper, together with bits of flesh found in the trunk of a car, as well as pieces of teeth, teeth caps, bones, and flesh found along the banks of the Housatonic, led forensic investigators to establish the fate of Helle Crafts.

On June 5, 1986, 52-year-old Bruce Nickell suddenly died after taking two tablets of Extra-Strength Excedrin. An autopsy indicated pulmonary emphysema as the cause of death. On June 11, Sue Snow, a 40-year-old bank manager, called 911 after taking two Extra-Strength Excedrin. She died at the hospital several hours later of an unknown cause. During an autopsy, an assisting physician noticed the faint odor of bitter almonds, suggesting cyanide poisoning.

Laboratory tests confirmed the presence of a fatal concentration of cyanide in her blood.

On June 17, Nickell's widow, Stella Nickell, called police to ask if her husband's death might be related to the contaminated Excedrin that had been reported in the Seattle area where the Nickells and Sue Snow lived. Although Bruce Nickell had been buried, he had volunteered his organs, and, therefore, a sample of his blood had been preserved. A test showed that his blood did indeed contain cyanide, as did the two bottles of Extra-Strength Excedrin capsules in the Nickells' medicine cabinet.

Analysis of the capsules also revealed the presence of an algicide used to destroy algae in fish tanks. Analysts concluded that whoever had mixed the cyanide and Excedrin had done so in a container that had held the algicide. Police had noticed a fish tank in the Nickells' home, and investigators began inquiring at local pet stores to see if Stella Nickell had purchased algicide recently. One clerk readily identified her as having made such a purchase. Police also discovered that while the Nickells were deeply in debt, Stella Nickell had bought an additional $40,000 policy on her husband's life should he die accidentally. Furthermore, as a state employee, his life was insured for $31,000, with an additional $105,000 should he die accidentally. Death by contaminated pills would be considered an accidental death; death from emphysema would not be accidental. Stella Nickell would receive an extra $145,000 if her husband's death was by accident. When she agreed to a polygraph test, she failed, but a failed polygraph test would not bring a conviction.

When Stella Nickell's estranged daughter contacted police, she told them that her mother had often talked of killing her husband, but a good defense attorney would attribute such testimony to a disgruntled daughter seeking revenge. However, the daughter also told police that her mother had done research on poisons at a number of libraries. When detectives went to the libraries, they learned that Stella Nickell had read a number of books on cyanide. In fact, she had never returned one entitled *Human Poisoning*. Another book she had read, *Deadly Harvest*, was sent to the FBI Crime Lab. There 84 of her fingerprints were lifted from the pages. The prints were primarily located on the pages that described cyanide poisoning.

On May 9, 1988, Stella Nickell was convicted and sentenced to 90 years in prison. She had been so intent on proving that her husband died "accidentally" that she was willing to take the lives of others as well as her husband's by adding cyanide to bottles of Excedrin capsules and planting them in a pharmacy in order to establish a cause of death that suited her purpose.

According to the coroner, the slaying of JonBenet Ramsey in December 1996 was the result of strangulation. There were deep cuts and bruises around her neck. However, a blow to the head that left a skull fracture 8½ inches long and 4 inches wide could also have been the cause of death. The person who murdered this child remains at large. *See also* Arsenic Poisoning; Carbon Monoxide; Coroner; Crime Scene; Cyanide; Drowning; Lung Flotation; Medical Examiner; Neutron Activation Analysis; Pathology; Poisons; Toxicologists; Toxicology.

References

Evans, Colin. *The Casebook of Forensic Detection: How Science Solved 100 of the World's Most Baffling Crimes.* New York: Wiley, 1996.

Lane, Brian. *The Encyclopedia of Forensic Science.* London: Headline, 1992.

Wilson, Colin. *Clues! A History of Forensic Detection.* New York: Warner Books, 1991.

Chaplin, Charlie: Paternity Case (1943–1946)

Movie star Charlie Chaplin, born in England but active in Hollywood, especially in silent films, was known not only for his box-office appeal as a comedian but also for his three failed marriages and a long list of relationships with young women. One such relationship led to a blood test to determine the paternity of an unwed woman's baby. The results of that test revealed that forensic science, even when it is conclusive, does not always lead to a just verdict.

The young woman who proclaimed that she was pregnant, out of wedlock, with Chaplin's child was an aspiring actress named Joan Barry. The announcement and ensuing newspaper frenzy led to a series of trials that went on from 1943 to 1946 and forced Chaplin into a self-imposed exile from the country where he enjoyed his greatest success.

Rich and successful, Chaplin made no secret of his appreciation of attractive women. The 54-year-old Chaplin met the 22-year-old Barry at a dinner party in June 1941. According to all accounts, Barry was very aggressive in her pursuit of the comedian before the two began a quiet love affair. By the fall of 1942, Chaplin felt that Barry had some promise as an actress, and he signed the would-be starlet to a $75-a-week contract, paid for some dental work, and sent her to acting school. As the year progressed, Chaplin was slow to recognize a mental deterioration in his protégé. Barry stopped attending acting classes and began to show up at Chaplin's home, where she displayed erratic and irritating behavior. On one occasion, she wrecked her Cadillac in the star's driveway. On another, she began smashing windows when Chaplin refused to open his door. Barry finally admitted that she had no desire to continue to pursue an acting career; she wanted to return to her home in Brooklyn. Chaplin obliged by releasing her from her contract, gave her $5,000, and provided tickets for both her and her mother to return home.

In December 1942, Barry returned to Hollywood to confront her estranged lover. She broke into Chaplin's home waving a gun and threatening to commit suicide while announcing that she was pregnant and demanding help. Chaplin responded by having her arrested for unlawful entry. Upon her release, the jilted woman told her story to Hollywood gossip columnist and longtime Chaplin adversary Hedda Hopper. Following Hopper's column about the incident, Chaplin was roasted by the press. Chaplin, involved in a relationship with 17-year-old Oona O'Neill, whom he planned to marry, denied any responsibility for Barry's child. He admitted that there had been a relationship, but he insisted that the child was not his.

In June 1943, Barry sued Chaplin, claiming that he had fathered the child. Because the child's blood could not be tested before it was four months old, a temporary settlement was reached in which Chaplin agreed to pay for Barry's living and medical expenses until blood tests could determine paternity. Should the tests prove that Chaplin was not the father, it was agreed that the suit would be dropped.

Meanwhile, Chaplin faced other legal issues and concerns about his reputation that could affect his ability to attract moviegoers. Chaplin, aside from his romantic dalliances, made an inviting target for the press. The actor was not an American citizen and expressed no interest in becoming one, despite his financial success in the United States. He was also considered sympathetic to leftist political causes, which only added to the distrust of the comedian.

In February 1944, the U.S. government indicted Chaplin for violating the Mann Act, a statute designed to fight commercial prostitution that made illegal the interstate transportation of women for immoral purposes. The government alleged that Chaplin paid for Barry's ticket to New York City with immoral purposes in mind. Barry testified that while Chaplin had been in New York to give a speech, he had renewed their relationship at a local hotel, a charge that could have sent Chaplin to prison for 25 years along with a fine of $25,000. While all this furor was going on, Barry gave birth to a baby girl she named Carol Ann.

At the Mann Act trial, Chaplin admitted meeting Barry in New York at her insistence. He stated that the meeting lasted about 30 minutes, that the two were with a mutual friend throughout the meeting, and that there had been no sexual contact. Chaplin's lawyer made the telling point that there was

no need for Chaplin to send Barry to New York to prostitute herself when the woman would have willingly given herself to Chaplin at any time or place. On April 4, 1944, the jury found the comedian not guilty. The acquittal, however, only intensified the growing public anti-Chaplin sentiment as the movie star entered the paternity trial.

The blood tests, which were monitored by a three-doctor panel, showed conclusively that Chaplin was not Carol Ann's father. Chaplin, who was type O, and Barry, who was type A, could not have conceived a child who was type B. Nonetheless, Barry's lawyers pursued their efforts to have Chaplin declared the father of Carol Ann. In 1953, California passed a law that upheld paternity blood tests and would have led to the dismissal of the case against Chaplin, but this was of no help to the accused in 1945. After the first trial in 1945 ended with a hung jury, the trial judge ordered another trial.

In the next trial, the plaintiff's counsel, lacking scientific evidence, resorted to an emotional bludgeoning of the jury. Barry sat in the front of the courtroom every day holding her baby, while her lawyers assaulted Chaplin to the point that he was once reduced to tears. He was portrayed as a libertine, a Svengali who seduced young girls, and a degenerate. As the trial drew to a close, Barry's defenders exhorted the jury to compare and acknowledge the common facial features of Chaplin and Carol Ann. The jury ignored the blood tests and found Chaplin responsible. Two appeals that extended the case for another year both failed to change the conviction. It would appear that Chaplin was found guilty of violating conventional morality, for the blood tests made it clear that he was not the father of Barry's child.

As the press continued its unrelenting attack, Chaplin's next film was widely boycotted, and the government renewed its attack by charging him with tax evasion. In September 1952, Chaplin, his wife Oona, and their child left the United States. The U.S. attorney general promptly rescinded the actor's reentry permit, and the Chaplins settled in Switzerland. In 1972, at the age of

82, Chaplin returned to the United States to be honored for his contributions to moving pictures as an actor and comedian. Five years later he died. In 1953, Joan Barry was diagnosed as a schizophrenic and was committed to a California state mental facility. Carol Ann disappeared, and what happened to her is unknown. *See also* Blood.

References

Chaplin, Charles. *My Autobiography.* New York: Simon & Schuster, 1964.

Conleir, Joseph R. "The Charlie Chaplin and Joan Barry Affair—1943." *The People's Almanac #2,* ed. David Wallechinsky and Irving Wallace. New York: Bantam Books, 1978.

Robinson, David. *Chaplin: His Life and Art.* New York: McGraw-Hill, 1985.

Cherrill, Frederick R. (1892–1964)

British detective Frederick R. Cherrill Scotland Yard's leading fingerprint expert in the 1930s and 1940s. In 1931 Cherrill investigated a case involving palmprint evidence. The criminal, faced with a palmprint that matched his own, pleaded guilty. As a result of his plea, the palmprint evidence was never used in court. Eleven years later, Cherrill provided palmprint evidence that led to the first conviction in Britain based on such evidence. Leonard Moules, an elderly pawnbroker, was fatally injured in his shop from severe blows to the head, probably from the butt of a gun. Cherrill found one palmprint inside the shop's safe that was not Moules's or his assistant's. A witness had seen two men in the area with a gun at the time of the crime; he knew them only as Sam and George. Investigators were able to find a George Silverosa in the area whose palmprint matched the one found in the safe. Silverosa turned in his accomplice Sam Dashwood, and the two men were hanged for the murder.

In another case, Cherrill's fingerprint evidence helped catch a murderer. In 1943, a naked woman's body was found in a sack. No physical evidence was found with the body, and her face had been beaten so badly that it was unrecognizable. Three months later, in early 1944, a piece of a black coat was found in some trash. The coat had a dry cleaner's tag that led the police to the home of Bertie Manton. Manton claimed that his wife had left him after an argument three months earlier but that he had heard from her in letters sent from Hampstead that he showed to the investigators. Cherrill took fingerprints from the apartment. He noted that all items had been carefully wiped down as if someone had tried to remove all traces of fingerprints. Finally, Cherrill found a pickle jar that had a single thumbprint that matched the murder victim's. Investigators looked more carefully at the letters supposedly written by Manton's wife. In the letters, Hampstead had been misspelled as "Hamstead." Asked to give a handwriting sample, Manton misspelled Hampstead in the same way. Confronted with these pieces of evidence, Manton confessed.

Colin Wilson notes that in Cherrill's autobiography, *Fingerprints Never Lie: The Autobiography of Fred Cherrill,* he talks about the similarity between fingerprints and handwriting. Cherrill states that both fingerprints and handwriting possess intricate details that can be detected and used to identify an individual. In the Manton case, both forensic areas were utilized to elicit a confession from the killer. *See also* Disputed Documents; Fingerprints.

References

Lane, Brian. *The Encyclopedia of Forensic Science.* London: Headline, 1992.

Wilson, Colin. *Clues! A History of Forensic Detection.* New York: Warner Books, 1991.

Chloral Hydrate

Chloral hydrate ($C_2Cl_3H_3O_2$), a crystalline solid made by the addition of water (H_2O) to chloral (C_2Cl_3HO), is popularly known as a "Mickey Finn," or "knockout drops," when it is mixed with alcohol. Although it is one of the oldest soporifics used in medicine and is often prescribed for insomnia or as a daytime sedative, it also can be fatal.

A Mickey Finn will cause anyone who drinks it to become unconscious. Although it is seldom referred to as chloral hydrate, it is a common theme in movies and stories involving crime. It has also played a role in real-life criminal activity. Criminals in bars have used it to drug and then rob their victims. In some cases, the victims died from an overdose or from adverse effects of the substance on someone who had heart, liver, or kidney ailments. It was also commonly used by bartenders to silence aggressive customers who they feared might start a fight.

Mickey Finns are sometimes used as a date-rape drug. However, Rohypnol, also known as "roofies," "rope," or "roach," a drug related to Valium, is more commonly slipped into a woman's drink prior to a date rape.

In Pasadena, California, a dentist, Dr. Drueceil Ford, faced felony child-abuse charges for allegedly over-sedating five children with a chloral hydrate syrup designed to quiet frightened youngsters undergoing dental procedures. According to Deputy District Attorney Albert H. MacKenzie, a 15-year-old girl went into cardiac arrest and stopped breathing following large doses of chloral hydrate. The girl survived but suffered brain damage. Based on numerous complaints, MacKenzie believes that Ford used chloral hydrate because it allowed her to perform more procedures in less time and, therefore, make more money. The death of a 4-year-old boy in a Santa Ana dentist's office in 1997 was instrumental in the passage of legislation in California requiring dentists to be certified before being allowed to use sedatives such as chloral hydrate on patients younger than 13.

References

The American Medical Association Encyclopedia of Medicine. Charles B. Clayman, ed. Heidi Hough, editor-in-chief. New York: Random House, 1989.

Lane, Brian. *The Encyclopedia of Forensic Science.* London: Headline, 1992.

Marqu, Julie. "Dentist Charged with Felonies in Use of Sedatives; Courts: Pasadena Woman Is Accused of Committing Child Abuse by Giving Overdoses of Chloral Hydrate." *Los Angeles Times*, Apr. 29, 2000.

Chromatography

Chromatography is the separation and tentative identification of a mixture of substances on the basis of the rate at which they are carried through a fixed medium by a moving medium. The method is frequently used by forensic scientists as a means of analyzing chemical substances found at the scene of a crime. For example, chromatographic techniques can detect remains of explosives used to start a fire, crack open a safe, or destroy a building.

Simple chromatography can be demonstrated by cutting a long inch-wide strip from a paper towel. Paint a thin line of food coloring across the strip about an inch from one end. Place the tip of the painted end of the strip in a container of water, being careful not to wet the food coloring. Tape the other end of the paper-towel strip to a chair or cabinet so that the strip hangs vertically with its lower tip immersed in water. Capillary action will cause water to move up the towel, carrying the food coloring with it. Because the food coloring is a mixture of pigments, some of them will move upward at a greater speed than others, causing the colors in the mixture to separate into distinct bands.

Thin-Layer Chromatography (TLC)

Thin-layer chromatography is similar to the paper-towel chromatography experiment. It consists of a stationary solid phase coupled with a moving liquid phase. The choice of which solid and liquid phases to use from a vast number of possibilities depends on the nature of the suspected material, the nature of the crime, and the experience of the forensic scientist. The process is fast and inexpensive, and generally requires a sample with a mass of no more than 0.0001 gram. But, like gas chromatography, discussed later, it is not an absolute means of identifying substances. Further definitive testing is required to obtain a positive identification.

A glass plate is coated with a thin layer of granular material such as a silica gel and a binder that adheres to the plate. Several spots of the unknown liquid or dissolved solid sample are placed at the bottom of the plate, which is placed upright with its lower end touching a liquid that will rise up the plate by capillary action. The rising liquid is the moving phase that carries substances in the spots upward at different rates of motion. As the moving liquid phase approaches the top of the plate, the plate is removed and dried.

When the plate is viewed in ultraviolet light, substances that fluoresce are seen as bright spots against a dark background. If the coating on the plate includes a fluorescent material, substances that do not fluoresce will be seen as dark spots on a bright background. If fluorescence is not involved, a reagent may be sprayed onto the plate. Separated substances will appear as distinct spots on the plate.

Gas Chromatography

Gas chromatography is so named because an inert gas stream, usually of nitrogen or helium, is used to carry a liquid mixture through a long, narrow column filled with particles of diatomaceous earth or a similar inert solid material. Some instruments use long capillary tubes whose interior surface is coated with a liquid. This method of analysis is widely used because it is fast and sensitive. It can detect quantities as small as a billionth of a gram.

After being heated, the vaporized sample of the liquid mixture is carried along the column by the inert gas stream. As the components of the mixture travel, they separate into the pure substances found in the injected sample. At the end of the column they enter the detector. The substances that travel fastest through the column enter the detector first; those that travel more slowly enter the detector one by one until the slowest-moving substance has traversed the column. Each substance produces an electrical signal that is triggered by a change such as a change in the thermal conductivity of the gas. The signal is recorded on a moving sheet of paper, producing a chromatogram. The horizontal axis represents the time for the substances to pass through the column and reach the detector and is referred to as the retention time. Retention time is a characteristic property of each pure substance. The vertical axis shows a series of peaks, each one characteristic of the retention time of a particular substance. The height of the peak provides an indication of the quantity involved.

By comparing the chromatogram of the sample being tested to chromatograms of known substances, the components of the mixture being analyzed can be identified. However, gas chromatography, like density, does not offer an absolute means of identification. Just as two substances may have the same density, so other substances not present in the mixture might have the same retention time as those revealed by the chromatogram. Tentative identifications obtained by gas chromatography must be confirmed by a more definitive test.

Pyrolysis Gas Chromatography

Some solid materials such as paint chips and fibers, which are often useful clues for forensic scientists, will not dissolve in a solvent that can be injected into a gas chromatograph. It may, however, be possible to vaporize these substances and then introduce the resulting mixture of substances into an inert gas stream. Such a technique is called pyrolysis gas chromatography. The resulting chromatogram, which usually consists of a large number of peaks, is called a pyrogram. The pyrogram, when it is compared with pyrograms of known samples, such as paint chips or fibers from a variety of different automobile models, can be used to identify the source or nature of a sample obtained at a crime scene.

High-Performance Liquid Chromatography (HPLC)

For substances that decompose when heated, such as organic explosives and some drugs, a moving liquid under pressure may replace the inert gas as the moving medium. The sample under investigation is then in-

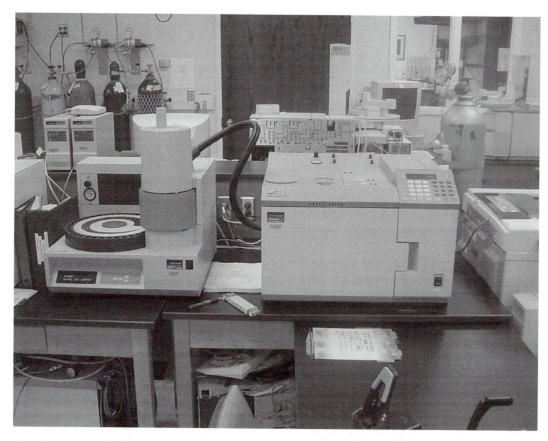

Gas chromatograph. Courtesy of the Florida Department of Law Enforcement.

jected as a liquid into the moving stream. HPLC allows the chromatographic process to take place at lower temperatures so that suspicious substances can be tentatively identified without being decomposed by the higher temperatures at which gas chromatography takes place.

Electrophoresis

Electrophoresis involves no moving phase, but the technique is similar to chromatography. An electric field is established across a starch or agar gel resting on a glass plate. Substances placed on the slide migrate in response to the electric field. The rate at which they move depends on their electric charge and their mass.

Electrophoresis is particularly useful in analyzing bloodstains. Many of the enzymes found in blood are proteins that will separate and move at different rates in an electric field.

Staining reveals the proteins as separate bands that can be compared with similar bands produced with known blood enzymes.

Athletes and Chromatography

In September 1988, Canadian sprinter Ben Johnson, Olympic winner of the 100-meter dash that summer, was stripped of his gold medal when he tested positive for anabolic steroids. Johnson is but one of many athletes who have been disciplined for using drugs believed to enhance their athletic prowess.

Laboratories that test athletes and other people suspected of drug use test in two stages. In the first part of the test, a urine sample is mixed with a substance that contains antibodies for a number of types of drugs as well as compounds made from the drugs themselves that are bonded to chemical enzymes. In a drug-free sample, the added antibodies and compounds combine, leaving the sample essentially free of enzyme-tagged drug compounds. In samples that contain metabolized drugs, most of the antibodies combine with the metabolized drugs, leaving many of the enzyme-tagged compounds in the fluid. By using spectrophotometry, the amount of light of a particular wavelength absorbed by the enzyme-tagged drugs in the sample can be determined. High absorption indicates that the antibodies have bonded with metabolized drugs that were in the urine, leaving a high concentration of the tagged drugs.

If the first test is positive, a second test can be performed using gas chromatography and mass spectrometry. The mass spectrometer is used to confirm the identification of substances detected by chromatography. It was this testing in 1988 that indicated that Ben Johnson had been using steroids.

Chromatography in an Arson Case

Following a fire on Highland Street in Revere, Massachusetts, on December 14, 1990, police collected samples that were tested using gas chromatography. Three samples indicated the presence of petroleum distillate, one sample was positive for gasoline, and another revealed no evidence of an accelerant. When three young men were charged with setting the blaze, their attorney called Michael Higgins, the president of K-CHEM Laboratories, a private forensic laboratory, to testify in an effort to have the charges dismissed. Higgins complained that no samples had been preserved for confirmatory testing. He also stated that no blank tests had been run to check for contamination and that no standard tests had been performed. Police officials responded that it was not common practice to save samples for long because they were seldom used for further testing and the space was needed for new samples. They further stated that blank tests were not necessary because the machine was cleaned between runs and heated to 240°C to vaporize any accelerants that might be present as contamination. They also claimed that standards were unnecessary because the parameters of the machine had not changed. After hearing

both sides, the court denied dismissal. The men were tried and convicted. *See also* Mass Spectrometry; Spectrophotometry.

References

Goodman, Brenda Dekoker. "Scene of the Crime: High-Tech Ways to See and Collect Evidence," *Scientific American*, Mar. 1998, pp. 35–36.

Grolier Multimedia Encyclopedia. Danbury, CT: Grolier Electronic Publishing, 1995.

Saferstein, Richard. *Criminalistics: An Introduction to Forensic Science*. Upper Saddle River, NJ: Prentice Hall, 1998.

Zonderman, Jon. *Beyond the Crime Lab: The New Science of Investigation*. New York: Wiley, 1990.

Cleveland Torso Murders

The Cleveland torso murders are considered by many to be one of the most gruesome and mysterious serial murders in American history. For forensic scientists, the case presented the nearly impossible task of identifying multiple victims based on body parts. An American Jack the Ripper, the Mad Butcher of Kingsbury Run murdered, mutilated, and dismembered 12 victims between 1934 and 1938, most of whom were never identified. Despite meticulously following thousands of leads, the police were continually stymied by dead-end investigations, and the Kingsbury Run murders remain unsolved.

Kingsbury Run was a ravine cutting across Cleveland's East Side to the Cuyahoga River that funneled the city's railroad traffic to Youngstown. Scattered along the scarlike gorge were the shanties and shacks of the transient and marginalized population of the depression era. On September 23, 1935, two boys slid down the steep side of Jackass Hill into the ravine. At the bottom of the hill they noticed something in the weeds and investigated. What they found—two headless male torsos—was the beginning of a reign of terror that gripped Cleveland until 1938, when the killings abruptly ended for no apparent reason.

Though these murders were more horrific than normal, crimes were not unusual in the

Kingsbury Run neighborhood, which was accustomed to violence and murder. Detectives found the initial two torsos neatly laid out and cleaned. There was no sign of a struggle, no blood on the ground, and decomposition had already begun. The officers concluded that the victims had been killed elsewhere and moved to Kingsbury Run. This was no simple feat given the steep incline into the cut. A search of the area uncovered two heads buried nearby but no other substantive clues. Authorities turned to the morgue for answers.

The coroner's office could not offer much information. The cause of death was recorded as decapitation. Death by decapitation is very rare in murders, and the coroner took particular notice of the professionalism of the cuts to the bodies. The bodies were so badly decomposed that victim number one could only be described in general physical terms of height, weight, and age. A fingerprint from victim number two identified him as 28-year-old Edward Andrassy, a local tough, brawler, and drunk. Inquiries ascertained that Andrassy was estranged from his family and had lived on the streets for several years. He was considered a womanizer, and there were also whispers of homosexuality.

No one in the police department immediately connected the double murder with a killing of nearly identical circumstances the previous year. An unemployed carpenter walking the shores of Lake Erie happened across the lower torso of a woman buried in the sand. Several days later, the rest of the body was located farther along the beach. Dubbed the "Lady of the Lake" by the press, she was never identified, and the case was closed. As the torso investigation progressed, the woman became an unofficial part of the case and was ultimately designated as victim zero.

Four months after the grisly discovery of the two torsos, the howling of a dog alerted neighbors to some half-bushel baskets near a factory wall. The police arrived to discover portions of a body neatly wrapped in burlap. The morgue obtained fingerprints that identified the remains as those of 42-year-old

Florence Polillo, like Andrassy, a shadowy figure of the Kingsbury neighborhood. Decapitation was again the cause of death. Despite the best efforts of the police, promising leads evaporated. One year later, the killer struck again. Labeled the "Tattooed Man" by law-enforcement authorities, this victim was also decapitated. Although a fingerprint search returned no identification, detectives were reasonably optimistic about getting a name due to six distinctive tattoos with either a name or initials in them. Photos were distributed and even a death mask was made. During the summers of 1936 and 1937, officials estimated that more than seven million people saw the victim's photo or mask at two fairlike Great Lakes expositions. There were no results. Even an aggressive search of missing-person files, tattoo parlors, and merchant-seamen hangouts failed to put a name to the body.

Two other murders, both decapitations and dismemberments, followed in rapid succession. No identification of the victims was possible. Eliot Ness, of Chicago fame, Cleveland's new public safety director, assigned 20 detectives to the case. Detectives checked the state mental facilities and followed up on recently released patients. Investigators went undercover in Kingsbury Run shanties hoping to lure the killer into action, and the *Cleveland News* and the city council offered $1,000 rewards for any information, all to no avail. Ness ordered a cleanup of the Kingsbury district. Every hobo was brought in for an interview, warned about the killer, and urged to find another place to live. On a professional level, Ness called for a meeting of the major players in the case that became known as the "Torso Clinic."

The "Clinic" was an attempt to profile the killer and coordinate all the information available. The county prosecutor, the chief of police, the pathologist who performed the autopsies, homicide detectives, and outside consultants attended the gathering. The group easily decided that the fiend was a psychopath, but most likely not insane. The culprit was definitely a man and very strong. There was no way a woman could have hauled those body parts into Kingsbury Run. Further, the manner of death by decapitation would have been physically very difficult for a woman. The killer was also likely from the Kingsbury area since he had successfully avoided police surveillance and his presence had not alerted suspicion in any of the residents. The skill of the dissection indicated that the "butcher" had an acute knowledge of anatomy, which always brought the inquiry back to the medical profession. This view was buttressed by the fact that the murderer needed a place of his own where he could work in confidence, undisturbed, to cut up the bodies and clean them. What better place than a doctor's office or maybe a butcher shop? Finally, based on the basis of the postmortem examinations of the stomach contents of the victims, officials felt the killer lured the indigents to their death with promises of food and shelter. Most were convinced that it was a local doctor, medical student, or male nurse who was responsible for the carnage. Police focused their ongoing efforts on anyone in the medical profession who had a history of drug or alcohol abuse, might have been involved in homosexual activities, or was generally considered eccentric.

Months went by with no activity, and investigators began to feel that the slaughter was over. Their hopes were dashed in February 1937 with the discovery of the remains of a young woman on the shores of Lake Erie. The site was very near where the "Lady of the Lake" had been found three years previously. June and July provided more unidentifiable torsos and remains, bringing the murder total to 10. The next year, on August 16, 1938, victims 11 and 12 were uncovered at a lakeside dump. After 12 brutal murders, three years of intense investigations, and countless man hours, Ness reacted. The day after the grisly discoveries, he raided the shanties and shacks and burned them down. Seeking a record of all the vagrants, he had each one arrested and fingerprinted. The killings ended.

As police morale bottomed out and public outrage heightened, law-enforcement officials did have a suspect. Ness never referred to the suspect publicly by name, preferring to call him "Gaylord Sondheim" and speaking of him only in very general terms. At first glance, Dr. Frank Sweeney appeared to be the embodiment of the American success story. Born and reared on the edge of the Kingsbury Run area, he had worked and put himself through medical school. Sweeney returned to Cleveland to take a surgical position at St. Alexis Hospital, the facility that served the Kingsbury Run community. Despite all this, Sweeney was a severe alcoholic who could turn violent when provoked. His family broke up and he separated from his wife in 1934. They ultimately divorced in 1936. At the same time, Sweeney was dismissed from his position at St. Alexis. Realizing his problem, Sweeney periodically checked himself into the Sandusky, Ohio, veterans hospital for rehabilitation. The hospital was his alibi. He always seemed to be out of town at the time a torso turned up in Cleveland.

Ness and his detectives put the all-out press on the suspect doctor. They checked his mail, searched his room, and even followed him on a regular basis. An enterprising detective checked the Sandusky records and realized that there was no accountability for patients; they could come and go from the hospital as they pleased. Sweeney was an outpatient. Ness called Sweeney in and confronted him with his suspicion that the doctor was the "butcher." Reputedly, Sweeney defiantly said, "Prove it," and left. Two days after the interview, Sweeney checked himself into the hospital. He remained in one type of facility or another for the rest of his life. He died in 1965. The case remained open for a time but it still remains unsolved. *See also* Serial Killers.

References

Bardsley, Marilyn. *The Kingsbury Run Murders*. Great Falls, VA: Dark Horse Multimedia, 1998. (Available on Internet Web http://crime library.com/kingsbury2/kingplog.htm.)

Nickel, Steven. *Torso: The Story of Eliot Ness and the Search for a Psychopathic Killer*. Winston Salem, NC: John F. Blair, 1989.

Purvis, James. *Great Unsolved Mysteries*. New York: Grosset & Dunlap, 1978.

Codeine. *See* Opium.

Coffee, Dr. Daniel Webster

Dr. Daniel Webster Coffee, a fictional character created by Lawrence Blochman, is chief pathologist and director of the laboratories at Pasteur Hospital in a fictional town named Northbank located somewhere in the midwestern part of the United States. Coffee uses his knowledge of forensic pathology to help his young detective friend, Max Ritter, utilize science and technology to solve very difficult criminal cases. The author's knowledge and experience with pathology enabled him to write stories that were true to life and thereby to make a portion of the reading public aware of the role of forensic science in solving crimes.

The Dr. Coffee stories were written in the 1950s and 1960s. One collection of stories about Dr. Coffee published in 1950 was called *Diagnosis: Homicide*. Another collection, published in 1964, was titled *Clues for Dr. Coffee*. Both collections have introductions written by well-known medical examiners of the era. Medical examiner Thomas A. Gonzales wrote the introduction for *Diagnosis: Homicide*, and Dr. Milton Helpern wrote the one for *Clues for Dr. Coffee. See also* Blochman, Lawrence; Gonzales, Thomas A.; Helpern, Milton.

References

Penzler, Otto, Chris Steinbrunner, and Marvin Lachman, eds. *Detectionary*. Woodstock, NY: Overlook Press, 1977.
Steinbrunner, Chris, and Otto Penzler, eds. *Encyclopedia of Mystery and Detection*. New York: McGraw-Hill, 1976.

Colt, Thatcher

Thatcher Colt was the creation of novelist C. Fulton Oursler, writing under the pseudonym Anthony Abbot. Oursler drew inspiration for the character from two former New York City police commissioners, Theodore Roosevelt and Grover A. Whalen. A big, strong man, Colt was a socially prominent New Yorker who joined the police force after service in World War I. As the series opens, Colt is already commissioner of police, ably assisted by Abbot, his Watson, who chronicles the hero's adventures.

The novels are noteworthy for their intriguing action and the smart police work of the commissioner. The first two novels were loosely based on real crimes. *About the Murder of Geraldine Foster* (1930) was a version of the Lizzie Borden affair, and *About the Murder of the Clergyman's Mistress* (1931) was inspired by the Hall-Mills case. The novels follow the course of Colt's career and beyond. In *About the Murder of a Man Afraid of Women* (1937), it is the day before Colt's wedding when he begins his investigation. The 1939 novel *The Creeps* follows a retired Abbot, Colt, and their wives on a Thanksgiving vacation to Cape Cod that turns into a murder case. Colt also enjoyed a brief life beyond novelization. Adolphe Menjou played the commissioner in two 1930s movies. A *Thatcher Colt* radio series also aired on BBC in 1936. *See also* Abbot, Anthony.

Comparison Microscope

A comparison microscope is used by forensic scientists to simultaneously compare two specimens such as the shells of two bullets. It was invented in the early 1920s by Philip O. Gravelle. The instrument consists of an optical bridge and two compound microscopes. The optical bridge—a series of mirrors and lenses mounted in a horizontal tube—connects the objective lenses of two compound microscopes to a single binocular viewer. Gazing into the binocular eyepieces of a comparison microscope, the viewer sees a field divided into two equal halves by a fine line. The object mounted under the microscope to the viewer's right is seen in the right half of the field. The object mounted under the microscope to the viewer's left is seen in the left half of the field. As a result, two dif-

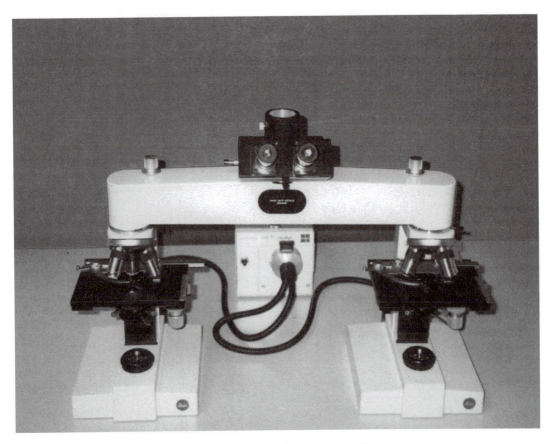

Comparison microscope. Courtesy of David Exline, ChemIcon Inc., Pittsburgh, PA.

ferent specimens can be viewed together and compared.

Shortly after it was invented, the comparison microscope was used to review the evidence used by the prosecution in the trial of Nicola Sacco and Bartolome Vanzetti. Sacco and Vanzetti were arrested for the murder of two security guards during a robbery in Braintree, Massachusetts, on April 15, 1920. Both men were immigrants and members of an anarchist group that supported the violent overthrow of the government. The men were convicted and sentenced to death.

Left-wing groups worldwide protested the conviction, claiming that Sacco and Vanzetti were the victims of an American "Red Scare" following the rise of communism in Russia. After Felix Frankfurter, who later became a U.S. Supreme Court justice, raised questions about the trial, Massachusetts governor Alvan T. Fuller, under pressure from Harvard

president A. Lawrence Lowell, appointed Lowell chairman of a committee to investigate the trial. As part of their investigation, the committee asked Calvin Goddard, an expert at the Bureau of Forensic Ballistics in New York, to review the evidence. Using a comparison microscope, which had not been invented at the time of the trial, Goddard could see that the striations on a bullet that had lodged in one of the security guard victims matched those on bullets fired from Sacco's gun. Despite the evidence, protests regarding the convictions continued, and in 1961 and 1983, ballistics tests were conducted again. Both tests confirmed Goddard's original analysis. Sacco was guilty. Whether or not Vanzetti was guilty as well remains unknown. Vanzetti maintained his innocence even during the moments preceding his electrocution.

The work involved in comparing bullets

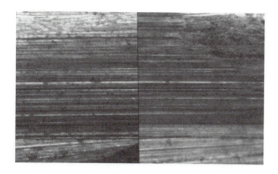

Photomicrograph of two bullets. Courtesy of the National Forensic Science Technology Center.

through a comparison microscope that used to take hours can now be done in seconds using a new high-tech system known as IBIS (Integrated Ballistic Identification System). An IBIS computer scans each bullet's unique markings and stores it as a video image in its memory. When a bullet from a crime scene is subjected to IBIS, it searches for a match with the thousands of other bullets and casings in its memory bank, each associated with a known gun. To date, markings from more than 100,000 gun barrels have been stored in IBIS. Eventually, as IBIS's database grows, law-enforcement authorities hope to be able to trace any bullet found at a crime scene to the gun that fired it.

During the last five years of the twentieth century, Boston police, using IBIS, learned that multiple killings were the result of the same gun. Bullets from the same 9-mm pistol, for example, were involved in more than a dozen different shootings in widely separated sections of the city. Police believe that shooters, knowing that a gun is "hot" following a crime, sell the gun, loan it, or give it to others. As a result, the weapon cannot be traced to a particular owner. *See also* Ballistics; Electron Microscope; Magnifiers; Spectrophotometry.

References

Evans, Colin. *The Casebook of Forensic Detection: How Science Solved 100 of the World's Most Baffling Crimes.* New York: Wiley, 1996.

Johnson, Alexander, and Aaron Brown. "Tracking Guns with IBIS Is Faster and Easier." *ABC World News Saturday*, Aug. 9, 1997.

Latour, Francie. "This Is the Gun." *Boston Globe*, Sept. 20, 2000, p. 1.

Saferstein, Richard. *Criminalistics: An Introduction to Forensic Science.* Upper Saddle River, NJ: Prentice Hall, 1998.

Computer Crime

Computer crimes are one of the major problems facing forensic science today. The crime has no simple definition, but commonly, computer crimes involve traditional crimes such as fraud, forgery, and general mischief carried out by use of a computer or computers. The people who engage in computer crime are represented by all age levels, from children to retired adults. Some are novices, some are professionals, but most (about 75 percent) are insiders, employees who, though they believe that it is immoral to steal from an individual, have no qualms about obtaining funds illegally from a corporation, usually the one that employs them (Fraser). The Robin Hood syndrome—taking from corporations to satisfy the needs of individuals—is the vogue among computer criminals. Computer crimes represent the downside of modern technology and have led legislatures to recognize that information as well as matter and energy constitutes property.

An employee with access to a company's electronically stored files and some understanding of computer science can quite easily channel money to an account that will automatically erase itself, leaving no evidence that it ever existed. Skimming a penny or two from each transaction can generate a sizable nest egg over time. Because computer crime is so difficult to detect, it often goes unnoticed, and even when it is detected, it is seldom reported. In addition to the embarrassment of having to admit being duped, companies fear that bad publicity will cost more in sales and goodwill than can ever be recovered financially through the courts. Usually, the only punishment the criminal receives is termination of employment.

Fraud is far easier to accomplish in the age of computers than it was when documents were kept in hard-copy files. Data in computer files can be instantly changed, modified, or erased without leaving fingerprints or detectable changes or erasures. Whenever cash is replaced by electronic transactions, fraud by the use of computers, not by breaking and entering, is prevalent. This is often accomplished by what are called "Trojan horses," in which someone with a detailed knowledge of computers and computer programming covertly adds instructions to a computer program. As a result, the program will generate unauthorized actions while carrying out the functions it was designed for. After the illegal actions, such as a transfer of funds, are carried out, the altered parts of the program self-destruct, leaving no evidence of illegal or unauthorized acts. In addition to fraud, forgery is readily accomplished by altering data in documents that are electronically stored. At the hard-copy level, color laser copiers can now generate fake documents that are so authentic in appearance that they are indistinguishable from those that are truly legitimate.

Computer programs and data can be sabotaged by what are called viruses, "worms," or logic bombs known also as "time bombs." Computer viruses are a series of program codes that attach themselves to a program and then, like those that invade human cells, propagate to other programs. Because many computers are connected through the Internet, a virus can become widespread rather quickly. Once a virus is in a program, it may do nothing more than leave a harmless message; however, more dangerous viruses can produce serious damage by destroying or altering data or programs.

If we consider a computer virus to be similar to a malignant tumor, a "worm" is the equivalent of a benign tumor. A worm, like a virus, holds codes that infiltrate and change or destroy data in a particular program, but it is unable to replicate and spread to other programs. For example, a worm that infiltrates a bank's computers might transfer money to an illegitimate account, but it will not spread to the programs of other banks.

A logic bomb is time related. Designed to destroy or modify data at some future date, such bombs are difficult to detect before they "explode." Often, a time bomb is used as a means of obtaining ransom. The criminal will invoke demands such as a terrorist might make. If the demands are not met, the "bomb" explodes and valuable data or programs are lost.

Antivirus programs have been developed, but computer criminals find ways to circumvent the "vaccine." Although most computer crime is carried out by insiders, there is a growing concern about organized groups of computer criminals who share information (stolen access numbers, passwords, and software) on electronic bulletin boards and through voice-mailbox systems.

Computer technology in the hands of the hacker "good guys" can be the undoing of virus writers. In March 1999, a computer virus known as Melissa, disguised as important e-mail, spread quickly around the globe and invaded 1.2 million computers in the United States. A number of independent hackers, America Online employees, and a high-tech FBI unit found that the code used in the virus resembled the work of a virus writer known as VicodinES. At the Vicodin Web site, Richard Smith, president of Phar Lap Software, downloaded the information found there. Within the files he found names, one of which appeared three times. Earlier, an automatic virus detector had found that the virus had entered through e-mail generated at skyrocket@aol.com. America Online discovered that someone had stolen that account and had then logged onto the Internet from New Jersey. With that information, the FBI traced the source back to a single telephone, and the suspect was arrested. Less than a week had passed between the appearance of the virus and the arrest of David Smith. In December 1999, Smith pleaded guilty to charges of creating the virus and causing $80 million of damage.

In that same month, Andrew Miffleton pleaded guilty to intent to defraud. He had, through his computer, illegally acquired a list of computer passwords from an Internet company (Verio, Inc.) that he had made available to other hackers. In the same manner, he had obtained mobile identification numbers for cellular-phone service, an AT&T calling-card number, and five credit-card numbers (Brenner). On July 24, 2000, Miffleton was sentenced to 21 months in prison and fined $3,000. He was also ordered to pay Verio $89,400.

In September 2000, Jonathan Lebed, 15, settled his case with the Securities and Exchange Commission (SEC) by agreeing to return illegal profits of $272,826 plus $12,174 interest. In return, Lebed was not charged with any wrongdoing related to his securities fraud. Sitting at his computer, Lebed would buy large amounts of a lightly traded stock on the NASDAQ or the over-the-counter bulletin board. The next morning he would send out hundreds of messages under different names over Internet chat sites extolling the stock. When the price rose, he would sell and reap a large profit. The scheme, according to the SEC, is a common one and is easy to trace. In fact, Lebed was a small part of schemes involving $10 million in illegal profits that were all prosecuted simultaneously. What was uncommon was for someone so young to be involved in such collusion. *See also* Disputed Documents; Forgery.

References

Brenner, Anita Susan. *Cyber-Rights and Criminal Justice: Legal Resources on the Web.* brenner@cyberspace.org.

Clark, Franklin, and Ken Diliberto. *Investigating Computer Crime.* Boca Raton, FL: CRC Press, 1996.

Fraser, Bruce T. "UN Manual on the Prevention and Control of Computer Related Crime." *Computer Crime Research Resources.* Florida State University, Tallahassee. btfl553@mailer.fsu.edu.

Icove, David J., Karl Seget, and William R. Von Storch. *Computer Crime: A Crimefighter's Handbook.* Sebastopol, CA: O'Reilly & Associates, 1995.

Kadlec, Daniel. "Crimes and Misdeminors: A Teenager Shows How Easily Stocks Can Be Manipulated and How Hard It Is to Get Away with It. So Why Are So Many Hailing Him as a Genius?" *Time,* Oct. 2, 2000, pp. 52–54.

Taylor, Chris. "How They Caught Him." *Time,* April 12, 1999, p. 66.

Contrecoup

Contrecoup is a medical term that refers to an injury to one point of an organ or part resulting from a blow at another site. It is also known as a counterstroke or counterblow. An example of a contrecoup injury is a blow to the back of the head causing an injury to the front parts of the brain. It is a frequent occurrence in victims of violent crimes.

Cornwell, Patricia (1956–   )

Patricia Cornwell is an American-born crime novelist whose main character, Dr. Kay Scarpetta, the chief medical examiner of the state of Virginia, uses scientific and medical knowledge to investigate crimes. Her first novel, *Postmortem* (1990), won the five most prestigious crime-writer awards, the only time the feat has been accomplished. She is the only American woman to win England's coveted Gold Dagger honor for her work *Cruel and Unusual* (1993). On March 1, 1996, six of the author's novels were on *USA Today's* list of the 25 all-time best-selling crime novels.

The hallmarks of Cornwell's novels are her intensive research and focus on science in criminal detection. Cornwell is justifiably proud of her meticulous research and has attended hundreds of autopsies and medical-school lectures. Each of her novels takes great pains to detail the importance of autopsies and forensic science in solving crimes. Scarpetta has to deal with the mistakes of an inexperienced medical examiner who misinterprets several crucial clues in *The Body Farm* (1994). In *Unnatural Exposure* (1997), Cornwell describes the necessary process of degreasing and defleshing bones to better study how a body was dismembered, and in *From Potter's Field* (1995), the author highlights the dangers of autopsies for the doctors and attendants from airborne infections. She describes their protective cloth-

Patricia Cornwell. © Irene Shulgin, Cornwell Productions.

ing in detail, likening it to armor. In each of the novels, Cornwell selects a specific problem of forensic medicine and investigation to highlight. To provide the authenticity necessary for *Point of Origin* (1998), a story about an attempt to cover up a murder by fire, Cornwell immersed herself in the study of fire science and worked closely for two years with the Bureau of Alcohol, Tobacco, and Firearms. *Unnatural Exposure* (1997) provides Scarpetta with a mutant death-dealing virus created by a colleague and former teacher envious of the medical examiner's success. The plot allows Cornwell to discuss the nation's infectious-disease facilities and investigative capabilities. Such realism is essential to Cornwell's novels. Unlike other writers who may emphasize analyzing the mind of the killer or determining the identity of the culprit, Cornwell is more interested in the excitement of detection through science and a quest for justice. She usually publishes one novel a year that is released in July to ensure summer readership. In 1998, the crime writer donated $1.5 million of her earnings to help Virginia create an institute to train forensic scientists and pathologists.

Between 1984 and 1986, Patricia Cornwell began to write in earnest, producing three novels featuring detective Joe Constable, involved in some unremarkable plots about poisonings and about Congo missionaries involved with diamonds. The novels were unceremoniously rejected. A confused author, searching for her voice, she contacted editor Sara Ann Freed at Mysterious Press for advice. Freed suggested developing and expanding a minor character, Kay Scarpetta, and further recommended that Cornwell base her writing on her own experience. The result was *Postmortem*, which Cornwell finally sold to Scribner's for $6,000 after seven rejections. *See also* Scarpetta, Dr. Kay.

References

Cantwell, Mary. "How to Make a Corpse Talk." *New York Times Magazine*, July 14, 1996.
Passero, Kathy. "Stranger Than Fiction: The True-life Drama of Novelist Patricia Cornwell." *Biography*, May 1998.

Schick, Elizabeth A., ed. "Patricia Cornwell." *Current Biography*. New York: H.W. Wilson Co., 1997.

Coroner

For many years, the coroner was the public official, often elected, who investigated untimely, violent, or unnatural deaths in order to determine the cause of death and whether criminal behavior was involved. Today the job usually is performed by a county or state medical examiner.

The coroner's office inquires into deaths that occur under suspicious circumstances, that are the result of unknown causes, or that have not been certified by a physician. Coroners also identify remains, notify survivors, and help settle estates and insurance claims. They also aid other law-enforcement officers in investigating deaths. The qualifications for the position are not very well established. There are no nationwide standards for the appointive or elective post. In some states, a sheriff, district attorney, or local mortician might hold the job. As late as 1946, the only requirement for the post in North Carolina was a belief in God and certification that the nominee had never participated in or arranged a duel. When necessary, these medical laymen may empower a local practitioner to conduct autopsies. The lack of training and professional medical requirements for an office that has increasingly come to demand specialized and technical knowledge has left the coroner's office open to criticism as the weak link in the investigative system for deaths.

As originally established in twelfth century England, the position was not medical in nature. The word *coroner* is a corruption of the original *crowner*. Coroners were elected county officials charged with protecting royal properties, fees, and interests within the bounds of their jurisdiction. The office was transplanted to America during the colonial period.

The two activities most often associated with the coroner's office are the autopsy and the inquest. If there is a question about the cause of death, the coroner can order an autopsy before issuing a death certificate. When it is essential to determine the cause of death, the coroner has investigative discretion, and his or her order supersedes the wishes of the family, the hospital, and even the attending physician. An autopsy can be ordered not only for criminal concerns but also for public health reasons, as in the case of a dangerous disease.

In suspicious or untimely deaths, the coroner can also conduct his or her own investigation, called an inquest. An inquest is not a trial but rather a preliminary fact-finding probe to uncover evidence that may be of use in a criminal investigation. The coroner's jury can issue subpoenas and warrants for the arrest of suspects implicated in a murder. The verdict of the inquest is not final but can be used as probable cause for an indictment. The proceedings and testimony at an inquest are generally inadmissible at a murder trial. Today, most inquests are usually limited to establishing the cause of death.

In 1877, Massachusetts realized that many coroners lacked the medical expertise and legal training for their position. The Bay State was the first to adopt the modern medical-examiner office that required professional medical training and experience in pathology. New York City introduced the medical-examiner system in 1918. The death of a prominent citizen and a coroner who refused to certify the death unless he selected the mortician led to the discovery of wide-ranging bribery and dishonesty in the coroner's office. The first statewide medical examiner's office was established in Maryland in 1939. By the 1980s, most states had replaced the old coroner's office with a county- or statewide medical examiner.

The medical-examiner post has two major advantages over the coroner's office. First, there is the professional requirement. The medical examiner's staff is usually composed of one or two pathologists, a chemist, a forensic toxicologist, various lab technicians, and investigators. The office may vary in size, but all involved are trained professionals. Second, the position is nonpolitical and independent. The medical examiner's job is not affected by political change. The medical examiner is not dismissed unless charged and found guilty of a crime. *See also* Medical Examiner.

Reference

Vorpagel, Russell. "Role of the Coroner." *Encyclopedia of Crime and Justice*. New York: Free Press, 1983.

Corrosive Sublimate

Corrosive sublimate is another name for mercuric chloride ($HgCl_2$), a colorless solid of deliquescent needles soluble in water, alcohol, and ether. The substance is extremely poisonous and was used in the past as a "murder weapon." Since the invention of guns, mercuric chloride, like other poisons, has been used very sparingly to poison people. (Guns and poisons are similar in that the killer never has to touch the victim.)

Corrosive sublimate was widely used in the past as a disinfectant, hypnotic, and antispasmodic, but it has been replaced by substances less irritating to the skin. When it is ingested, it damages the tissues of the mouth, stomach, intestine, and kidneys, causing bloody diarrhea, shutdown of urine production, and, possibly, uremic death.

During the reign of England's King James I, Sir Thomas Overbury was reportedly murdered by poisoning. Overbury had helped Robert Carr, who became Viscount Rochester, gain favor with the king. However, when Overbury counseled Carr to discontinue his adulterous relationship with Lady Frances Howard, wife of the earl of Essex, Carr and Howard swore vengeance against him. Through intrigue, Carr was able to put Overbury in a situation that led the king to imprison him. As Overbury languished in the Tower, Carr, pretending to be his friend, sent him food laced with arsenic and various other poisons. Although Overbury became ill, he seemed to be able to withstand the poisons. Consequently, Carr resorted to corrosive sublimate, which killed Overbury within hours. *See also* Poisons.

References

Memoirs of Popular Delusions, Vol. 2—The Slow Poisoners. http://www.student.dtu.dk/c973572/drugs/chemistry/data/e.
Turkington, Carol. *The Home Health Guide to Poisons and Antidotes.* New York: Facts on File, 1994.

Counterfeiting

Counterfeiting is the forging of government obligations such as paper money or bonds. Formed originally to track down counterfeiters, the Secret Service was so successful that counterfeiting was greatly reduced, but recent advances in copying techniques have made counterfeiting a more common crime.

Counterfeiting of U.S. currency extends well beyond its shores. The General Accounting Office (GAO) in a February 26, 1996, report indicated that approximately $29 million of counterfeit U.S. currency was in circulation throughout the world (*Massive Dollar Counterfeiting Activities*). The phony paper money is believed to have been produced in North Korea or Italy. It is believed that counterfeit money has been generated in North Korea in an effort to surmount a shortage of foreign currency. A month after the GAO report was issued, a North Korean diplomat and a Japanese hijacker were arrested in Japan and found to be carrying more than $36,000 of counterfeit U.S. money. The Department of the Treasury's Bureau of Alcohol, Tobacco, and Firearms laboratories provide the facilities for investigation of counterfeit materials and maintain a large information base on paper and ink.

Counterfeiting in the United States

Counterfeiting in North America is older than the United States. In colonial times, each colony issued paper currency, which was often printed on but one side and was made using simple woodcut or engraved designs that were easily copied. It is estimated that in those times about half the paper money in circulation was counterfeit.

After the U.S. government was established, paper money was issued by individual banks. Because there were so many forms of currency in circulation, it was difficult to distinguish real from counterfeit, and the crime flourished. It was thought that the establishment of a national currency in 1863 with intricate designs and fine engraving would eliminate counterfeiting. With the creation of a national currency, the Secret Service was established by Abraham Lincoln in 1865 at his last cabinet meeting. It was conceived as a general law-enforcement agency designed to combat forgery and counterfeiting of government checks and bonds as well as currency. Apparently counterfeiters rose to the challenge, because the number of counterfeit bills in circulation actually increased.

Not surprisingly, counterfeiting grew during the depression years following the stock-market crash in 1929. In an effort to combat the crime, the Secret Service and the Treasury Department provided banks with detailed information about how to detect fake bills and bonds. The public, however, was not informed for fear that this would encourage the unemployed to turn to crime. This approach changed when Frank J. Wilson became the director of the Secret Service. Wilson provided the public with information about counterfeiting, including a movie, *Know Your Money*, that showed people how to detect fake currency. Counterfeiting declined during World War II as the economy improved, but it increased again during the 1950s with the development of photocopying technology. The development of copiers with color capability led to another outbreak of counterfeiting during the 1990s.

A number of companies have developed devices for detecting counterfeit currency. For example, there is a pen containing a chemical substance that leaves an amber color on real bills and dark markings on fake currency. A magnetic-ink detector will locate magnetic ink on bills and produce different sounds or light to indicate whether or not they are counterfeit. There are also ultraviolet-light detectors that can identify fluorescent fibers.

The Treasury Department is constantly devising methods to detect counterfeit bills and to make the work of counterfeiters more difficult. In an effort to discourage counterfeiters, the government is changing its currency. It started with the $100 note, the largest denomination paper bill that is printed, because counterfeiters generally print the larger denominations. In the new currency, the portrait on the front of a bill is more detailed and off-center to provide room for a watermark. Within the portrait is a series of concentric lines that are very difficult to replicate. A watermark of the portrait is visible on the right-hand side of the note when it is held up to bright light. The number denoting the denomination of the note in the lower right corner appears green when it is viewed head-on but darkens when it is viewed at an angle. Microprinting of "USA 100," "USA 50," and so on, which is very difficult to duplicate, can be found within the number of the denomination located in the lower left corner of the note. Similar printing of "United States of America" can be found on the lower part of the portrait's "frame." A vertical security polymer thread is located on the left-hand side of each note. Within the thread are the words "USA" and the note's denomination, which can be read from either side of the paper. The thread contains a fluorescent substance that emits a reddish glow under ultraviolet light. Despite the efforts of the Treasury Department, counterfeiters continue their efforts to make money the easy way.

Pueblo Counterfeiting Scheme Uncovered

Forensic scientists are often aided by tips they receive from so-called crime stoppers. This was the case in June 2000 when narcotics officers in Pueblo, Colorado, together with a detective investigating counterfeit bills and the Secret Service, arrested 20-year-old Jerimiah Hale and charged him with forgery. A search of the young man's home and that of a second home implicated by the initial search led police to confiscate a number of counterfeit bills. Investigators believe that the fake $1, $5, $10, $50, and $100 bills were made using a computer and a color printer. The public was alerted to the fact that counterfeit bills were being circulated, that each denomination had identical serial numbers, and that the paper had a different "feel" than legitimate bills. *See also* Disputed Documents; Forgery.

References

A Brief Counterfeiting History in U.S. http://www.servicemart.com/money/histr1.htm. Accessed August 23, 2001.

Massive Dollar Counterfeiting Activities. http://www.koreascope.org/ktext/english/sub/2/1/nk9_4.htm. Accessed August 23, 2001.

Perez, Gayle. "Pueblo, Colo., Counterfeiting Scheme Exposed." *Pueblo Chieftain* (Pueblo, Colorado), June 22, 2000.

Crafts, Richard. *See* Woodchipper Murder.

Crime Laboratories

Crime laboratories are the facilities where forensic scientists perform various tests on physical evidence to help solve crimes. The French criminologist Edmond Locard, who was inspired as a teenager by Sherlock Holmes mysteries, is credited with forming the world's first crime laboratory in 1910. His lab consisted of a microscope and a spectroscope in a small room above the courthouse in Lyon, but it helped him solve crimes and established the value of science in criminal investigations. August Vollmer is credited with initiating the first crime laboratory in the United States 13 years later.

Today, the United States and France remain in the forefront of forensic investigations. France has a national police force under the supervision of the Ministry of the Interior and has five forensic laboratories located in Paris, Lille, Lyon, Marseilles, and Toulouse. In the United States, there is no national system of crime labs, but there are more than 2,250 public labs employing the services of approximately 3,500 scientists who examine crime-scene evidence. Individual states have a system of labs, but each state has its own organization. Some states have regional labs that work with local police departments. Other states have labs in each county, and many large cities, such as New York and Los Angeles, have their own labs.

At least 80 countries around the world have one or more crime laboratories. England, like France, has a national system of crime laboratories. It maintains regional labs that are supervised by a central Home Office. However, London does have its own lab within the Metropolitan Police Department. England also has the Central Research Establishment, which was created in 1966. It is the first laboratory that performs only research in the area of forensic science, nothing else. In Canada, the laboratories are affiliated with three government-funded organizations. The Royal Canadian Mounted Police maintains eight regional labs. In Toronto, there is the Centre of Forensic Science, and in Montreal, the Institute of Legal Medicine and Police Science. It is likely that computer technology will eventually allow all nations to merge and form a single cohesive system so that data and evidence can be shared by police worldwide.

Various crime labs analyze blood, hair, fibers, glass, paint, soil, and other materials found at a crime scene. They work on latent-fingerprint removal and fingerprint identification, weapon markings, and the authenticity of documents. Some crime laboratories specialize in one area, while others perform many types of analysis. Some are attached to the medical examiner's office, others work as a consulting partner to police departments, and a few are affiliated with universities, such as the Northwestern University Crime Laboratory, which was formed as a result of the St. Valentine's Day massacre of 1929. Members of a gang ruled by Al Capone disguised themselves as police officers and killed seven people in a garage. Two jurors from the court case were so disturbed by this crime that they financed the crime laboratory.

In the United States, the federal government has four major agencies with crime laboratories. One is the FBI Crime Laboratory, which is part of the Department of Justice. It was first set up under the direction of J.

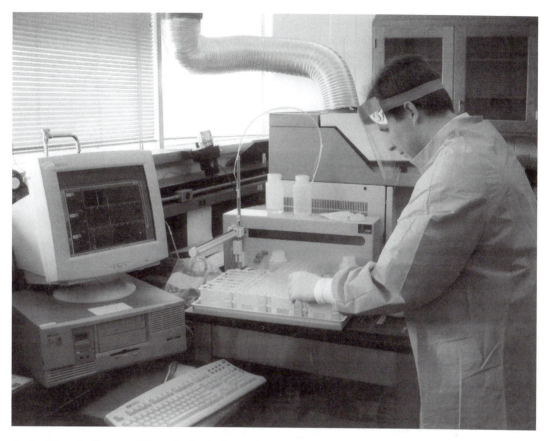

At the Georgia Bureau of Investigation's Crime Laboratory, a scientist prepares a sample run on an inductively coupled plasma/mass spectrometer for the elemental analysis of biological samples. Courtesy of the Georgia Bureau of Investigation, Division of Forensic Sciences.

Edgar Hoover in November 1932 with only a microscope and some ultraviolet-light equipment, but it has grown to be the largest and best-equipped crime lab in the world. Also under the Department of Justice are the Drug Enforcement Administration laboratories that deal with drug-related crimes. The U.S. Postal Inspection Service maintains labs for investigating crimes involving the violation of federal laws related to the postal service. Under the Department of the Treasury are the Bureau of Alcohol, Tobacco, and Firearms laboratories, which investigate bombs, weapons, and threats as well as maintaining a large information base on paper and ink since this is the agency that investigates crimes involving counterfeit money. Local law-enforcement agencies may ask any federal crime laboratory for assistance in investigating crimes related to that lab's area of expertise. There is no charge for such service.

Crime laboratories are not flawless. Even the FBI Crime Laboratory, which is held in high regard around the world, can make mistakes. Following the Oklahoma City bombing of a federal building, there were allegations of mishandling of evidence and altering of lab findings to support the prosecution. The Los Angeles Police Department Crime Laboratory was also involved in a major controversy. During the trial of O.J. Simpson for the murder of his ex-wife and her friend, the Los Angeles Lab was accused of not handling DNA evidence properly.

In an effort to reduce errors made by crime laboratories, the American Society of Crime Laboratory Directors (ASCLD) in 1981 approved a program of laboratory accreditation and established the American Society of Crime Laboratory Directors Laboratory Accreditation Board (ASCLD/LAB). In 1982, the eight crime laboratories of the Illinois State Police were the first to receive accreditation. Since then, many other crime labs have undergone the board's evaluation procedures. By June 1999, there were 182 ASCLD/LAB-accredited crime laboratories, including labs in Canada, Australia, New Zealand, Singapore, and Hong Kong, as well as the United States.

Accreditation is a voluntary procedure. Any crime laboratory may ask to be accredited by ASCLD/LAB. To gain accreditation, a crime laboratory has to demonstrate that its management, personnel, operating procedures, equipment, plant, security, and safety procedures meet ASCLD/LAB standards in eight areas—toxicology, trace evidence, serology, DNA, disputed documents, firearms, tool marks, and latent prints. The accreditation process ensures that a crime laboratory provides proficiency testing, continuing education, and other programs that enable the laboratory to provide high-quality service to the criminal justice system.

The Paul Coverdell National Forensic Improvement Act of 2000 is an incentive for crime laboratories to seek accreditation. The law authorizes that $482 million be made available for improving the quality, timeliness, and credibility of forensic science for criminal justice purposes. Grants will be available to accredited crime laboratories and medical examiners' offices and to those seeking accreditation. Seventy-five percent of the money will be given as grants to states on the basis of population. The remaining funds will be distributed in the form of discretionary grants. The act states that it is the "sense of Congress regarding the obligation of grantee states to ensure access to post-conviction DNA testing and competent counsel in capital cases." *See also* FBI Crime Laboratory; Locard, Edmond; St. Valentine's Day Massacre; Vollmer, August.

References

The American Society of Crime Laboratory Directors (ASCLD). http://www.ascld.org.

Cyriax, Oliver. *Crime: An Encyclopedia*. North Pomfret, VT: Trafalgar Square, 1996.

Evans, Colin. *The Casebook of Forensic Detection: How Science Solved 100 of the World's Most Baffling Crimes*. New York: Wiley, 1996.

Lane, Brian. *The Encyclopedia of Forensic Science*. London: Headline, 1992.

Saferstein, Richard. *Criminalistics: An Introduction to Forensic Science*. Englewood Cliffs, NJ: Prentice Hall, 1990.

Sifakis, Carl. *The Encyclopedia of American Crime*. New York: Smithmark, 1992.

Theoharis, Athan G., ed. *The FBI: A Comprehensive Reference Guide*. Phoenix, AZ: Oryx Press, 1999.

Crime Scene

The crime scene is the place where evidence leading to the solution of that crime may be found. Consequently, police will try to preserve it in an undisturbed state for trained investigators who will examine the area and search for evidence. The first duty of police who reach a crime scene is to help and provide medical care for any injured parties. Once any persons needing medical attention are cared for, police seal off the area with the familiar wide yellow band and establish a guard to be certain that unauthorized personnel do not enter the established boundaries.

When investigators arrive, they photograph the scene and any physical evidence from several angles. To provide estimates of magnitude, items are often photographed with a ruler beside the object whose image is to be recorded. Increasingly, police are using video camcorders to photograph crime scenes.

Because of the increased incidence of AIDS and hepatitis B, crime-scene investigators are at risk at sites where body fluids may exist. Most crime investigators have been inoculated for hepatitis B and don latex

A police officer carefully taking evidence. In this case, Jim Bell of the Salt Lake City police department gathers forensic evidence on the sixth floor of the Judge Building near a door blown off its hinges by a blast on October 16, 1985. The bombing was the second of three to occur during the "Mormon Forgery Murders" case. The first bomb, exploded the previous day, killed two people. © *Deseret News*/Don Grayston.

gloves before touching any evidence at a crime scene.

Investigators also make detailed notes describing the scene and the location of pieces of evidence relative to other fixed points at the location and write thorough descriptions of weapons, wounds, bodies, and any objects of significance. The notes must be detailed because crimes are often not solved for months or years and memories of the details of such scenes quickly fade from the minds of detectives who investigate many crimes in the course of a year. In addition to photographs and notes, investigators make detailed sketches of the scene that include measure-

ments and dimensions to reveal the location of bodies, weapons, and other pieces of evidence relative to fixed points such as doors, windows, posts, and so on.

Investigators conduct a careful search for evidence. Weapons used to kill or injure a victim may be recovered at the scene. If there is a victim, evidence may have been transferred from criminal to victim or left nearby. Such things as hair, fibers, blood, semen, saliva, bite marks, tool marks, and fingerprints may be discovered by a careful search. For example, at the scene of a hit-and-run accident, the victim's clothing may contain particles of paint or glass from the car that struck him or her. The car, in turn, may have carried away drops of the victim's blood, pieces of skin, or fibers from his or her clothing. Materials that require laboratory analysis, such as bloodstains, hair, bullets, and so on, should be carefully bagged for transportation to a forensic laboratory. If possible, such evidence should not be separated from the surface in or on which it is found.

A sweeping or vacuuming of a crime scene may reveal microscopic evidence not visible to the naked eye that will be seen under a laboratory microscope. Depending on the nature of the crime, other items that may be taken to the laboratory for careful examination might include fingernail scrapings from the victim and any suspects, the clothes of the victim or suspects, hair, blood, bullets, bullet casings, and vaginal, anal, and oral swabs, as well as hand swabs (if a suspect is believed to have discharged a gun).

New Technology

New technology promises to reduce the drudgery of the manual process associated with crime-scene investigation. With computer-aided-design (CAD) software similar to that employed by architects and structural engineers, audio and videotaped images can be used to construct three-dimensional drawings of the crime scene as well as chronologically arranging, recording, and integrating information provided by investigators. Under evaluation is a laptop computer and associated digital video and still-picture cameras, as well as laser range finders and a global-positioning system, that will allow investigators to transmit information from a crime scene directly to a forensic laboratory. Research is also progressing on a truly portable laboratory that will allow detectives to process evidence, including DNA evidence, at the crime scene, thereby reducing the possibility of contamination.

As an example of how modern technology can affect criminal investigations, consider a drug investigation that took place in Kennewick, Washington, in 1996. Complaints about unpleasant odors from a house led police to suspect that methamphetamines were being prepared there. Unable to obtain a warrant on the basis of visual evidence, authorities employed a vapor-collecting system from Pacific Northwest Lab. Analysis of the vapors provided the evidence needed to obtain a warrant. When police raided the residence, they found the evidence they needed to convict the occupants of making illegal drugs.

Legal Precautions

The investigators in Kennewick took no action until they were able to obtain a warrant. No prosecutor likes to lose a case because police violated the constitutional rights of the accused. The Fourth Amendment of the U.S. Constitution states, "The right of the people to be secure in their persons, houses, papers, and effects, against unreasonable searches and seizures, shall not be violated, and no warrants shall issue, but upon probable cause, supported by oath or affirmation, and particularly describing the place to be searched, and the persons or things to be seized."

The U.S. Supreme Court has been strict in its interpretation of this amendment, and a number of accused persons who might otherwise have been convicted have gone free because police did not obtain a warrant before conducting a search. Any search conducted without a warrant must be done only in an emergency situation, a need to prevent the loss or destruction of evidence, in connection with a lawful arrest, or with the consent of the person being searched. Unless one of these conditions exists, police should first obtain a warrant before conducting any search of a crime scene.

In cases involving arson, timing is a key factor. The investigation must be started as soon as possible because an accelerant that may have been used to start the fire will evaporate quickly. However, in *Michigan v. Tyler* (1978), the U.S. Supreme Court ruled that "entry to a fire requires no warrant, and once in the building, officials may remain there for a reasonable time to investigate the cause of a blaze. Thereafter, additional entries to investigate the cause of a fire must be made pursuant to the warrant procedures. . . . Consequently, if police are called to investigate a fire after others have left the scene, they must obtain a search warrant or risk having a court deny the admission of any evidence uncovered during the investigation."

Tiny Pieces of Crime-Scene Evidence

The tiny bits of evidence surrounding the scene of Helle Crafts's murder in 1986 (the Woodchipper murder) masked the gruesome nature of the act. A number of circumstances finally led police to conclude that something was wrong. Richard Crafts's explanations about his wife's disappearance kept changing. His credit-card receipts showed that five days before Helle Crafts disappeared, he made a down payment on a large freezer, which was delivered on the day she returned to Newtown, Connecticut, and was returned several days later. He had rented a woodchipper in Darien, a good hour's drive from his home. Why? There was no evidence of wood chips or tree work around his home.

The Craftses's nanny/housekeeper had discovered a brownish, grapefruit-sized stain on the rug in the Craftes's bedroom. Crafts said that it was a kerosene spill from a portable heater he had used during a recent storm. He pulled the carpet up and took it to the dump. It was never recovered.

Once the case was placed in the hands of the state police, forensic experts discovered bloodstains on one side of the mattress in the master bedroom. Henry Lee, a blood-spatter expert, determined that a blow from a blunt object produced the stain. Tests showed that the blood came from someone with type-O blood—the same type as Helle Crafts's. In the trunk of Richard Crafts's car, Lee found human hairs and tiny pieces of flesh and bones. Convinced that Helle Crafts was dead, police had to ask, where was the body?

Police began questioning people. One person recalled seeing someone operating what might have been a woodchipper beside the Housatonic River in Southbury at night during a storm. During an intensive search of the area along the riverbank, police found traces of blonde hair, two caps from teeth, a fingernail, bone fragments, a small amount of blood, and a toenail, none of which could be conclusively identified as the victim's. On the basis of the initial results, the state police established a field lab at the site by the river and called in the state's forensic dental expert in the hope of finding definitive evidence.

Another person recalled seeing a woodchipper on a bridge above the Housatonic River in mid-November. Divers searched for clues under the bridge and recovered a chain saw. Between the teeth of the saw, forensic scientists found fragments of bleached, dyed human hair that were consistent with hair found on the river's bank as well as strands from the victim's own brush. Inside the chain saw's housing were bits of blue-green fabric and traces of flesh and blood. Tests indicated that the blood was human and type O, the same type as Helle Crafts's. Though the chain saw's serial numbers had been filed off, the scientists were able to restore the numbers and trace the machine's ownership. It belonged to Richard Crafts.

The discovery of a tooth on the bank of the river enabled forensic dental experts to demonstrate that it definitively came from the mouth of Helle Crafts. From the condition of the tooth, they deduced that a tremendous force, such as a saw or woodchipper, had broken the tooth from its seat. The tooth, together with skin and hair evidence, allowed the medical examiner to determine that Helle Crafts was indeed dead two months after her disappearance. The ev-

idence was convincing. Even without an intact body and with less than six ounces of bodily remains, prosecutors were able to make a convincing case. Richard Crafts was arrested, indicted, and finally convicted of his wife's murder.

A Toronto Murder Scene

In April 1998, the bodies of two teenage girls were found behind a factory in Toronto, Canada. As he walked toward the body of one of the victims, Detective Mike Davis knew that her throat had been slit before she ran to the point where she collapsed. There were pools of dried blood every three to four feet from the parking lot to the wall where the victim lay, curled in the fetal position. Every heartbeat would send blood gushing from her throat. The second victim lay on her back, arms and legs outstretched. She, too, had been cut across the throat. A knife blade was found near the first victim. Some distance away lay the handle of a broken knife, presumably from the same weapon.

Descriptions of the two girls broadcast on local television led to a mother's phone call and a photograph of her missing daughter. The photograph enabled police to identify the first victim, 16-year-old Isha Cleverdon. Cleverdon's mother suggested that police talk to Kayla McAlmont, who had been planning to go out with Isha the night she disappeared. McAlmont told police that Isha had left a message on another friend's answering machine saying that she had gone out with a friend named Cheri but would return soon. The number from which she had called was on the friend's call display, and McAlmonts had tried unsuccessfully to reach Isha Cleverdon at that number.

Cheri had been staying with Isha, and police were able to identify her as Cheri Doucette from signed papers she had left in Isha's room. The number left on the call display was that of a cell phone owned by Carol Edwards. Detectives found that Carol Edwards had a home phone, which led them to an address and an apartment. When they knocked and entered the apartment, they discovered that Carol Edwards was a 33-year-old man. He denied any knowledge of the two girls, but police discovered blood on his car and obtained a search warrant. In his apartment, they found a cream-colored suit, recently washed but stained, a cell phone with a number matching the one left on the call display, and a knife block with an empty space. Edwards claimed to have no memory of the night in question and said that he suffered from memory loss as a result of an industrial accident two years earlier. However, a marijuana cigarette with both Edwards's and Doucette's DNA was found in the ashtray of Edwards's car and Edwards's DNA was found in Doucette's mouth. A jury found Edwards guilty, and he was sentenced to two concurrent life sentences.

Ignoring Crime-Scene Evidence

In the case of Roy Criner, prosecutors tended to ignore evidence collected at a crime scene. Criner was convicted of the rape and murder of a 16-year-old girl in 1986. The evidence presented to the jury consisted of Criner's bragging to a few friends that he had picked up a woman who was hitchhiking and had had sex with her. There was no physical evidence that the girl Criner mentioned to friends was the same girl who had been raped and murdered. Recognizing that such minimal evidence lacked the substance needed for a murder conviction, the prosecutor sought and obtained a conviction for rape.

In 1997, persistent journalists were able to convince a court that DNA tests not available in 1986 should be used to test the semen collected at the crime scene. The tests revealed that the semen found in the girl's body was not Criner's. Still, prosecutors maintained that the semen's DNA only proved that someone else had had sex with the girl; it did not prove that Criner had not raped her. Perhaps, they argued, Criner used a condom or did not ejaculate. As a result, the conviction was upheld by the state's highest court.

Later, Bob Burtman, a journalist with the *Houston Press*, and a private investigator employed by Criner's family had a cigarette butt found at the crime scene subjected to DNA

testing. The DNA on the butt matched the DNA found in the semen, indicating that the rapist, who could not have been Criner, had been at the crime scene. In July 2000, 14 years after the crime, the district attorney, the sheriff, and a Montgomery County District Court judge all recommended that Criner be pardoned and he was. *See also* Automobile Accidents; Ballistics; Bite Marks; Blood; Bruises; Carbon Monoxide; Cause of Death; Drowning; Fiber Evidence; Fingerprints; Forensic Photography; Glass Evidence; Gunshot Wounds; Hair Evidence; Rape; Soil Evidence; Strangulation; Time of Death; Tool Marks; Trace Evidence.

References

Fisher, Barry A.J. *Techniques of Crime Scene Investigation.* 6th ed. Boca Raton, FL: CRC Press, 2000.

Ragle, L. *Crime Scene.* New York: Avon Books, 1995.

Raspberry, William. "Lone-Star Lessons." *Washington Post*, Aug. 4, 2000, p. A29.

Raymond, Joan. "Forget the Pipe, Sherlock: Gear for Tomorrow's Detectives." *Newsweek*, June 22, 1998, p. 12.

Shepard, Michelle. "A Case of Murder." *Toronto Star*, Sept. 30, 2000.

"Technology Unlocks the Architecture of a Crime." *American City and County*, Aug. 1996, p. 58.

Crime Statistics

Crime statistics during the last decade of the twentieth century reveal a declining crime rate. According to the U.S. Department of Justice Bureau of Justice Statistics, in 1991, there were 3.7 million violent crimes (murders, rapes, robberies, and assaults) in the United States. By 1998, that number had declined to 2.8 million. During the same decade, despite population growth, the total number of crimes of all types declined from 14.9 million in 1991 to 12.5 million in 1998.

Explaining why the crime rate diminished is not a simple matter. As with most social phenomena, there is no way to control the variables. Some attribute the reduction in

Dr. Hawley Harvey Crippen. © Bettmann/CORBIS.

crime to the booming economy and low un-employment that characterized the 1990s. Others argue that it is the result of the increase in police officers on the streets and growing cooperation among law-enforcement agencies. Because younger people are most crime prone, demographers maintain that the aging of baby boomers has led to a reduction in the number of young people in the population and a consequent reduction in crimes. Many conservatives credit the decline in crime to legislation that mandated longer prison terms.

The murder rate reached a peak of 10.3 per 100,000 in 1980. In 1991, it was still 9.8 per 100,000, but by 1998 it had declined to 6.3 per 100,000, which was the lowest in 30 years. The fact that the use of firearms in murders dropped from 13,101 in 1997 to 8,816 in 1998 led many to conclude that efforts to limit access to guns were beginning to have a positive effect, but others argue that economic well-being and longer prison terms are responsible.

During a two-year period from 1994 to 1996, the number of felony convictions increased by 14 percent in state courts and 11 percent in federal courts. Furthermore, between 1988 and 1996, the number of felony convictions grew at a faster rate than the number of felony arrests. Both statistics suggest improvement in forensic science. *See also* Arson; Homicide.

References

Gardner, Robert, and Edward A. Shore. *Math and Society.* New York: Watts, 1995.

Sniffen, Michael J. "Crime in the U.S." Washington, DC: Associated Press, Oct. 17, 1999.

U.S. Department of Justice: Bureau of Justice Statistics. http://www.ojp.usdoj.gov/bjs/.

The World Almanac and Book of Facts, 1999. Mahwah, NJ: World Almanac Books, 1998.

Criminal Anthropology. *See* Forensic Anthropology.

Crippen Murder Case (1910)

The Crippen murder case and the subsequent trial were touted by the London press as a landmark in criminology. Those prosecuting the case possessed some startling forensic evidence making the crime an early British submission for the "Crime of the Century." The sensational murder had everything necessary to capture the popular imagination: an unfamiliar and exotic poison, a long-term love affair, a grisly dismemberment, a transatlantic escape attempt terminated by one of the technological marvels of the age, and the central role of forensic science in providing a final conviction.

Hawley Harvey Crippen was an American living in London at the time of the murder. He arrived there by a circuitous route due to a dubious medical background. Although Crippen held the title of doctor, his homeopathic degree labeled him a quack in the eyes of the medical establishment. The Michigan native moved to New York City after a failed first marriage and promptly fell in love with a 19-year-old Polish girl, Cora Turner. The new Mrs. Crippen, who preferred her stage name of Belle Elmore, was convinced that she was destined for theatrical fame. The vain, plump, and demanding woman also expected a lifestyle suitable for a doctor's wife, an entitlement that Crippen found difficult to maintain.

To meet his financial obligations, Crippen left his medical practice and became a patent-medicine salesman. He moved to London with his company and simultaneously managed his wife's "career." As a manager, he frequently paid theaters to book his wife and even hired a composer to create songs specifically for the aspiring songbird. He was ultimately fired from his sales position due to his dance-hall activities and drifted to a number of other positions, even losing a malpractice suit on one occasion.

As a partner in a dental firm, Crippen met Ethel Le Neve, a 17-year-old bookkeeper/secretary, and was immediately smitten. Crippen began a lengthy affair, first as a confidant and then as a lover. Because he was beset by the temper tantrums of a disappointed wife who flaunted her numerous affairs, most neighbors felt sorry for the unassuming American doctor.

Cora Crippen, alias Belle Elmore, was last seen alive after a dinner party at her home in early January 1910. The following day a letter from the would-be songstress appeared at the Music Hall Ladies Guild saying that she was going to the United States to nurse a sick relative. At the same time, Dr. Crippen was pawning his wife's diamond ring and earrings. Just before Easter, Crippen informed his wife's friends that she was ill in Los Angeles. A short time later the doctor announced Cora Crippen's death.

Cora Crippen's music-hall friends were less than convinced by Crippen's reports, and their suspicions were only heightened when Ethel Le Neve moved into the Crippens' home and began appearing regularly in public with the doctor. Friends checked the ship lines and discovered that no one listed as Crippen or Elmore had sailed to America. Others verified, through friends in New York City, that no one by those names had passed through American customs. By June 1910, people were concerned enough to go to Scotland Yard with their fears.

Inspector Walter Dew was assigned to the inquiry and went to interview Crippen. A frightened Crippen readily admitted to lying. The truth was that Cora Crippen had run off with a more affluent lover, and Crippen had lied about her death to avoid a scandal. A search of the house uncovered nothing, and Dew left the scene feeling that there was nothing unusual about Crippen's explanation.

Crippen might have gotten away with his crime if he had not wilted under pressure. Two days after the initial interview, Dew returned to the Crippen home to find the doctor and his mistress gone. Convinced that something was wrong, Scotland Yard began an intensive three-day search of the house. In the coal cellar, buried in quicklime, were the gruesome remains of a human being. The remains were headless and limbless, and all the bones had been removed to further hinder identification. Any indication of the victim's sex had been destroyed. It was clear that someone with an extensive knowledge of anatomy was the murderer. Further exploration revealed that the torso was wrapped in a man's pajama top.

While the newspaper headlines screamed about murder and mutilation complete with lurid stories and pictures of the vanished suspect, the two lovers, traveling as father and son under the name of Robinson, went to Antwerp and boarded the SS *Montrose* bound for Canada. Henry Kendall, the captain of the ship, quickly realized that the Robinson boy was really a woman in disguise. Further, Mr. Robinson was strikingly similar to the photos of the "cellar murderer" Crippen, albeit without the glasses and moustache. Kendall radioed his superiors. Inspector Dew immediately boarded the liner *Laurentic* and arrived in Montreal ahead of the suspects. On July 31, 1910, Dew, disguised as a pilot, boarded the *Montrose*. He discovered Crippen walking the deck and approached him, saying, "Good morning, Dr. Crippen." Crippen quietly replied, "Good morning, Mr. Dew," and both fugitives were taken into custody.

The five-day trial dominated English headlines. Crippen's defense steadily denied that the remains were those of Cora Crippen. She had left her husband, and all the police had was circumstantial evidence. The defense's alternative scenario was that the unfortunate victim had been buried in the cellar prior to Crippen's occupancy of the house. The pajamas were explained away as an attempt to implicate the innocent doctor.

As soon as the remains were discovered, a team of prominent forensic experts was brought in. The most significant member of the scientific authorities turned out to be the most junior—Bernard Spilsbury, an expert on scars. The prosecution's forensic case centered on a small 5½" × 7" patch of skin. The defense asserted that this was merely a fold of skin from a thigh. Spilsbury, however, proved that the skin was from the stomach, and the wrinkle was not a fold but a scar from an old operation. Cora Crippen's friends confirmed that she had a similar scar. Other experts found traces of the poison hyoscine in the torso during the autopsy. The lawyers went on to prove that Crippen had purchased an amount of this poison days before his wife disappeared. Finally, the forensic experts dismissed the defense's claim of an earlier crime by placing the time of the burial at four to eight months prior to the discovery of the remains.

The final piece of the Crown's case was provided by Crippen himself. He was caught in a direct lie. The middle-aged doctor insisted that his missing pajama top had been lost three or four years earlier. The prosecutor was able to tell the court that the material had not been made until 1908 and that Crippen had purchased the pajamas in 1909. On October 27, 1910, the jury returned a verdict of guilty, and Crippen was hanged on November 23, 1910. The case was so well known that Crippen was eventually made a character in Madame Tussaud's Wax Museum in the Chamber of Horrors.

Ethel Le Neve was tried as an accessory. The defense portrayed her as an innocent, young girl seduced and manipulated by the Svengali-like Crippen. After only a 20-minute deliberation, the jury found her innocent. Le Neve migrated to Toronto, Canada, but eventually returned to England. She married an accountant and reared two children in the obscurity of a London suburb. She died in 1967 at the age of 84. *See also* Poisons; Spilsbury, Sir Bernard H.

References

Cullen, Tom. *Crippen: The Mild Murder.* London: Bodley Head, 1977.

Goodman, Jonathan. *The Crippen File.* London: Allison and Busby, 1985.

Gordon, Richard. *A Question of Guilt: The Curious Case of Dr. Crippen.* New York: Atheneum, 1981.

CSI: Crime Scene Investigation

CSI: Crime Scene Investigation is a CBS television drama that premiered in the fall of 2000. The program follows a small forensic team on the graveyard shift in Las Vegas. The theme of the program is that the analysis of evidence collected at the crime scene is far more useful in solving crimes than eyewitness and verbal accounts of a crime. Fibers, blood, stains, position of corpses, and even toenails provide the grist the criminologists work with as they seek to re-create what happened at a crime scene.

The primary characters are Gil Grissom (portrayed by William Peterson), who heads the crew, and Catherine Willows (played by

Marg Helgenberger). Grissom set the tone of the series in the initial show by pulling a maggot from a dead body and announcing that the victim had been dead for seven days. Grissom frequently admonishes his team to focus on what cannot lie—the evidence.

Each week, a number of stories occupy the cast and pose forensic dilemmas to solve. In the premiere episode, "the nerd squad," so described by a detective, reviews what appears to be a home invasion and self-defense murder. However, an oddly tied shoelace triggers suspicion that is reinforced when hairs from the crime scene are closely examined. A microscopic view of the hairs reveals that they were pulled from the victim's head. There are "seeds" or "bulbs" at the ends of the follicles. Normally, hairs lost from a person's head do not include the seed. The entire case ultimately hinges on a small segment of a toenail found in the dead man's shoe.

In another episode, the squad encounters a man who has been robbed by a prostitute but remembers nothing about the experience. Here, the important clue is the victim's discoloration, which an investigator attributes to scopolamine spray that rendered the naïve out-of-towner unconscious.

In yet another episode, the jumping death of a casino jackpot winner raises the question: was it murder or suicide? When Grissom reaches the scene, he begins his evaluation by "letting the body talk to him." He believes the cause of death to be murder since people who commit suicide seldom wear their glasses when they jump. To test his hypothesis, the investigator uses a dummy simulator and camera to make comparison studies on body position. A colleague throws, pushes, and re-creates a jump using the dummy while Grissom uses a camera to photograph body positions after each trial. In examining the victim's possessions, Grissom finds rug fibers caught in the jumper's watch stem that match those in the hotel room's carpet. This bit of evidence leads the forensic expert to deduce that the victim was dragged across the room before being thrown out the window.

In the same episode, the murder of a rookie detective is the occasion for showing how ballistics tests are conducted. Her death is also used to demonstrate how DNA samples derived from scrapings under the victim's fingernails can be used as a way to establish guilt and make an arrest.

The series provides interesting examples of how forensic science is used to solve crime. The fact that CBS introduced a program with forensic science as its theme, and the show's apparent popularity, as evidenced by its high ratings, reveal the American public's enthusiasm for and interest in this subject.

Cyanide

Cyanide, the layman's term for poisonous compounds such as potassium cyanide (KCN) and sodium cyanide (NaCN), has played a role in crimes, both real and fictional. For example, a 1983 Agatha Christie made-for-television movie revealed a component of the plot in its title, *Sparkling Cyanide*. In the all-too-real world, the Nazis used Zyklon B during the Holocaust to kill Jews at Auschwitz (*Chapter Nine: The "Final Solution"*). Zyklon B was another name for hydrogen cyanide (HCN).

More recently, in August 1998, police in the Los Angeles area arrested Kathryn Schoonover for attempting to commit murder by mail. An observant person saw Schoonover placing small bags of powder into envelopes and, on a counter nearby, a container with a poison symbol. The observer reported his concern to postal authorities, who called the police. Schoonover, who was 50 and a cancer patient, had planned to send 100 envelopes containing the powder, which was labeled as a free nutritional supplement, to doctors and police officers. Forensic chemists analyzed the powder and found that it was a cyanide salt. Further investigation revealed that a few people in the New York area had received previously mailed envelopes containing the cyanide, but no one was believed to have ingested any of the powder. To explain her behavior, Schoonover, who had been in a New York psychiatric hospital in 1994, claimed that doctors had given her drugs that caused her to abort a fetus. Doctors, however, said that she had not been pregnant at the time she was treated.

Cyanide is poisonous because by reacting with an enzyme, cytochrome oxidase, it removes it from the respiratory cycle. Without this enzyme, cells are unable to make use of the oxygen that is available in the blood.

A fatal dose of cyanide ions requires about 250 milligrams (mg). Symptoms of cyanide poisoning include nausea and dizziness leading to loss of consciousness. An odor of bitter almonds can often be detected on the victim's breath or emanating from the body. The odor comes from hydrogen cyanide gas (HCN), which forms when cyanide compounds come in contact with an acid such as the acids normally found in human stomachs. Treatment involves inhaling amyl nitrate together with injections of sodium nitrite ($NaNO_2$) and sodium thiosulfate ($Na_2S_2O_3 \cdot 5H_2O$). However, the effects of cyanide are so rapid that treatment must begin very soon after exposure.

Ingestion of more than 50 milligram mg of hydrogen cyanide, which dissolves in water to form hydrocyanic acid, is usually fatal. With a boiling point of 26° C (79° F), hydrogen cyanide, which can be obtained from the pits of wild cherries, is very volatile. Its odor of bitter almonds is related to the fact that hydrogen cyanide is a component of bitter-almond water. Breathing hydrogen cyanide vapors containing less than 1 percent of the gas can cause death within minutes. The gas can also enter the blood through the skin, but the effects are generally less pronounced than when it is inhaled.

Nazi executioners probably made hydrogen cyanide by adding hydrochloric acid to a cyanide salt such as potassium cyanide. A similar process is used in many states that carry out capital punishment. The first state to use this method of execution was Nevada, which introduced it in 1924. Sodium cyanide pellets are dropped into an acid and react to form hydrogen cyanide. The prisoner then breathes the hydrogen cyanide gas that is produced. Because its action is so rapid, many regard it as the most humane way to carry out an execution. However, the Nazis used it because they found it more efficient than carbon monoxide. *See also* Arsenic Poisoning; Atropine; Opium; Poisons; Strychnine.

References

Chang, Raymond. *Chemistry* 2nd ed. New York: Random House, 1984.

Channel 3000 News. *Woman Arrested in Cyanide Mail Scare.* http://wisctv.com/news/stories/news-980824–081245.html. Accessed August 29, 2001.

D

Date Rape. *See* Rape.

Dead-Body Evidence. *See* Arsenic Poisoning; Ballistics; Bite Marks; Blood; Bruises; Cause of Death; DNA Evidence; Drowning; Forensic Anthropology; Gunshot Wounds; Lung Flotation; Odontology; Pathology; Poison; Rape; Strangulation; Strychnine; Time of Death; Trace Evidence.

Deaver, Jeffery (1950–)

American novelist Jeffery Deaver created Lincoln Rhyme, one of the most remarkable characters in forensic detective fiction. Rhyme, a New York Police Department (NYPD) forensic-team leader and nationally recognized criminalist, was tragically paralyzed when a beam crushed his spinal cord at a construction site crime scene. The quadriplegic investigator is able to move only a single finger, his head, and his shoulders. Rhyme relies on his considerable deductive abilities, technology, and the physical skills of his protégée, fiery Amelia Sachs, to carry out inquiries. Rhyme's specialty is the minute scientific analysis of the physical evidence from a crime scene. His trace-evidence expertise and analysis provide the fabric of Deaver's contorted plots. Rhyme cannot be the stereotypical action hero, so he uses his brain, supplemented by Sachs's legwork and a two-way radio to talk her through decoding crime scenes. In Deaver's most widely recognized novel, *The Bone Collector* (1997), the writer even includes a glossary of forensic terms to aid the reader. In 1999, Universal Pictures made *The Bone Collector* into a feature-length film starring Denzel Washington.

The Chicago-born Deaver, a Fordham law-school graduate who began writing full-time in 1990, builds his plots from extensive research and the work of real forensic investigators. He has written 14 suspense novels, of which the Rhyme series is the best known. Deaver's forensic clues are always played out in the context of what he describes as condensed time. A quick resolution is essential to save lives and foil the next crime. For example, *The Coffin Dancer* (1998) unfolds in a 48-hour period as Rhyme races to stop a hired killer, and *The Empty Chair* (2000) takes place over a weekend. *The Devil's Teardrop* (1999) introduces a new forensic hero, Patrick Kinkaid, a document expert, who has to find a killer in 12 hours before he massacres a crowd of millennium New Year's Eve revelers.

The Bone Collector, the first of the Rhyme thrillers, introduces a depressed Rhyme, on the verge of suicide, and Amelia Sachs, an ex-model turned cop. Deaver uses Sachs to establish his emphasis on crime-scene investigation. The discovery of a body near a New York City railroad track leads Sachs to intuitively preserve and photograph the scene. She buys an inexpensive cardboard camera to photograph the area before it is compromised and even has the presence of mind to use a dollar bill to establish the scale of a footprint left at the site. Intrigued by the murders and the officer's native skill, Rhyme agrees to lead the investigative team, thereby creating a tense collegial association between the wheelchair-bound scientist and the second-guessing Sachs that eventually leads to romance in *The Coffin Dancer.*

The murderer in *The Bone Collector*, who is obviously familiar with forensic crime-solving procedures, baits Rhyme by leaving clues indicating the location of the next murder. *The Coffin Dancer* is a good example of the breadth of Deaver's forensic content. The novel opens with Rhyme studying a soil sample, a favorite trace element, as it relates to the disappearance of a federal undercover agent. Deaver, through his protagonist, launches into a discussion about how a good criminalist should know his local soils and regions. Rhyme even drops some professional secrets, discussing emergency vehicles as a great source of good soil evidence. They are usually at the scene before it is contaminated, ordinarily have good tires with deep grooves that hold trace evidence, and drive from the scene directly to the hospital, eliminating some conflicting soil evidence. The treads of shoes, described by Rhyme as spongelike, as well as the pockets in clothes, provide all sorts of forensic clues.

In *The Empty Chair*, Deaver returns to soil as a valuable clue. He meticulously details Rhyme's painstaking review of the trace evidence to track down a suspect in the wilds of North Carolina. The author also comments on bad forensic science in the novel. To Rhyme's dismay, local police use spray paint, notorious for running, to outline a body and fail to make any attempt to evaluate the footprints left behind at the crime scene. In order not to compromise the scene of a crime Sachs uses elastic bands around her shoes to

distinguish her prints from those involved in the inquiry.

In fact, Deaver's novels are full of nuggets of forensic lore such as the fact that 90 percent of a bomb survives a blast, leaving enough residue to type the device. At another point, Rhyme informs readers that unclaimed airline luggage is frequently donated by the airlines to the FBI, where it is used to establish the effects of explosions on baggage. Deaver explains, too, that strangulation victims bite their tongues, that fellow cops are among the greatest crime-scene contaminators, and that the FBI's Explosives Reference Collection is the most extensive database on explosive devices in the world. In *The Devil's Teardrop*, Deaver explores forensic facts on paper, ink, and handwriting. Blood spatter, tool marks, and fingerprints are other topics that come under Deaver's microscope in creating the Rhyme novels.

References

"Author Jeffery Deaver: The Empty Chair." http://chat.yahoo.com/c/events/transcript/authors/051000deaver.html.

Bickel, Bill. "A Chat with Jeffery Deaver." http://crime.about.co . . . /crime/library/weekly/aa090299.htm.

Deaver, Jeffery. "Jeffery Deaver—The Official Website." http://www.jefferydeaver.com/.

Gallagher, Cathy. "Lincoln Rhyme Returns." http://mysterybooks.a . . . books/library/weekly/aa000525a.htm.

Deptford Murders (1905)

The Deptford murders, of an elderly couple in the London suburb of Deptford, were the first case in England in which the perpetrators were convicted on the basis of fingerprint evidence. Earlier, fingerprinting had been used as a way to check criminals' aliases and as ancillary evidence in a number of legal cases, but this case was the first true test of fingerprinting as evidence.

As 7:15 A.M. on Monday, March 27, 1905, the local milkman saw two men emerge from the paint shop at 34 High Street. The business was managed by 70-year-old Thomas Farrow and his wife Anne.

Early that morning, a painter, waiting for friends, saw an old man with what appeared to be blood on his face, hands, and shirt come out of the store. He looked up and down the street and went back inside. The witness looked for a policeman, but when his friends arrived, he left. At 8:30, the shop boy, William Jones, came to work and found the store locked. This was very unusual since Farrow had a habit of opening early to catch tradesmen on the way to their jobs. Unable to rouse anyone inside the store, Jones went to tell the owner, George Chapman. Chapman sent an assistant with the boy, and the two broke into the business via the kitchen window. There they discovered Farrow dead on the floor due to extensive head injuries. Farrow's wife was in her bed, barely alive, but suffering from injuries similar to her husband's. Mrs. Farrow died within days without giving any information to law-enforcement officials.

In reconstructing the crime, police believed that it was a robbery gone bad. Farrow usually turned over the weekly receipts to Chapman on Monday morning. When police searched the premises, they found no money, but did discover two black stocking masks and an empty cash box under the bed. Police speculated that criminals broke into the house. Hearing the disturbance, Farrow got up to investigate. The culprits overcame the manager, beating him severely and leaving him for dead. Then they went upstairs and did the same to Mrs. Farrow, discovering the cash box in the process. Investigators believed that it had been the wounded Farrow that the painter and a young girl had seen at the door of the business. The discovery of the box would change criminal investigation forever.

A clear thumbprint was found on a tray inside the box. The local authorities and Scotland Yard printed anyone, including the dead couple, who had access to the evidence and found no match. A comparison with known criminals in Scotland Yard's files also failed to produce a suspect. The police concluded that the print was probably the murderer's.

On the basis of a witness's partial description and local gossip, the police picked up two neighborhood toughs, the Stratton brothers, for questioning. The brothers, Alfred and Albert, were known to be prone to violence and had been questioned in previous criminal investigations but never arrested. In the course of the interviews, the much-amused suspects were fingerprinted. Scotland Yard's experts established to their satisfaction that the print on the box was the right thumb of the older Stratton, Alfred. The police were convinced that they had their murderers, but the question remained, would the fingerprint be admissible as the major piece of evidence in an otherwise-circumstantial case?

When the trial opened on May 5, 1905, it was not just the Strattons who were looking at a legal judgment. The future of fingerprinting as evidence was also before the court. Neither the judge nor the jury was familiar with the fingerprinting process. Prosecutor Sir Richard Muir, with the aid of Scotland Yard detective Charles Collins, painstakingly laid out the science of fingerprinting. The detective used two photographic enlargements of the box print and the one taken from Alfred Stratton to present the case. Collins pointed out 11 identical characteristics of the two prints. Since no two people's fingerprints are identical, Collins said that he was persuaded enough by the similarities to swear that they had been made by the same individual, in this case, Alfred Stratton.

Not to be outdone, the defense called Dr. Henry Faulds, one of the key developers of fingerprints as an identity tool, as its chief expert. Faulds's testimony may have been tainted by the bitterness he felt toward the government for its failure to adequately recognize his role as a fingerprinting pioneer. The defense expert disagreed with Scotland Yard's evaluation. Faulds asserted that the prints were different because of discrepancies in the lines of the print. The jury asked for an explanation about why the government print was so clear and the one on the box blurred. Collins, using a jury member, showed that the amount of pressure put on the finger during the process created blurring but did not change the lines that established the print pattern. The jury was clearly impressed and in just under three and a half hours convicted the Strattons of murder. The "Deptford murderers" were condemned to death by hanging, and fingerprints as evidence were established in English law. *See also* Fingerprints.

References

Howe, Sir Ronald. *The Story of Scotland Yard.* New York: Horizon Press, 1966.

Thorwald, Jürgen. *The Century of the Detective.* New York: Harcourt, Brace & World, 1965.

DeSalvo, Albert (1931–1973)

Albert DeSalvo, known as the Boston Strangler, was never arrested for his alleged actions as a serial killer. He confessed to being the strangler to a fellow inmate after being incarcerated for rape. In September 2000, nearly 30 years after his death, some members of his family and some of the family of Mary Sullivan, one of the victims he claimed to have strangled, sued local and state authorities seeking data they believe will prove that DeSalvo was not the Boston Strangler. They suggested that both DeSalvo's and Sullivan's bodies be exhumed so that tissues needed for DNA tests could be obtained. In October 2000, Mary Sullivan's body was exhumed, and more than 60 samples of DNA were collected. In addition, Massachusetts attorney general Thomas Reilly's office found missing crime files, which included evidence from the Sullivan murder. In February 2001, federal court judge William Young let the suit stand. However, there was no resolution as to whether investigators from the attorney general's office will share evidence with the two families or with private investigators. Reilly had argued that because no one was ever charged with the crimes, the petitioners were not entitled to evidence collected in the case.

DeSalvo's relatives believe that he made up the confession so that he could earn money from a book and a movie based on the murders. They contend that his confession contains errors that reveal that he did not commit the murders attributed to him. For example, in his confession, he said that he raped Mary Sullivan prior to killing her, but no semen was found on Mary's body. He also claimed to have gagged her, but no gag was ever found. Police had claimed earlier that DeSalvo's confession contained details that only the strangler could know, including the fact that he would often race home after an attack so he could play with his two children. His family and others maintain that he could have obtained the information from detailed newspaper accounts of the crimes.

In 1956, at the age of 25, Albert DeSalvo settled in Boston, where he was a construction worker with a wife and two children. When time allowed, he posed as a talent scout for a modeling photographer. He would knock on a door and explain to any young woman who might open it that she might have a career as a model if her measurements were right. Molestation commonly accompanied the taking of vital statistics and frequently there would be consensual sex based on a promise of a job. In 1960, this activity led to an arrest that cost him a year in jail.

A killing spree began on June 14, 1962, shortly after DeSalvo was released from prison. During the next 18 months, 12 women in the Boston area were strangled after being raped repeatedly. Police could identify the strangler because his victims were always found with an article of their clothing that had been used to strangle them wrapped about their neck and tied in a uniquely identifiable bow knot.

After January 4, 1964, the murders ceased. However, a rapist, albeit a gentler man than the Boston Strangler, continued to create fear among women in Boston. Descriptions of this rapist reminded police of the "Measuring Man" who had been sent to jail in 1961, and DeSalvo was arrested again. Following his arrest, DeSalvo was identified by one of the women he had raped. He was judged to be mentally ill and was sent to the

Bridgewater Mental Institute outside Boston. Meanwhile, police continued to search for the Boston Strangler.

Asked to establish a profile of the strangler, New York psychiatrist James Brussel proposed that the killer was of southern European heritage because strangling is a common means of killing in that area. He was an impotent homosexual and suffered from an Oedipal complex because of a dominating mother, which explained his killing of older women. The murders of five young women reflected his struggle with impotency.

In truth, unlike most serial killers, DeSalvo seemed truly motivated by sex. His wife told police that he demanded sex four or five times a day. Impotence, which often drives serial killers to fulfill their sexual needs by killing, was not a factor with DeSalvo. He had been hypersexual all his life. Led astray by the profile, police searched for months in Boston's homosexual community without success.

At Bridgewater, DeSalvo conversed at length with a fellow patient, George Nasser, who reported that DeSalvo confessed to being the long-sought-after Boston Strangler. There was no physical evidence linking DeSalvo to the murders, and police were dubious. Nasser might be the strangler. Perhaps he had convinced DeSalvo to confess so that Nasser could pass on part of the reward money to DeSalvo's wife.

Recognizing that defense lawyers would advise him to deny his confession if he were tried for murder, prosecutors charged him only with rape, and he was convicted because of the testimony of victims who readily identified him. He was sentenced to life in prison and sent back to Walpole State Prison. Seven years later, in 1973, the 42-year-old DeSalvo was stabbed to death by another inmate. *See also* Serial Killers.

References

"Accord Urged in Strangler Case; U.S. Wants an End to Evidence Fight." *Boston Globe*, March 1, 2001, p. B4.

"The Boston Strangler, Albert DeSalvo." http://www.crimelibrary.com/boston/bostonmain.htm. Accessed August 21, 2001.

Nash, Jay Robert. *Almanac of World Crime*. New York: Bonanza Books, 1986.

Newton, Michael. *Hunting Humans: An Encyclopedia of Modern Serial Killers*. Port Townsend, WA: Loompanics, 1990.

Zonderman, Jon. *Beyond the Crime Lab: The New Science of Investigation*. New York: Wiley, 1990.

Disputed Documents

The analysis of disputed documents is one of the major tasks confronting forensic scientists. Were the words on a ransom note written by the suspect in custody? Is the signature on a check authentic or forged? Can the words on a charred sheet of paper be made visible? These are the kind of questions that document experts try to answer.

Just by holding two paper samples up to a light, one can readily determine whether they are from different sources. Microscopic examination of the fibers, fillers, sizers, starches, and resins in the paper can also prove useful in comparing a disputed document with a possible source. Quality paper will probably reveal a company watermark impressed on the paper during its manufacture. If the watermark is genuine, that portion of the paper will be thinner and less dense than the rest because fibers are brushed away as the mark is imprinted. A forged watermark, imprinted after the paper has been made, will not be less dense. Density differences can be detected by beta rays (electrons emitted from the nuclei of radioactive atoms), which are deflected by the fibers in paper. The test, known as beta radiography, involves placing the paper in question between a sheet of plastic containing carbon 14 (C-14), a beta emitter, and a film sensitive to beta rays. If the watermark is genuine, its image will appear on the film after a period of 12–24 hours, because the low-density paper in the watermark will deflect the electrons less than the rest of the paper. Beta radiography was used to test a map of Vinland (northeastern North America) supposedly made by Norsemen before Columbus's voyage to the New World. The watermark was proven authentic, indicating that the paper was made circa 1450; however, thin-layer chromatography revealed that ink in the map was not made before 1930.

Handwriting

Handwriting, like fingerprints, is unique, but is far more difficult to identify. Even a handwriting expert cannot determine with certainty that a document was written by a suspect. To provide their best opinion, experts need to examine as many documents as possible, preferably extensive ones, to determine the suspect's natural style. Documents that are crime related are often written crudely in an unnatural style or in a deliberately different manner to avoid the subconscious movements that characterize a person's normal writing. Furthermore, an individual's handwriting changes with age and under the influence of alcohol or other drugs.

In procuring a handwriting sample, the suspect, using a pen or pencil similar to one used to write the document in question, should be required to write several pages of dictated material without receiving any help regarding punctuation, spelling, or the use of upper- or lowercase letters. The dictation should include words and phrases found in the document being used as evidence and should be repeated several times. A suspect will generally have difficulty repeating deliberate attempts to change his or her writing style if he or she is required to write lengthy documents several times.

If the signature on a check is believed to have been forged, the suspect should be asked to write and sign a number of checks for comparison with numerous authentic signatures. A handwriting expert can then examine the style and offer an opinion. Since a person's signatures are never exactly the same, one that is identical with another can be viewed as evidence of forgery. It was probably copied using tracing paper.

To compare a suspect's handwriting with a writing sample taken as evidence, an expert will cover the sheets with tracing paper to avoid contact with evidence. He or she will then mark the high point of each letter along the written lines with a dot. Connecting the dots will produce a pattern of zigzag lines. The patterns can be compared with those of the disputed documents. A similar procedure can be carried out for the low points of each letter and for spaces between letters (if any exist) and words. The angles of the writing can be compared by drawing diagonal lines through letters both above and below the line (real or imaginary) on which the words are written. Finally, the expert will look for similarities and differences in the letters, such as loops in the *f*'s, *g*'s, *h*'s, *j*'s, *k*'s, *l*'s, *q*'s, and *y*'s.

The legality of requiring suspects to provide samples of their handwriting was not established until 1967. In *Gilbert v. California* (1967), the U.S. Supreme Court ruled that asking a suspect to submit handwriting samples before the appointment of counsel was legal and did not violate protection provided by the Fifth Amendment. In *United States v. Mara* (1973), the Court maintained that demands for writing samples did not constitute unreasonable search and were not, therefore, in violation of the Fourth Amendment.

Indented Writing

After one writes on a pad of paper, the sheet beneath the one on which the message has been written may have indentations that match the words written on the top page. (If there had been carbon paper between the pages, the indented writing would have appeared as a copy of the upper sheet.) Such indented writing may be a valuable clue if the indented words match those on such evidence as a ransom note or a bookmaker's betting records.

Indented writing can be seen more clearly when it is viewed with side lighting, when light is passed over the document at an oblique angle. Even very faint indentations can be seen by using electrostatic detection apparatus (ESDA). The reason is not clear, but the electrical properties of paper are changed by compression. Consequently, indented words can be detected using ESDA.

A scientist viewing results from the electrostatic detector apparatus (ESDA) that processes paper items for indented handwriting. Courtesy of the Georgia Bureau of Investigation, Division of Forensic Sciences.

The document in question is placed on a metal plate and covered with a sheet of mylar held firmly in place by a vacuum. An electric field is applied to the apparatus, which is then sprayed with a mixture of toner and fine glass particles that stick to the charged portions, making the indented words visible.

The Secret Service used ESDA in 1988 while investigating scientific fraud for a congressional committee. Its tests showed that laboratory data recorded in notebooks, supposedly in sequence, actually came from experiments that had been conducted two years apart. For example, page 25 from a 1986 notebook had been written while it was directly above page 30 in a 1984 notebook.

Typed Documents

Although computers and word processors have nearly eliminated the use of typewriters, a few disputed documents written on a typewriter still appear, and many earlier typed documents are often examined to determine the date and machine on which they were written. Different makes and models of typewriters and printers produce different print, which can often be identified. Evidence of forgery has sometimes been based on dates. A document supposedly written in 1941 obviously could not have been typed on a machine that was not manufactured until 1950. Occasionally, investigators have found a copy of the original print on discarded carbon paper or on a used typewriter ribbon.

With wear, typewriters become unique and identifiable. Some characters appear heavier or lighter, thicker or thinner, or possibly reveal chips in the type ball or bar that presses the letters against the paper. Disputed documents can be compared with a sample typed or printed using the suspected machine. Comparisons can be made by placing the documents side by side or by projecting and superimposing images of the two documents on a screen so that similarities or differences become apparent.

Charred and Hard-to-Read Documents

Even when criminals attempt to destroy evidence by burning, forensic scientists are frequently able to restore enough print or written words to provide valuable evidence. The charred bits of paper found at a crime scene are sprayed with a dilute solution of polyvinyl acetate in acetone. The pieces of paper can then be flattened and placed on cotton in a box before being carried to a laboratory, where they are placed in a mixture of water, glycerin, and alcohol in which they float. By photographing the fragments in oblique or infrared light, the words or numbers may become visible.

Digital image processing may enhance the visual quality of a hard-to-read document. The document is scanned with a television camera while a computer converts the image into a series of digital intensity values called pixels. The image may be improved by darkening, adding light, or changing the contrast between pixels.

One killer who thought that a bomb would blow any evidence to undetectable smithereens was thwarted by persistent forensic scientists. On December 27, 1922, Clementine Chapman of Marshfield, Wisconsin, opened a package mailed to her husband James. The package contained a bomb, which exploded, killing her and crippling James Chapman. Scraps of the packaging paper were collected and examined by John Tyrell, a forensic expert on disputed documents, who found that the address had been written awkwardly by a person who misspelled Marshfield as Marsflld. From that information alone, Tyrell was quite certain that the killer was foreign and probably Swedish. The postmark, however, led investigators to the mailbox of Thorval Moen, who lived in a rural area outside Marshfield. Moen, who was not Swedish, denied mailing the package. The only Swede in the area was John Magnuson, who could have easily mailed the package from Moen's mailbox. It was known that Chapman and Magnuson had been engaged in a land dispute.

Police obtained a warrant and searched Magnuson's farm, where they found pieces of wood similar to those used in making the bomb. A writing sample revealed that Magnuson's handwriting was similar to that on the package and Marshfield was again misspelled, this time as Marsfld. Handwriting expert Albert Osborn agreed that the writing was so unique that it must have come from the hand of Magnuson. Professor of the Swedish language at the University of Minnesota Jill Stromberg explained that the *sh* sound does not occur in Swedish and neither the *ie* or *ei* combination of letters appears in Swedish. Consequently, *marsh* would be pronounced and spelled *mars*, and field would be written as *fld* or *flld*.

In addition, another forensic expert testified that wood fragments from the bomb matched shavings found in Magnuson's workshop, and still another expert found that pieces of metal from the bomb matched steel found in the same workshop. On the basis of all this forensic evidence, Magnuson was

Handwriting on paper with ink spilled over it. Courtesy of the Georgia Bureau of Investigation, Division of Forensic Sciences.

found guilty and was sentenced to life in prison.

Ink Evidence

The use of microspectrophotometry makes it possible to compare inked lines without destroying them. Inks from pens and paper can be compared using thin-layer chromatography. The Secret Service controls the International Ink Library, which has information on the chemical composition of more than 6,000 inks, and the Bureau of Alcohol, Tobacco, and Firearms Laboratory has chromatogram patterns from more than 3,000 different inks that can be used for purposes of comparison. Many manufacturers, at the bureau's request, now tag their products so that the year they were made can be readily determined. In a number of cases, the data stored at the bureau have shown that dated documents were prepared using ink made well after the date on the document (backdating). In the Magnuson case discussed ear-

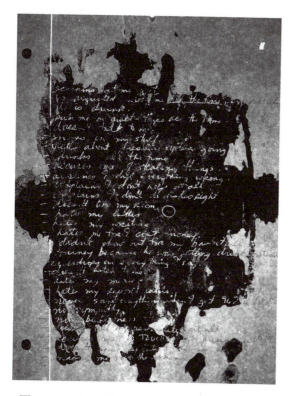

The same piece of paper, photographed using infrared radiation. Courtesy of the Georgia Bureau of Investigation, Division of Forensic Sciences.

lier, chromatography was used to show that the ink used to address the bomb was the same peculiar mixture of inks found in a pen available to Magnuson. Forensic experts were also able to determine that the pen tip used to write the address was the same type as the pen that contained the ink.

The Hitler Diaries

The forged Hitler Diaries illustrate the role that forensic science can play in cases involving disputed documents. In 1981, Gerd Heidemann, a staff journalist for the German publisher Gruner und Jahr, claimed to have located a man who had acquired the long-sought-after diaries of Adolf Hitler. The company agreed to pay this source, who wished to remain anonymous, the equivalent of two million dollars for the documents. Gruner und Jahr hired several handwriting experts who compared the documents with

known samples of Hitler's writing and agreed that the diaries were authentic.

As Heidemann provided increasing numbers of the 27 volumes, the work progressed at a rapid pace until German forensic scientists reported that the documents were a hoax. Using ultraviolet light, the scientists found that the paper on which the diaries were written contained a paper whitener known as blankophor, which had not come into use until after World War II. Furthermore, the inks used to pen the documents were also post-1945, and the binder threads were made of viscor and polyester, both of which were not available during Hitler's lifetime. Finally, measurements of the evaporation rate of the chloride in the ink revealed that all the diaries had been penned within the last year.

Heidemann, who had diverted much of the company's money into his own accounts and was facing prison, now revealed his source. It was Konrad Kujau, whom police identified as a known forger of art and Nazi memorabilia. Both Heidemann and Kujau received jail sentences, and Gruner und Jahr became painfully aware of the difficulty experts face in comparing handwriting samples.

The Gregory Breeden Case and an Issue

Typically, people who are arrested for passing bad checks are pressured by prosecuting attorneys to make restitution before charges are filed, but the case of Gregory Breeden is quite different. Breeden, who was sentenced in Kansas City, Missouri, on July 5, 1995, is serving two consecutive 5-year sentences and one concurrent 4.5-year sentence for writing bad checks totaling $1,800. In August 2000, he was denied parole despite his recognized eligibility.

Authorities are reluctant to release Breeden because he is a prime suspect in the murders of seven people beginning in 1982. He was, in fact, charged with one of those murders in 1996, and there is reason to believe that he is a serial killer. After his third marriage ended, he was seen on the strip in Independence, Missouri, where prostitutes are available. Many of these women told their probation officer that they feared Breeden, and his neighbors in Kansas City had seen him with one of the murdered prostitutes. For these and, perhaps, other reasons, Breeden became the main suspect for the murders (by stabbing) of six Independence prostitutes and the shooting of a 13-year-old girl. After Breeden's arrest for a bad check in 1994, police searched his home and found clothes and other items that had belonged to some of the murdered prostitutes. However, convincing evidence such as a murder weapon, blood, or bloodstains was never discovered. As a result, the murder charge was dropped in 1999.

Some lawyers are disturbed by the sentences given to Breeden. Ron Hall, a defense attorney responding to questions about Breeden, said, "It should bother us. The law assumes a man will be punished for the crime he did, not the crime that we think he did" (Canon). John Fougere, spokesman for the board that denied Breeden parole, stated, "It is legal to consider his status as a suspect in considering parole. They certainly have the power to revisit that. . . . But they never planned to set [Breeden] free" (Canon).

The FBI Crime Laboratory's Questioned Documents Unit

The Questioned Documents Unit (QDU) at the FBI Crime Laboratory supports federal, state, local, and international law-enforcement agencies by providing advanced technical and forensic support with examinations, reports, and testimony; technical support of field investigators; training provided to FBI and other federal and state examiners; and research and evaluation of newly developed technology. To aid in physical and comparison examinations, the QDU maintains many files, including ones of anonymous letters and bank-robbery notes, check writers, national fraudulent checks, national motor-vehicle certificates of title, office equipment, shoeprints, tire treads, and watermarks. *See also* Chromatography; FBI Crime Laboratory; Forgery; Lindbergh Kidnapping and Trial; Mad Bomber, The; Spectrophotometry.

References

Canon, Scott. "Uncharged Murder Suspect Stays in Jail." *Boston Globe*, Sept. 20, 2000, p. 1.

Crimes and Punishment: The Illustrated Crime Encyclopedia. Westport, CT: H.S. Stuttman, 1994.

Evans, Colin. *The Casebook of Forensic Detection: How Science Solved 100 of the World's Most Baffling Crimes.* New York: Wiley, 1996.

Held, Dorothy-Anne E. "Handwriting, Typewriting, Shoeprints, and Tire Treads: FBI Laboratory's Questioned Documents Unit." *Forensic Science Communications,* April 2001. http://www.fbi.gov/hq/lab/fsc/backissue/april2001/held.htm. Accessed August 24, 2001.

Lane, Brian. *The Encyclopedia of Forensic Science.* London: Headline, 1992.

District Attorney

In the United States, a district attorney is a public official responsible for prosecuting criminal cases locally for the state or federal government. The district attorney decides which cases go to trial based on the basis of available evidence. In some state jurisdictions, the district attorney is called the prosecuting attorney or county attorney. In the federal court system, there is a district attorney for each of the federal judicial districts, and these district attorneys can also be referred to as U.S. attorneys.

Most states have an elected district attorney for each of their counties. Also, in almost every state, the state attorney general has no control over district attorneys. This means that a district attorney is a very powerful person in the criminal justice system. There is no requirement that all district attorneys in a state have the same policies or priorities. As elected officials, district attorneys can be affected by local public sentiment and influenced in their enforcement priorities. This is particularly true if the district attorney wants to pursue a political career leading to a higher office.

The district attorney is at the center of the criminal justice process. Law-enforcement agencies arrest suspected criminals, but it is the district attorney who controls which cases will be brought to court. He or she must carefully consider the evidence and decide whether or not the documentation is sufficiently convincing to obtain a conviction. It is also the district attorney's responsibility to decide with what crime the person should be charged, and this makes a difference as to the sentence or punishment. The district attorney is also involved in plea-agreement negotiations and makes recommendations to the judge on pretrial release and sentences. Many county district attorneys have responsibility not only for criminal matters but for civil cases as well. This is also true at the federal level, as U.S. attorneys are local trial lawyers for civil and criminal cases.

Reference

Renstrom, Peter G. *The American Law Dictionary.* Santa Barbara, CA: ABC-CLIO, 1991.

DNA Evidence

DNA (deoxyribonucleic acid) evidence is as unique as fingerprints in identifying a suspect. Consequently, its use by forensic scientists has grown dramatically in recent years. DNA is a polymer found in the cells of living organisms. It makes up the chromosomes that are passed from one generation to the next in the egg and sperm cells that unite to form a zygote, which, through millions of cell divisions, becomes an embryo, a fetus, and finally a baby. Humans have 23 pairs of chromosomes in most cells. Egg and sperm cells contain only one member of each of the 23 pairs; therefore, each parent contributes one member of each pair of chromosomes to the zygote. Because most cells of the human body contain a nucleus with chromosomes, DNA can be obtained from blood, bone, semen, saliva, skin, hair follicles, and muscle cells.

The structure of DNA molecules was first deduced by James Watson and Francis Crick in the 1950s. They found that a DNA molecule is similar to a long spiral ladder or double helix. The sides of the ladder are made of deoxyribose sugar and phosphate groups.

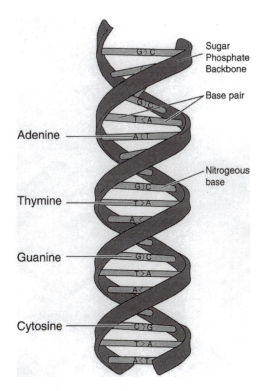

Deoxyribonucleic acid (DNA). DNA molecules are polymers made of many nucleotide units. The rungs are base pairs; the sides are sugar and phosphate units arranged one after another. National Institutes of Health, National Genome Research Institute.

The rungs consist of pairs of bases (adenine [A], cytosine [C], guanine [G], and thymine [T]). Nucleotides, which differ only in the bases they contain, are the fundamental units of the polymeric DNA molecules. Each nucleotide consists of a deoxyribose sugar, a phosphate, and a base. A pair of bases (A and T or C and G), one from each nucleotide, constitutes the rungs that hold opposite sides of the sugar-phosphate ladder together. The average human chromosome contains 100 million base pairs. The total number of base pairs in the chromosomes of a human cell is about 3 billion. By a series of rather complicated reactions, the nucleotides in DNA code the development and life processes of living organisms. For that reason, DNA is often referred to as the genetic blueprint for life.

The base pairs that join the two sides of the double helix are the result of chemical bonds between either A and T or C and G. Adenine and thymine will bond, and so will guanine and cytosine. However, A-C, A-G, T-C, and T-G bonds do not form. Consequently, if the bases that form half-rungs attached to one side of a portion of the ladder are AATCGGA, the bases forming half-rungs on the opposite side of the ladder will be TTAGCCT.

Most human DNA is the same for all humans, but a few sections differ. Some of these sections have more or fewer identical sequences than others and make each person's DNA unique. These are the sections that are used to identify people. What are known as restriction enzymes act like scissors that can be used to cut the DNA molecules at specific sites. A great many of these enzymes are available; consequently, forensic scientists can choose where they want to snip the DNA.

DNA in Forensic Science

A testing technique known as RFLP is used to carry out DNA typing. In this process, restriction enzymes are used to cut portions of the molecules that differ slightly among humans into fragments of different lengths. Length differences of DNA strands are referred to as restriction fragment length polymorphisms (RFLPs); hence the name of the test. Fragments from different sources, such as a victim, a suspect, and a sample obtained from a crime scene are placed side by side in a gel and subjected to electrophoresis. The smaller particles will move through the gel at a faster rate than the larger ones. Once the electrophoresis is completed, the separated fragments are transferred to a nylon membrane in much the same way that ink is transferred to a blotter. Next, radioactive DNA probes (pieces of DNA that have been exposed to radiation) that bind with specific fragments are added to the pattern on the membrane. The fragments with which the probes unite are known as targets. X-ray film is then placed over the nylon for several days. After the film is developed, bands can be seen on it. The bands are the result of radiation released by the probes that stuck to specific fragments.

By comparing the parallel patterns of bands from the different sources, it is possible to determine whether two samples of DNA could or could not have come from the same individual. A probe producing a match might reveal a DNA target that occurs in 1 of 10 members of a population, but 9 such probes, each with an occurrence of 1/10, would establish a frequency of one in a billion ($0.1^9 = 0.000000001$). FBI rules for sampling require 13 matches, and the bureau claims that 200 cases have been solved using DNA evidence. At the same time, 75 people on death row have been released when DNA evidence proved their innocence.

To carry out RFLP testing, thousands of DNA molecules are required. Because DNA samples are often very limited in quantity, RFLP testing of DNA would be of little use in forensic science were it not for DNA polymerases—enzymes that will assemble new DNA strands in sequence so as to make copies of the original. Furthermore, these enzymes can act outside the cell so that billions of copies of a DNA sample can be prepared in the laboratory. By the use of a technique known as PCR (polymerase chain reaction), as few as 50 molecules of DNA from a crime scene can be copied repeatedly to prepare enough DNA for RFLP testing. Each PCR cycle doubles the amount of DNA, so within a few hours 30 cycles can increase the quantity of DNA a billionfold (2^{30}). As a result, even a sample of DNA as small as the saliva on a cigarette butt can provide sufficient material for comparison with other DNA.

In England, more than 350,000 genetic "fingerprints" are available to police, and up to 500 matches between crime-scene samples and gene prints on file are made each week. English police often conduct mass screening in an effort to find DNA matches. Such screenings have provided a 70 percent success rate. Anyone can decline to provide a DNA sample, but that person is then regarded with suspicion.

Despite the reliability of DNA evidence and the fact that police look for it first, it plays a role in less than 1 percent of all crimes. Even in cases involving rape, less than

Burned pelvis bones. If bones are not reduced to ash, they can be used either to obtain nuclear DNA results from the marrow, if any of it survives, or mitochondrial DNA from the burned bones. Courtesy of the Division of Forensic Science, Commonwealth of Virginia, Department of Criminal Justice Services.

50 percent result in DNA evidence that can be used. Technicians contaminate vaginal swabs; the rapist uses a condom; the victim bathes before reporting the crime. Even if DNA evidence is available from the crime scene, the suspect must be found and DNA evidence collected from him or her. But DNA evidence is so significant that police will do almost anything to obtain it. In 1998, a detective tailing a man suspected of robbery and rape used a paper towel to sop up saliva when the suspect spit on a sidewalk. Forensic scientists found a match between the DNA in the saliva and that in semen collected from one of the victims.

More often there is no suspect. A database with DNA typing of every citizen would make it possible to scan the data and obtain a match. Such a proposal has been made, but many believe that such action would be a violation of an individual's civil rights and that DNA data could be used for political as well as forensic purposes.

Cases

In 1984, researchers at England's Leicester University invented the RFLP technique discussed earlier for recording DNA segments in a bar-code pattern. Three years later, police seeking the person guilty of raping and killing two teenage girls took blood samples from every man between 13 and 30 years of age who lived in three villages near Leicester. Testing revealed that the DNA in blood from 27-year-old Colin Pitchfork, a baker from Leicester, matched the DNA in the semen recovered from the girls' bodies. Pitchfork was the first person convicted of a crime on the basis of DNA testing. At the same time, 17-year-old Rodney Buckland, who had been a suspect, was cleared on the basis of the DNA evidence.

The first conviction resulting from DNA evidence in the United States occurred in 1987. Tommie Lee Andrews had left his fingerprints on the screen door of one of his rape victims, but police believed that he was a serial rapist who had terrorized the Orange County area of Florida, and they sought to convict him of other rapes and robberies. On the basis of blood samples from Andrews and semen collected 17 months earlier from Nancy Hodge, a rape victim who had identified Andrews as the man who had raped her, forensic scientists were certain that Andrews was guilty. However, both Andrews's sister and girlfriend testified that he had been at home on the night Hodge was attacked. The first trial resulted in a jury deadlocked at 11 for conviction to 1. Defense strategy prevented the experts for the prosecution from explaining how they arrived at a figure of 10 billion to 1 as the probable frequency of the DNA used as evidence. In two trials that followed, the prosecution was able to explain the statistics associated with the DNA evidence, and Andrews was sentenced to a total of 115 years.

DNA "fingerprints" do not always lead to conviction. O.J. Simpson, the famous football player accused of murdering his ex-wife and her friend, was acquitted in 1995 because the defense successfully accused the police of sloppy blood work and possible tampering with evidence.

In the fall of 1998, a woman who had been a comatose patient in a Lawrence, Massachusetts, nursing home since 1995 gave birth to a premature baby. To determine who had raped the woman, police obtained voluntary blood samples from more than 20 men who had access to the victim. Police later arrested a nurse's assistant whose DNA indicated that he had fathered the child. He was later tried and convicted.

The Ronald Cotton Case

In July 1984, Jennifer Thompson retained her rationality even as she was raped at knifepoint. She claimed she carefully observed the facial features of her attacker and was, within hours after being attacked, able to work with a detective to produce a composite sketch of the rapist. A week later, she identified the man who had raped her from a lineup of six men. During the court case that followed, she pointed to Ronald Cotton and stated that he was definitely the man who had raped her. The jury, swayed by Thompson's convincing certainty about her attacker, found Cotton guilty, and on January 17, 1985, he was sentenced to life in prison.

About a year later, a rapist named Bobby Poole arrived at the prison where Cotton was serving his sentence. Poole bragged to other inmates that Cotton was serving some of the time he (Poole) should have received. Poole's loose tongue, together with the knowledge that a rape similar to Thompson's had occurred an hour later in the same North Carolina town, enabled Cotton's lawyer to obtain a second trial. However, Jennifer Thompson testified and again identified Cotton, not Poole, as the man who had raped her.

In 1995, still professing his innocence, Cotton convinced University of North Carolina law professor Richard Rosen to ask the courts to conduct DNA tests on him and Poole. Rosen was concerned that a man had received a life sentence on the basis of eyewitness testimony, which he regarded as unreliable. Rosen's concern was justified. The DNA results revealed that Poole, not Cotton, was Thompson's rapist. (Thompson later apologized to Cotton and became an ardent opponent of the death penalty.)

The Clyde Charles Case

Early on the morning of March 12, 1981, brothers Clyde and Marlo Charles left a Louisiana bar and started hitchhiking in opposite directions for their respective homes. At 3:50 A.M. that same day, police found a young battered woman on her hands and knees at the side of a road not far from the bar the Charles brothers had left somewhat earlier. The woman, who was hysterical, said that she had been raped by a black man. The deputy sheriff who had discovered the woman radioed other deputies to be on the lookout for a suspect. At 4:05 A.M., police found Clyde

Charles hitchhiking and took him into custody. At 4:35 A.M., the victim identified Charles, the only black man in a lineup, as the rapist.

At the trial, Charles denied the charge, but the victim's identification and claim that the rapist had told her his name was Clyde together with trace evidence—two light brown hairs on Clyde Charles's shirt that were similar to the victim's—led to a conviction and a sentence of life in prison. During the trial, defense lawyer Thomas Divens filed a motion naming Marlo Charles as an alternative suspect, but the motion was never pursued. A rape kit containing a sample of semen from the victim's vagina was placed in a refrigerated locker; however, DNA testing was not available at the time.

In 1991, the imprisoned Clyde Charles learned of DNA testing and requested that his DNA be compared with the DNA stored in the rape kit. His request was denied by Louisiana courts for seven years. In 1998, Clyde Charles's sister, Lois Hill, learned of the Innocence Project at Yeshiva University's Benjamin Cardozo School of Law. Barry Scheck, a lawyer with the project, filed a federal civil rights lawsuit in May 1999. On November 19 of that year, Scheck informed Charles that the DNA test had shown that he was innocent. In December, Charles was released from prison; however, the courts then pursued Divens's alternative-suspect motion and required Marlo Charles to undergo DNA testing. On April 6, 2000, the results showed that Marlo had been the rapist, and he was arrested the next day. He was convicted of the crime on March 1, 2002.

More DNA Cases

Often, DNA evidence has been used to prove the innocence of people wrongly charged with or convicted of crimes they did not commit. In 1994, Derrick Coleman, a New Jersey Nets forward, was cleared of raping a young Detroit woman. DNA typing revealed that semen collected from the woman could not have come from Coleman.

The charges against Newfoundland resident Gregory Parsons, who had been convicted in 1994 of murdering his mother in 1991 were stayed in 1998 when new DNA testing showed that evidence collected at the crime scene could not have been left by Parsons. Parsons received legal help from James Lockyer, a Toronto lawyer who has used DNA evidence to free several people accused of murder.

In 1992, Diane Burkhard, in an effort to exonerate her mother and arrive at the truth about her own paternity, contested the will of John Brooks, claiming that she was his illegitimate daughter. Burkhard had learned that she had probably been conceived at the time Brooks raped her mother in 1943. Burkhard's mother, who was attacked on her 18th birthday by her uncle's brother-in-law, had never revealed the circumstances surrounding Burkhard's conception and birth until recently. A state court ordered that Brooks's body be exhumed and tested for paternity. A pathologist removed ribs and muscle tissue. DNA from Brooks's tissue was compared with that from the blood of Burkhard and her mother. The test indicated that Brooks was very likely Burkhard's father. Burkhard settled for half the estate ($90,000).

In 1998, geneticists tested descendants of Thomas Jefferson in an effort to determine whether or not he had sired offspring with Sally Hemings, one of his slaves. The evidence indicated that Jefferson was probably the father of Hemings's youngest son. More definite results cannot be obtained without exhuming Jefferson's body.

That useful DNA might be obtained from Jefferson's remains suggests the longevity of DNA. Concrete evidence of the durability of DNA was demonstrated in 1997 by a team of scientists led by Dr. Svante Pääbo (University of Munich), who extracted and analyzed DNA from the arm bone of the first Neanderthal skeleton to be discovered in 1856. (Neanderthals disappeared about 30,000 years ago.) Of the 379 DNA sites they examined, Neanderthal DNA differed, on the average, in 25.6 places from a similar section of modern human DNA. Comparable DNA from modern humans differs in only 8 places on the average between individuals. On the other hand, equivalent DNA from chimpanzees differs from that of humans in about 55 places. The scientists concluded that Neanderthals were a different species from *homo sapiens.*

Shortly after her son was murdered in 1997 Joyce McField was approached by a woman who claimed McField's son was the father of the fetus she was carrying. After the child was born, McField began to have doubts and hired a Houston company known as Identigene to conduct a paternity test. Police had kept a sample of her son's blood for evidence. When Identigene compared DNA from the baby with that obtained from the blood of McField's son, it was clear that the baby was not her granddaughter.

In 2001, efforts are under way to perform DNA testing in conjunction with the Boston Strangler case, in which victims were murdered between 1962 and 1964. Some members of the family of Albert DeSalvo (the murder suspect) and some family members of one of the victims in the case, Mary Sullivan, do not believe that DeSalvo was the murderer. Consequently, they have teamed in a lawsuit to have DNA evidence collected from Sullivan's exhumed body and from Richard DeSalvo, the brother of Albert DeSalvo, because Albert DeSalvo was killed in a fight in prison in 1973, and because some parts of DNA from brothers will be the same.

DNA Evidence and the Future

In the future, portable computers that can process DNA will allow investigators to process evidence at the scene of a crime. Eventually, it may be possible to provide a description of the suspect based on DNA evidence. The FBI has already opened a national computer system that enables law-enforcement agencies throughout the country to compare DNA evidence from a crime scene with DNA from convicted felons and that collected at the scenes of unsolved crimes. These FBI computer files contain genetic "fingerprints" from a quarter of a million convicted felons and evidence from 4,600 unsolved cases.

In late 1998, The *New York Times* reported that New York City police commissioner Howard Safir was considering taking saliva samples as well as fingerprints from everyone who is arrested. (The DNA record would be destroyed if the person arrested were found innocent.) Currently, DNA records are kept only for people who are convicted of major crimes such as murder or rape. Safir's proposal was opposed by the New York Civil Liberties Union because it viewed this as a violation of the Fourth Amendment's protection against unreasonable search and seizure. Attorney Barry Scheck, an expert on the legal aspects of DNA evidence, believes that such a plan, which is in effect in Louisiana and South Dakota, would be impractical.

Safir's suggestion, which is supported by the International Association of Chiefs of Police (IACP), has received less publicity since he resigned in August 2000 to take an executive position with an Atlanta security firm, Choice Point, where he will concentrate on the use of DNA testing in law enforcement. However, the New York state legislature passed a bill in 1999 that mandates DNA testing for those who commit violent felonies, and New York's governor, George Pataki, has proposed expanding the legislation to include testing of those guilty of misdemeanors. Such testing will become far easier when portable DNA labs being developed by the Whitehead Institute for Biomedical Research become available, possibly as early as 2002. *See also* Blood; Chromatography; Fingerprints; Forensic Anthropology; Lineup; Paternity Testing; Rape.

References

Adler, Jerry, and John McCormick. "The DNA Detectives." *Newsweek*, Nov. 16, 1998, p. 66.

Buckley, William F., Jr. "O.J. on Our Mind." *National Review*, June 26, 1995, p. 71.

"DNA Tests Clear Prisoner of Rape, But Implicate His Younger Brother." *NewsTimes* (Danbury, CT), November 26, 2000.

Ellis, David. "Sin of the Father: A Child Conceived in Rape 51 Years Ago Proves Her Mother Was Telling the Truth." *People Weekly*, July 25, 1994, p. 38ff.

Evett, I.W., and B.S. Weir. *Interpreting DNA Evidence.* Sunderland, MA: Sinauer Associates, 1998.

Gwynne, S.C. "Genes and Money." *Time*, Apr. 12, 1999, p. 69.

Kluger, Jeffrey. "DNA Detectives." *Time*, Jan. 11, 1999, pp. 62–63.

Mathis, Ayana. "Stop, Drop, and Swab." *Village Voice*, May 31–June 6, 2000.

O'Neill, Helen (Associated Press). "A Rape Victim Regrets Role as Witness." *The Boston Globe*, Oct. 1, 2000, pp. 18–19.

Pan, Esther Begley Sharon. "Jefferson's DNA Trail." *Newsweek*, Nov. 9, 1998, p. 66.

Riley, Donald E. "DNA Testing: An Introduction for Non-Scientists. An Illustrated Explanation." *Scientific Testimony—An Online Journal.* http://www.scientific.org/tutorials/articles/riley/riley.html.

Rosenberg, Debra. "A Murder Case That Will Not Die." *Newsweek*, March 5, 2001, pp. 50–51.

Dobkin, Harry (c. 1893–1943)

Harry Dobkin was a British murderer whose case was instrumental in establishing odontology (forensic dentistry) as a forensic tool in criminal investigations. The Russian-born Dobkin married Rachel Dubinski in September 1920. After three days, the two realized that the marriage was a mistake and separated. The brief liaison, however, led to the birth of a child nine months later. Following the birth, there began a series of confrontations between the estranged couple. Over the next 20 years, Rachel Dobkin regularly pursued her ex-husband for his failure to pay court-ordered child support. There were a number of violent meetings, and Dobkin spent some time in jail for his negligence and angry responses to his ex-wife's requests.

In 1941, Dobkin was working as a fire spotter for an insurance company in war-torn London. On April 15 of that year, there was a fire in the bombed-out shell of a Baptist chapel in Dobkin's area. Due to Nazi air raids, such fires were common, and it was extinguished with little fanfare. Fifteen months later, workmen clearing debris unearthed a badly burned human skeleton. Initially the death was attributed to the bombing. However, a closer examination at the scene and later at the mortuary produced a different conclusion.

Pathologist Keith Simpson, a pioneer in forensic dentistry, performed the autopsy and quickly confirmed his field observations. This was not another victim of war; the cause of death was murder. In a crude attempt to cover up the murder and prevent identification, the killer, with no anatomical experience, had severed the head and legs of the victim and then set the fire. To hasten decomposition, the murderer had spread lime over the remains. Despite these obstructions, Simpson was able to draw a number of conclusions based on his examination of the skeleton. The victim was a female, and by measuring bones, Simpson concluded that she was about 5'1". Based upon the sutures in the skull, as well as remnants of dark hair that was just turning gray, the pathologist placed her age between 40 and 50. The state of the remains led Simpson to believe that the woman had been dead for about a year to a year and a half. The fracture of a small bone in the throat established the cause of death as manual strangulation. The autopsy established two other facts that were pivotal in the case. First, this woman had an intrauterine fibroid tumor, and second, and of greater significance, the victim had extensive dental work, which Simpson knew offered the best chance for a positive identification.

A check of local missing persons turned up a prime possibility in Rachel Dobkin. The 49-year-old woman had been reported missing just days before the April fire by her sister. In a police interview, the sister reported that Rachel Dobkin had been treated for a fibroid tumor, which was confirmed by her doctor. She also supplied the name of Rachel Dobkin's dentist.

Although the lower jaw was missing, her upper jaw was intact and Simpson noticed not only the extensive dental work but the unusual dentition of the victim. Dr. Barnett Kopkin, a meticulous record keeper, sketched a chart of Rachel Dobkin's dental work. The doctor's work matched the murdered women's dentistry. Armed with the dental information he needed, Simpson went one step further in developing evidence. He successfully superimposed a photo of Rachel Dobkin over the skull of the murdered woman. The result was conclusive; there remained no doubt; the murdered woman was definitely Rachel Dobkin.

Harry Dobkin went to trial for the murder of his ex-wife in November 1942. The defense tried to dismiss Simpson's findings as mere speculation. Failing that, Dobkin's lawyers tried to convince the jury that if this woman was Rachel Dobkin, she had been killed in an air raid. This proved a futile defense, and Dobkin was convicted of murder. He was executed in January 1943. *See also* Odontology.

References

Shew, E. Spencer. *A Companion to Murder*. New York: Knopf, 1961.

Wilson, Colin and Patricia Pitnam. *Encyclopedia of Murder*. New York: G.P. Putnam & Sons, 1962.

Doyle, Sir Arthur Conan (1859–1930)

Scottish-born Sir Arthur Conan Doyle was the physician and novelist who created the well-known fictional detective Sherlock Holmes, whose methods are often cited as the basis for the scientific techniques used in crime detection today. At the time Doyle began writing his Sherlock Holmes detective stories, real-life detectives were usually uneducated men who used their underworld criminal acquaintances to do their jobs. But people in the 1880s had developed a great respect for science, so when Doyle had Holmes use deductive reasoning and scientific methods such as fingerprinting and firearm identification, the public loved it, and real detectives began using the same methods. Sherlock Holmes cases became required reading in many European police forces. Some believe that it was Doyle's detective stories that motivated Edmond Locard to set up the world's first forensic-science laboratory in France in 1910. Doyle himself used the same methods he wrote about in the Sherlock Holmes stories to prove the innocence of some men accused of crimes they did not commit. One such man was Oscar Slater, who Doyle proved had been wrongly identified as the killer of an elderly woman in 1908. Doyle was able to show that the witnesses against Slater had been coached by the police.

Doyle was educated at Stonyhurst College and the University of Edinburgh, where he developed his interest in chemistry and laboratory analysis. Before receiving his degree, he worked on two ship voyages that would later provide him with material for his adventure stories. While he was studying at the university, Doyle had been greatly influenced by Joseph Bell, one of his professors at Edinburgh. Bell was an expert in the use of deductive reasoning to diagnose disease, and Doyle, in his stories, had Holmes use these same principles to solve complex crimes through ingenious deductive reasoning. The guilty party was often determined by careful examination of muddy footprints or cigarette ashes left at the scene of the crime.

After graduating in 1885, Doyle set up a small medical practice as an eye specialist in Southsea, England. His lack of patients gave him time to write. The first Sherlock Holmes story, *A Study in Scarlet*, was published in *Beeton's Christmas Annual* in 1887. Besides Holmes, there was the narrator of the stories, Dr. Watson, who was Holmes's good-natured, inquisitive friend. In many of the stories, there was also the evil genius, Professor Moriarty.

Doyle was so successful as an author that he gave up his medical practice to devote himself entirely to writing after the publication of the second Holmes novel, *The Sign of the Four*, in *Lippincott's Magazine* in 1890. The *Strand Magazine* also published a series of short stories, *The Adventures of Sherlock Holmes*.

Doyle wanted to spend more time writing historical novels and science fiction. Consequently, in *The Final Problem* published in *Strand Magazine* in 1893, Holmes and Moriarity, while struggling, fall onto the rocks at

Reichenbach Falls. For several years, Doyle stopped writing Sherlock Holmes stories, but the public outcry was so great that in 1902 he published *The Hound of the Baskervilles* as a posthumous memoir of Sherlock Holmes. Then, in 1905, he began writing a new series of Holmes stories that began with *The Adventure of the Empty House* in which he reveals that Holmes was not killed at Reichenbach Falls but had gone to Tibet to carry out research and study with the Dalai Lama.

Doyle served in the Boer War as a physician, and when he returned to England, he wrote *The Great Boer War* (1900) and *The War in South Africa: Its Cause and Conduct* (1902). For these works justifying England's participation in the war, Doyle was knighted. Doyle's oldest son, Kingsley, was severely injured in World War I and later died from pneumonia. After his son's death, Doyle dedicated much time to spiritualism. He gave lectures and wrote 30 books on the subject, including *The Vital Spirit* and *The History of Spiritualism*. Doyle's autobiography, *Memories and Adventures*, was published in 1924. His last Sherlock Holmes book, *The Casebook of Sherlock Holmes*, was published in 1927, three years prior to his death. *See also* Holmes, Sherlock.

References

Arthur Conan Doyle Society. http://www.ashtree.bc.ca/acdsocy.html. Accessed August 21, 2001.

"Biography of Arthur Conan Doyle." http://pantheon.yale.edu/-yoder/mystery/doyle-bio.html.

Huntington, Tom. "Sherlock Holmes." *Historic Traveler.* Feb. 1997.

Redmond, Chris. Sherlockian.Net. http://www.sherlockian.net.

Saferstein, Richard. *Criminalistics: An Introduction to Forensic Science.* Englewood Cliffs, NJ: Prentice Hall, 1990.

Zonderman, Jon. *Beyond the Crime Lab: The New Science of Investigation.* New York: Wiley, 1990.

Drowning

Drowning is the third most common cause of death in the United States after automobile accidents and falls. Often forensic scientists face the task of determining whether the drowning was accidental or a case of homicide. In some cases, they find evidence that the victim was murdered before being dumped in water and that the cause of death was not drowning.

A person under water will hold his breath for as long as possible. When, at last, the first involuntary inhalation occurs, water enters the mouth and throat. At this point, one of two things may happen. In 80 percent of drownings, the victim suffocates when the lungs become filled with water. In the other 20 percent, a phenomenon known as "dry drowning" occurs. The moment water strikes the vocal cords, a strong laryngeal reflex contracts the muscles surrounding the larynx. This prevents air from entering the lungs and squeezes the vagus nerve, stopping the heart. These victims will be found with water in their stomachs but very little water in their lungs.

In the usual "wet drowning," the presence of fluids in the lungs prevents the oxygen in air, which normally fills the lungs, from reaching the blood. When a person breathes, air fills the approximately 700,000,000 alveoli—the tiny air sacs at the end of the tubes leading from the trachea (windpipe). The alveoli are surrounded by capillaries, and it is there that the hemoglobin in red blood cells unites with oxygen. This vital element is then carried by the bloodstream to the various cells of the body. The presence of water in the alveoli prevents oxygen from diffusing into the blood. Since the brain is most susceptible to a deficient supply of oxygen, a drowning person soon loses consciousness.

If the cause of death is wet drowning, the coroner will find a fine frothy foam throughout the respiratory system. The lungs will be soggy; the stomach will contain water that the victim swallowed. Some of the water that enters the lungs during the course of drowning will seep into the left side of the heart, where it will dilute the blood there. If a murder victim was dead and thrown into the water to make death appear to be caused by drowning, the blood in the heart will not be diluted.

Attempts to breathe by a person who is drowning will draw water containing a multitude of diatoms (microscopic plants) into the lungs and/or stomach and from there to the blood and the various organs of the body. An autopsy will reveal these tiny organisms in the body tissues of a drowning victim. In the case of a corpse dumped into water to make death appear to be caused by drowning, there will be diatoms in the lungs but not in other body tissues. Careful analysis of the diatoms can reveal the location in which the drowning occurred because these tiny shelled organisms can be used to identify the area from which they came.

In the case *Grace Dolan, Administrix v. Commonwealth* (1988), a wrongful-death action was entered against a metropolitan district commission by Grace Dolan, administrix, on behalf of a drowning victim. She claimed that the young woman had drowned because lifeguards employed by the commission were negligent in delaying an organized search for the victim after she was reported missing by her brother. She also claimed that hospital entries indicating that the victim had consumed angel dust, marijuana, and alcohol prior to entering the waist-deep water should have been excised because they were based on hearsay from an unreliable source. A forensic pathologist testified that the time for vital functions to cease was less than the time taken by lifeguards to organize a search. Twenty-five minutes elapsed from the time the victim was reported missing until her body was discovered by a volunteer, not a lifeguard. A screening by a hospital toxicologist indicated only that the victim had ingested Valium, a widely used tranquilizer, but in a concentration well below that needed to cause her to become unconscious. The case was settled out of court. *See also* Cause of Death; Crime Scene.

References

The American Medical Association Encyclopedia of Medicine. Ed. Charles B. Clayman, M.D. New York: Random House, 1989.

Cyriax, Oliver. *Crime: An Encyclopedia.* North Pomfret, VT: Trafalgar Square, 1996.

Grace Dolan, Administrix v. Commonwealth. Appeals Court of Massachusetts, Suffolk, 1988.

Junger, Sebastian. *The Perfect Storm.* New York: W.W. Norton, 1997.

Lane, Brian. *The Encyclopedia of Forensic Science.* London: Headline, 1992.

Drugs

Drugs are natural or synthetic substances used to produce physical or psychological effects. One of the challenges faced by forensic scientists is the identification of substances obtained by police officials that are believed to be illegal drugs.

The use of drugs by humans is not new. Sumerians were using opium to relieve pain at least 6,000 years ago, and marijuana was prevalent in Chinese society a millennium later. During the nineteenth century, there were virtually no laws governing drugs. Opium and its derivatives, morphine, heroin, and codeine, were the ingredients of patent medicines that could be purchased without a prescription. Although these medicines were useful in relieving pain, they were advertised as cures for almost every human ailment. Because of the addictive nature and ready availability of these drugs, the percentage of people addicted to narcotics was higher in 1900 than it is today.

People who use drugs for a prolonged period can become drug dependent. Dependence frequently develops among those who use alcohol, heroin, cocaine, and barbiturates. Dependence is less common with codeine and marijuana. The dependence can be either physical or psychological. If one becomes physically dependent, attempts to stop using the drug will cause chills, nausea and vomiting, cramps, pain, insomnia, and convulsions. Psychological dependence occurs when a person wants to use the drug on a regular basis because of the pleasure and false sense of well-being that it provides. Withdrawal when one is psychologically dependent can lead to depression and anxiety, but the physical effects are not present.

Narcotic Drugs

Narcotics are used in medicine to relieve pain and induce sleep. Most narcotics are derived from opium, which contains between 4 and 21 percent morphine. Heroin, which is derived from morphine, is preferred by drug addicts because it is soluble in water and can be injected as a liquid into the bloodstream, a practice known as "mainlining." Codeine is weaker than heroin and is not widely used by those addicted to narcotics. Synthetic drugs with opiatelike effects include methadone and pethidine. Methadone has been used to treat drug addicts because it seems to inhibit the craving for heroin without producing significant side effects.

Hallucinogenic Drugs

Hallucinogenic drugs produce changes in normal thought processes, perceptions, and moods. The most widely used hallucinogen is marijuana, which was introduced into the United States from Mexico in 1920. The drug, which is smoked, is a mixture of the crushed leaves, flowers, stems, and seeds of the marijuana plant (*Cannabis sativa*). A resinous extract of the plant is hashish, which is more potent than marijuana. By 1937, 46 states had passed laws making marijuana illegal. It is often mistakenly classified as a narcotic and designated as such by law. The chemical responsible for marijuana's hallucinatory effects is tetrahydrocannabinal (THC), which was isolated by chemists in 1964. Most of the THC is found in the plant's resin, flowers, and leaves; very little is present in the stem, roots, or seeds.

Other common hallucinogens are LSD (lysergic acid diethylamide), PCP (phencyclidine), STP (dimethyloxymethylamphetamine), DMT (dimethyltryptamine), mescaline, psilocybin, and Ecstasy. LSD, a very potent hallucinogen, is derived from the lysergic acid found in ergot, a poisonous fungus that grows on some grasses and grains. Just 25 micrograms can produce hallucinations for half a day. Ecstasy, also known as MDMA because its chemical name is methylenedioxymethamphetamine, was patented as an appetite suppressant. Later, it was found to produce feelings of happiness and a relaxed body state. However, adverse reactions, some of them fatal, have been reported. The hallucinations produced by these drugs range from a sense of space and time expansion (marijuana) to nightmarish horrors (LSD). Both LSD and PCP can lead to depression, violent behavior, hallucinatory flashbacks, and psychotic reactions.

Depressant Drugs

In small doses, depressants, such as alcohol and barbiturates, often called "downers," produce a mild euphoria followed by drowsiness. Larger doses can cause loss of muscular coordination as well as one's mental faculties. Prolonged use can lead to both psychological and physical dependence.

Tranquilizers, such as Valium, Librium, and Miltown, can reduce anxiety and feelings of stress without inducing drowsiness. Prolonged use of high doses of tranquilizers can lead to both psychological and physical dependence.

Sniffing glue, popular among children and among adults who cannot afford the usual drugs, produces a sense of euphoria that is followed by the effects of typical depressants. Prolonged use can lead to liver, heart, and brain damage.

Stimulants

Stimulants such as amphetamines ("uppers" or "speed") at prescribed dosages of 5–20 milligrams per day provide users with increased alertness and a sense of well-being and confidence. Undesirable side effects include loss of appetite, insomnia, itching, and paranoia. "Speed freaks" who inject amphetamines or methamphetamines experience a rush of pleasure that is often followed, to their dismay, by depression.

Cocaine, another stimulant and the second most commonly used drug in the United States, is obtained from coca plants grown in South America. Users typically sniff the white powder and may become psychologically dependent. Crack, which is obtained by heating a mixture of cocaine, baking soda, and water,

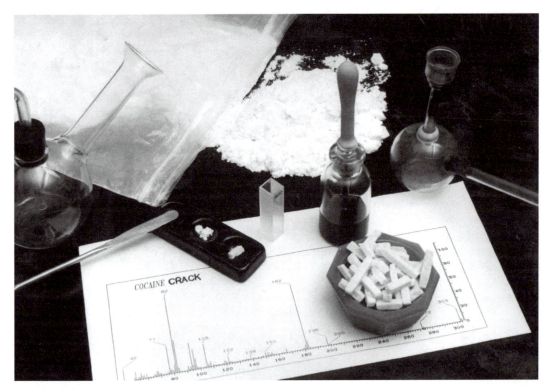

Cocaine and crack cocaine. Courtesy of the Georgia Bureau of Investigation, Division of Forensic Sciences.

is far more addictive than cocaine. After drying, the drug is broken into small pieces that can be smoked. It produces an instant high that is followed by depression and anxiety.

Anabolic Steroids

Anabolic steroids are synthetic compounds that are chemically similar to testosterone, a male sex hormone. Because steroids promote muscle growth, they have been used by athletes and body builders. Severe side effects such as liver cancer and other liver disorders, along with infertility, masculinization of females, depression, and personality changes, including increased anger and destructive behavior, have been observed. Use by teenagers can prematurely halt bone growth. These side effects led the federal government to regulate steroids and to classify them as controlled dangerous substances.

The Legal Response to Drug Use

In 1914, the widespread use of narcotic medicines led Congress to pass the Harrison Act. This law made it illegal to use narcotics without a prescription. Five years later, the Eighteenth Amendment was ratified, making the manufacture, sale, or transportation of intoxicating liquors illegal in the United States. The nation's experiment with Prohibition led to bootlegging, speakeasies, and widespread violation of the law. The recognition that Prohibition had been a failure resulted in the repeal of the amendment in 1933.

It was not until the 1960s that the youth rebellion against authority and the Vietnam War led to the misuse of drugs on a major scale. In response to the growing use of drugs, Congress in 1970 passed the Comprehensive Drug Abuse Prevention and Control Act, which established five schedules of classification based on a drug's potential for abuse and dependence as well as its medical value if prescribed by doctors. The severest penalties for illegal use or possession of drugs were for those in schedules I or II, under which an individual may be imprisoned for up to 20 years and/or fined as much as a million dollars for a first offense.

The drugs in schedule I have a high potential for abuse, have no medical use, and are limited to controlled research. These include heroin, LSD, marijuana, and methaqualone. Recently, a number of states have approved marijuana for medical use in treating glaucoma and the nausea associated with chemotherapy. The federal government has not followed the lead of the states because there are medicines that can be used to treat both glaucoma and nausea.

Drugs in schedule II—cocaine, methadone, PCP, and most amphetamines and barbiturates—are also regarded as having a high potential for abuse and dependency, but do have medical value and are available through carefully controlled prescriptions. The drugs in schedules III, IV, and V have a lower potential for abuse and dependency, are widely used in medicine, and draw less serious punishment when they are used illegally. These drugs include some barbiturates, tranquilizers, many codeine medicines, phenobarbital, and opiate mixtures that contain nonnarcotic substances. Since 1989, the Office of National Drug Control Policy has issued a National Drug Control Strategy that is designed to reduce the use of drugs through education, treatment, research, interdiction, law enforcement, and crop reduction and to assess the success of the current strategy.

Identification of Drugs

Drug suspects cannot be convicted unless police can prove that the substances they possessed, manufactured, or sold are illegal substances. Marijuana is probably the easiest drug to identify. Its botanical features are readily evident under a microscope. Confirmation can be achieved with the Duquenois-Levine color test. A mixture of 2 percent vanillin and 1 percent acetaldehyde in ethyl alcohol is added to the suspected substance. Concentrated hydrochloric acid is then added, followed by chloroform. The appearance of a purple color in the chloroform layer indicates marijuana.

Other suspect materials are first subjected to screening tests. These tests usually involve a color change characteristic of common illicit drugs. These preliminary tests must be followed by testing schemes devised by forensic chemists that will positively identify the drug. The microcrystalline test involves placing a sample of the suspected drug on a slide and then adding a reagent to see if there is a reaction that produces a crystalline precipitate. If a precipitate forms, the crystals can be examined under a microscope for characteristic shapes.

Both thin-layer and gas chromatography can provide tentative identification and eliminate many drugs from consideration. Ultraviolet spectrophotometry can establish high probability of a drug's identity. Ultraviolet spectrophotometry of a substance is often followed by infrared spectrophotometry because it provides unique patterns for many substances and, therefore, can often establish positive identification. For many drugs, gas chromatography combined with mass spectrometry will provide proof of the drug's identity.

Drugs and Crime

Drug-abuse violations accounted for 10 percent of all arrests in 1996. In addition to the crimes associated with the possession, sale, and use of drugs that became illegal when legislatures declared them so, there are crimes that stem from drug use. In 1997, according to the U.S. Department of Justice, 19 percent of state inmates and 16 percent of federal inmates said that they committed crimes in order to obtain drugs. This represents an increase from 17 percent and 10 percent, respectively, in 1991. On the other hand, drug-related homicides declined from 21,676 in 1991 to 15,289 in 1997.

Of the men arrested for assault in Chicago in 1996, 84 percent tested positive for drugs. In Philadelphia, 94 percent of the prostitutes arrested tested positive for drugs. Drug offenses accounted for three-quarters of the growth in the prison population between 1985 and 1995. A strong correlation exists between violence and psychoactive drugs, and many people convicted of theft do so to support their drug addictions. What fraction of the 1,486,300 arrests for theft (another 10 percent of all arrests) in 1996 was related to drug addiction is not known, but it was significant. The same is true of many other crimes such as robbery, assault, burglary, possession of stolen property, drunken driving, vagrancy, domestic disputes, and others.

The University of Michigan's 1997 annual survey of students in the 8th, 10th, and 12th grades indicated that drug use showed signs of leveling off after five years of growth. Similar results were obtained in 1998 according to the National Household Survey on Drug Abuse, which showed that 9.9 percent of 12- to 17-year-olds said that they were using illegal drugs, compared with 11.4 percent in the previous year. However, overall use of drugs did not change because 16.1 percent of 18- to 25-year-olds were using drugs, compared with 14.7 percent in 1997.

Marijuana remained the most widely used illegal drug, but alcohol is the most commonly used drug. About 50 percent of 12th graders drink it. Its use by adults is legal, but most students are not of legal age. However, alcohol use is significantly lower than it was in 1980 when 72 percent of 12th graders reported using it. Marijuana use among 8th graders fell to slightly less than 18 percent, but increased among 12th graders to 39 percent. Daily use is significantly lower for both age groups (about 1.1 percent for 8th graders and 5.8 percent for 12th graders).

Use of hallucinogens (LSD or PCP), inhalants, and stimulants remained at about the same level. Roughly one-third as many reported using these drugs in 1997 as reported using marijuana. A small fraction (2.1 percent) reported using heroin at least once, use of cocaine (once or more) was up 1.5 percent to 8.7 percent, crack use (once or more) rose from 3.3 percent to 3.9 percent, and there seems to be a slow but steady increase in cigarette smoking since 1992.

Cases

One of the most common methods used by police to arrest drug peddlers is to use undercover agents in sting operations. On September 23, 2000, police in Hackensack, New Jersey, seized two pounds of marijuana and arrested five men connected with a Patterson, New Jersey, street gang. An undercover police officer ordered a pound of marijuana from Juan Quinones. Quinones ordered the drug from an agent in Patterson. At 6:30 P.M. on September 23, two men from Patterson delivered a pound of marijuana and six plastic bags of cocaine. One of the men carried a loaded 9-mm semiautomatic handgun. Both men were arrested and charged with drug possession, possession with intent to sell, possession within 1,000 feet of a school, and unlawful possession of a handgun. Several hours later, two other men from Patterson arrived with a pound of marijuana and were also arrested and charged. The only charge against Quinones, who was held on $150,000 bail, was endangering the welfare of a child because he was allegedly packaging drugs while his girlfriend's child was in his apartment. Police said that other drug charges were pending further investigation. Quinones is a member of the Latin Kings, an organized-crime group estimated to contain 100,000 members throughout the country. Members are believed to range in age from 15 to 40.

On October 5, 2000, following a 16-month undercover operation, 60 police officers, armed with warrants, raided homes and apartments in Falmouth, Massachusetts, and arrested 17 suspects. All those arrested were males ranging in age from 16 to 55. The charges were based on 32 drug purchases of marijuana, cocaine, heroin, and Valium by undercover officers. Falmouth police Captain Roman Medeiros told a reporter that the police were led to investigate the suspects by anonymous tips and by complaints from people who lived in the neighborhood of those arrested. *See also* Chromatography; Opium.

References

Gaudiano, Nicole. "5 Arrested in Undercover Drug Sting." *Record* (Bergen County, NJ), Sept. 26, 2000, p. 101.

Lane, Brian. *The Encyclopedia of Forensic Science.* London: Headline, 1992.

1998 National Drug Control Strategy. http://www.whitehousedrugpolicy.gov/publications/policy/98nd/contents.html.

Peters, Paula. "Drug Bust Nets 17: Arrests Follow Sales to Undercover Officers." *Cape Cod Times,* Oct. 5, 2000, p. A3.

U.S. Department of Justice: Bureau of Justice Statistics. http://www.ojp.usdoj.gov/bjs/. Accessed August 24, 2001.

The World Almanac and Book of Facts, 1999. Mahwah, NJ: World Almanac Books, 1998.

E

Electron Microscope

Electron microscopes provide millionfold magnifications of objects, magnifications that are a thousand times greater than those seen under ordinary light microscopes. The scanning electron microscope, a more sophisticated version of the electron microscope, is of great value to forensic scientists. It can provide detailed comparisons of such trace evidence as tiny bits of paint, fiber, paper, wood, cloth, hair, and other materials as small as a wavelength of visible light (~0.0001 mm).

The idea for an electron microscope followed the work of Louis de Broglie, a French physicist who in 1924 first suggested that electrons and, indeed, all particles have a wavelength that depends on their momentum (mass times velocity). The greater an electron's momentum, the shorter its wavelength. Thus high-energy electrons, which move at very high speeds, have short wavelengths.

Ordinary light microscopes are limited by the wavelength of visible light. Objects comparable in size to the wavelength of visible light cannot be seen in an ordinary microscope because the light waves simply diffract around the object. The effect is similar to what one sees when water waves pass by a cork floating on a pond. The waves move right around the cork and provide no evidence of its presence. The same wave striking a larger object such as the hull of a boat will be seen to reflect off the boat's surface.

An ordinary light microscope produces an image by sending a beam of light through, or reflecting the light from, an object and then passing that light through a pair of lenses that refract (bend) the light to form an enlarged image of the object. An electron microscope does the same thing with a beam of electrons. The electron beam passes through coils of wire called a condensing lens. The wires in the coil carry an electric current (a stream of electrons confined to the wire). An electric current is surrounded by a magnetic field, and electrons are deflected by a magnetic field perpendicular to their motion. The strength of the magnetic field can be controlled by regulating the electric current through the coils. The coil is wound so that it acts as a convex lens would on a diverging beam of light; that is, it diverts the electrons just enough to produce a parallel beam.

Once the beam is parallel, it is allowed to pass through or be reflected from the sample being examined. In a transmission electron microscope, the electrons pass through the specimen being viewed. In a scanning electron microscope, the electrons are reflected by the specimen. After leaving the specimen, the electrons pass through two more mag-

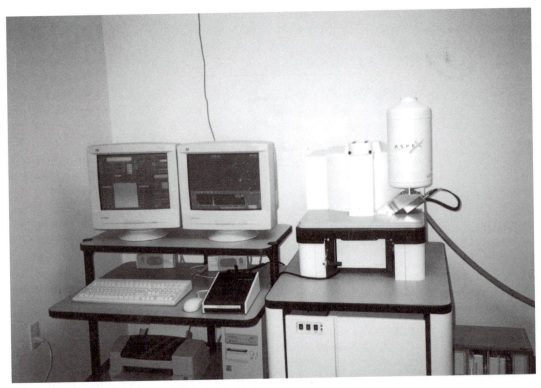

Electron scanning microscope. Courtesy of David Exline, ChemIcon Inc., Pittsburgh, PA.

netic fields produced by coils of current-carrying wire. These two fields are called the objective and the projection lenses. They have the same function as the objective and eyepiece lenses in a light microscope: they bend the beam of electrons to form a magnified image of the specimen. That image falls on either a photographic film or a fluorescent screen where it can be recorded or seen by an observer.

The source of the beam in an electron microscope is the electrons that "boil" off a hot tungsten-wire cathode in a vacuum, a process very similar to what is done to produce the electron beam in a television set. The electrons are then accelerated across a potential difference (voltage) of 10,000 to 100,000 volts to increase their momentum to a point where their wavelength is about 0.004 to 0.01 nanometer (nm) or 4 to 10 billionths of a millimeter. About 40,000 to 100,000 of these wavelengths could fit into a single wavelength of visible light. Because of their tiny size, these waves do not diffract around

very small objects. In fact, they can provide images of parts as small as half a nanometer (half a millionth of a millimeter).

With the scanning electron microscope, the smallest particles that can be discerned as clear images are about 10 nm (0.00001 mm) in size. However, the images provide a far superior three-dimensional view than the transmission electron microscope and since forensic science seldom requires a magnification of more than 10,000, the scanning electron microscope is the tool preferred by forensic scientists. Just as a television set sweeps a beam of electrons across the screen at high frequency, so the scanning electron microscope sweeps a fine-line beam of electrons across the specimen. The reflected electrons are captured by a positively charged anode, amplified, and swept across a fluorescent screen where the greatly magnified image can be viewed. In many cases in which trace evidence may hold the key to the solution of a crime, forensic scientists rely on the scanning electron microscope.

During the post mortem examination of Bulgarian defector Georgi Markov, doctors extracted a tiny pellet, about the size of a pinhead, from a wound in his thigh. When the sphere was examined under an electron microscope, it was seen to be made of a platinum-iridium alloy that would not be rejected by the body. Further analysis showed that two small holes, about 0.4 mm in diameter, had been bored into the pellet, which had been coated with sugar.

Forensic scientists concluded that the pellet had contained a sophisticated biotoxin held in place by the sugar coating. The poison had dissolved in the body and caused Markov's death. The scientists estimated that only half a milligram had been used. Based on the symptoms and the dosage, they determined that the poison was ricin. A pig injected with a comparable amount of the poison died within 24 hours.

On May 26, 1999, 19-year-old Katie Poirier disappeared from Moose Lake, Minnesota. That autumn, 51-year-old Donald Blom, a convicted kidnapper and sex offender, confessed to kidnapping the young woman, choking her, and burning her body. Later, claiming that he had been suicidal and pressured, he recanted the confession and sought a trial. In order to convict Blom, the prosecution had to prove that Poirier was dead. A crucial piece of evidence in the trial was a burned tooth found in a fire pit where a burned skeleton was found. A physical anthropologist told the court that the bones were those of a young woman; however, the fire had destroyed any DNA that might have been used for a positive identification.

Forensic scientist Donald Melandar used a scanning electron microscope to establish that the burned tooth contained zirconium and silicon, two elements found in a new type of cement used by Poirier's dentist in filling one of her teeth two weeks before she disappeared. Defense witnesses claimed that these two elements were too common in dentistry to be used to identify a tooth. Nevertheless, a jury convicted Blom in August 2000, and he was sentenced to life in prison.

See also Ballistics; Magnifiers; Woodchipper Murder.

References

Evans, Colin. *The Casebook of Forensic Detection: How Science Solved 100 of the World's Most Baffling Crimes.* New York: Wiley, 1996.

Macaulay, David. *The Way Things Work.* Boston: Houghton Mifflin, 1988.

Oakes, Larry, "Blom Defense Rests after Presenting Dental Expert." *Minneapolis Star Tribune,* Aug. 12, 2000.

Saferstein, Richard. *Criminalistics: An Introduction to Forensic Science.* Upper Saddle River, NJ: Prentice Hall, 1998.

Electrophoresis. *See* Chromatography.

Entomology. *See* Forensic Entomology.

Expert Testimony

Expert testimony is provided by people with competence in some aspect of forensic science or knowledge pertinent to a case. Such people are frequently called to testify in court by both prosecuting and defense attorneys. A trial judge must be satisfied that a witness called to give an expert opinion on a matter before the court is qualified to be heard. The credentials of such a witness are assessed on the basis of his or her education, experience, and training as reflected by the individual's advanced degrees, publications, membership in professional organizations, years of experience, and reputation. Before a ruling on the qualifications of a proposed expert witness is made, opposing attorneys will be given the opportunity to cross-examine the person.

To provide effective expert testimony, one must be a good teacher as well as an authority in the field relevant to the case. A judge or jury will find information provided by an expert valuable only if the pertinent data can be explained clearly in terms a layman can understand. Finally, an expert testifying in a trial should be concerned only with the truth based on the scientific or other information that is related to the case. He or she is there to provide valuable knowledge, not to serve as an advocate for either the defendant or the prosecutor.

It was the testimony of Harvey Wiley and others who investigated the harmful effects of food preservatives as well as the descriptions of the lack of sanitation in the meat-packing industry found in Upton Sinclair's novel *The Jungle* that led Congress to establish the Food and Drug Administration in 1906. Similarly, testimony by psychiatrists led to laws that established sex offenders as distinct from other criminals.

In defending O.J. Simpson, who was accused of murdering his ex-wife Nicole Brown Simpson and her friend Ron Goldman, his lawyers called on a blood expert, Herbert MacDonnell, to testify. MacDonnell told the court that the blood found on a sock in Simpson's bedroom had not been splattered. He said that the blood had been transferred in the form of a compression stain. In other words, someone had pressed Nicole Simpson's blood into the sock, which indicated that the evidence had been planted. Such expert testimony made it difficult for a jury to find Simpson guilty of murder beyond a shadow of a doubt.

One expert who is also an author of books that involve forensic science is Kathy Reichs. Reichs, who is a professor of anthropology at the University of North Carolina, serves as a consultant to the North Carolina Office of the Chief Medical Examiner and to the Laboratoire de Sciences Judiciaires et de Médecine Légale for Quebec, Canada. She is often called upon to give expert testimony in cases in which ordinary autopsies are not effective because of mutilated bones, decomposed and burned bodies, and so on. Her experiences with the police as a forensic expert have served as the basis for her novels, which include *Déjà Dead*, *Death du Jour*, and *Deadly Decisions*.

Experts, above all, should be honest in presenting information and in providing their credentials. In *Trapp v. American Trading and Production Corporation* in Kings County, NY, on March 5, 1998, the trial was set aside because the plaintiff's expert witness had falsified his qualifications. He held none of the academic degrees he had listed; in fact, he had completed only one year of college. While he had served in the U.S. Navy, he had been a chief warrant officer, not a captain. Furthermore, he held no license for unlimited tonnage in the merchant marine, as he had claimed. In view of the "expert's" deceit, the original verdict was set aside and a new trial was ordered.

References

Brice, Chris. "Expert Witness." *Advertiser* (Adelaide, Australia), Aug. 12, 2000, p. M23.

Conklin, John E. *Criminology*. 3rd ed. New York: Macmillan, 1989.

Saferstein, Richard. *Criminalistics: An Introduction to Forensic Science*. Upper Saddle River, NJ: Prentice Hall, 1998.

Eye Prints

Eye prints, the pattern of pigments on the irises of the eyes or blood vessels on the retina are as unique as fingerprints and, like fingerprints, can be used as a means of identification. Some police departments, in addition to taking fingerprints and facial photos, photograph a suspect's retinas, but more commonly retinal patterns are used at places such as nuclear weapons plants, missile-launching sites, and other areas where national security is involved. Recently some banks have begun using eye prints in place of PIN numbers as a means of identifying customers. The person to be identified looks into a pair of lenses while a scanner records the pigment patterns on the iris and feeds the data to a computer that compares the patterns with the iris patterns of persons stored in its memory. As is true of fingerprints, no two people, including identical twins, who are genetically identical, have the same iris pattern. In fact, the iris pattern of a person's right eye is different from that of his or her left eye. *See also* Fingerprints.

References

Eyedentify, Inc. http://www.eyedentify.com.

Flatow, Ira. "Analysis: Biometric Identification, Using Physical Characteristics Such as Voice Sample or Handprint to Identify a Person." *Talk of the Nation Science Friday*, Sept. 1, 2000.

Gardner, Robert. *Crime Lab 101: Experimenting with Crime Detection*. New York: Walker, 1992.

Eyewitness Account

An eyewitness account is the report of what an eyewitness saw or heard during or after a crime. Eyewitness accounts can convict a defendant without any other supporting evidence. Courts and jurors tend to place more faith in eyewitness accounts than in any other type of evidence. Unfortunately, forensic-science research has shown that eyewitness accounts are less reliable than people believe. They are based on a person's memory of an event, and memory can be influenced by the witness's thoughts, feelings, and reactions during the incident, or by information introduced after the event, such as that included in suggestive questions, newspaper or television reports, or overheard conversations.

There are ways to minimize the possibility of a false identification or suggestive influences. Witnesses should give a free narration before being asked specific questions. Questions should be asked as soon as possible so that misleading information from other sources does not become part of the remembered event. Witnesses can be asked to identify a suspect from photographs such as mug shots or a lineup. A lineup is usually a single suspect standing in a line with a group of others who are known to be innocent. The reliability of lineup identification is dependent on the length of time between witnessing the crime and attempting to identify the criminal.

Other factors that can influence the ability of a witness to make a correct identification are the amount of illumination and the duration of the event. Crimes frequently take place at night or in poor lighting and can be committed in a matter of seconds. Short observation periods make it difficult for the witness to take in a lot of information and poor lighting will make it difficult for the witness to get a good look at the criminal. If the witness has any vision or hearing disability, this will affect his or her ability to make an accurate identification. Also, contrary to popular belief, perception and memory under stress are not "highly tuned." When someone is in a life-threatening situation or is frightened, his or her attention is more likely focused on looking for an escape route or threatening weapons, not on the criminal's face or physical appearance.

Another way a person may be asked to identify a criminal is as an "earwitness" instead of an eyewitness. A witness may have not seen the criminal but heard him or her. This could happen if the crime occurred in a dark location or the criminal was wearing a mask, or the crime included threatening phone calls. Courts will accept voice identification as permissible evidence, as happened in the famous Lindbergh kidnapping case, in which Charles Lindbergh identified the voice he heard in the ransom exchange as Bruno Richard Hauptmann's. Mistakes can occur in earwitness accounts just as in eyewitness accounts. The witness can make a mistake because of a time delay between the crime and the identification, lack of attention to details, or other outside influences. The criminal can also hinder identification by changing his appearance or voice.

One example of a mistaken eyewitness identification occured in the Ronald Cotton Case. In July 1984, Jennifer Thompson was raped at knifepoint. Despite the emotion connected with the brutal attack, Thompson was able to carefully observe the facial features of her attacker. Not only was she able to work with a detective to produce a composite sketch of the rapist within hours after the crime, but also a week later, she identified the man who had raped her, picking him from a lineup of six men. During the court case, she pointed to Ronald Cotton as definitely the man who had raped her. Cotton, who had served an 18-month sentence for attempted sexual assault, was nervous, had no alibis, and, on January 17, 1985, was sentenced to life in prison.

About a year later, a rapist named Bobby Poole arrived at the prison where Cotton was serving his sentence. When Poole bragged to other inmates that Cotton was serving some of the time he (Poole) should have received,

Cotton's lawyer was able to obtain a second trial. However, when Jennifer Thompson testified, she still identified Cotton, not Poole, as the man who had raped her. Disconsolate, Cotton returned to prison. In 1995, Cotton convinced University of North Carolina law professor Richard Rosen to ask the courts to conduct DNA tests on him and Poole. Rosen agreed because Cotton had received a life sentence on the basis of eyewitness testimony, which Rosen regarded as unreliable. The DNA results finally vindicated Cotton by revealing that Bobby Poole was Thompson's rapist. Thompson sought Cotton out and apologized for her mistake. She has since become an ardent opponent of the death penalty.

Despite the problems associated with eyewitnesses, they are often helpful in locating suspects. At a July 23, 2000, party held in Wellfleet, Massachusetts, a young man was severely beaten by two men whom he did not know. On September 26, police arrested two suspects—Andrew Hoagland and Will Elliot—in nearby Eastham, Massachusetts. The arrests were based on reports of eyewitnesses who had attended the party. After Boston Celtics star Paul Pierce was repeatedly stabbed in a Boston nightclub in September 2000, eyewitness reports were used to identify and arrest suspects. *See also* Lindbergh Kidnapping and Trial; Rogues' Gallery.

References

Grau, Joseph J., ed. *Criminal and Civil Investigation Handbook*. New York: McGraw-Hill, 1981.

Loftus, E.F. *Eyewitness Testimony*. Cambridge, MA: Harvard University Press, 1996.

Marcus, Paul. "Eyewitness Identification: Constitutional Aspects." *Encyclopedia of Crime and Justice*, ed. Sanford H. Kadish. Vol. 2. New York: Free Press, 1983.

Yarmey, Daniel. "Eyewitness Identification: Psychological Aspects." *Encyclopedia of Crime and Justice*, ed. Sanford H. Kadish. Vol. 2. New York: Free Press, 1983.

F

Faurot, Joseph A. (1872–1942)

In 1904, Joseph A. Faurot was the first American to travel to London to study the new fingerprinting technique at Scotland Yard and, soon after, the first detective in the United States to get a conviction based on fingerprint evidence. He had convinced his boss, the commissioner of the New York City Police Department, William McAdoo, to let him make the trip. However, when Faurot returned to New York, McAdoo was no longer commissioner, and his successor did not believe that fingerprinting was of any value. Faurot realized that he had to convince the new commissioner and the news media of its worth.

In 1906, a prominent citizen had been robbed while staying at the Waldorf-Astoria Hotel in New York City. Faurot happened to see a man come out of someone else's suite at the hotel and arrested him. He fingerprinted the suspect and sent the prints to Scotland Yard. The suspect was identified as Daniel Nolan, alias Henry Johnson, an international jewel thief. Confronted with the report from Scotland Yard, he confessed.

In 1908, a nurse, Nellie Quinn, had been choked to death in her room. Since fingerprinting was still not an accepted practice, policemen and reporters at the crime scene had touched many items in the room. Faurot arrived and discovered a whiskey bottle under the bed that had not been touched by others. He found the nurse's prints on the bottle and one other set. He fingerprinted friends and neighbors of the nurse. The prints of one man, George W. Cramer, matched the other set of prints on the bottle. Cramer confessed after being confronted with the evidence.

The case that brought the most attention to the value of fingerprinting in solving crimes, which Faurot was advocating, was a 1911 burglary. Someone had broken into a millinery shop. The suspect, Caesar Cella, a notorious burglar, had an alibi for his whereabouts at the time of the burglary. He had been with friends at the Hippodrome that evening. He had then gone home with his wife and had not left home until the next morning. But Faurot had found dirty fingerprints that matched Cella's on the window through which the burglar had made his entrance. This was the first time that fingerprints were the sole incriminating evidence against a man who appeared to have a watertight alibi. Fingerprinting was very new in the United States, and the judge and jury were not convinced of its value until Faurot performed a demonstration. Faurot left the room, and 15 different people in the courtroom put their prints on a window. One of those people also put his prints on a glass

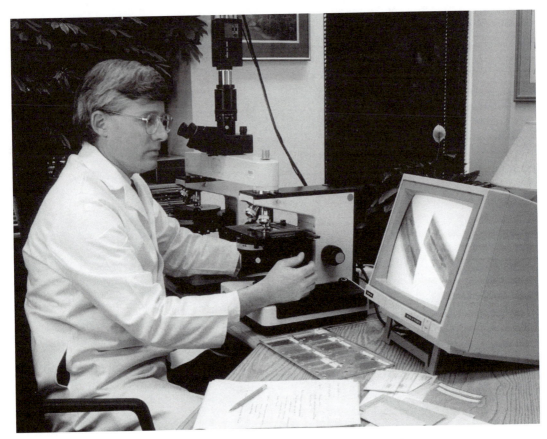

Criminologist examining trace evidence under comparison microscope. FBI photo.

desktop. Faurot was able to match the prints on the glass desktop with the correct ones on the window. The judge and jury were impressed and convicted the defendant based on the fingerprint evidence. Cella later admitted that he had waited for his wife to fall asleep and had then left their home to commit the burglary. *See also* Fingerprints.

References

Sifakis, Carl. *The Encyclopedia of American Crime.* New York: Smithmark, 1992.

Thorwald, Jürgen. *The Century of the Detective.* New York: Harcourt, Brace & World, 1965.

FBI Crime Laboratory

The FBI Crime Laboratory was established in November 1932 under the directorship of J. Edgar Hoover. It was a national laboratory that aimed to offer forensic services to all law-enforcement agencies in the United States. It is now the world's largest forensic laboratory.

The initial lab was simply a room in the Old Southern Railway Building at 13th Street and Pennsylvania Avenue NW in Washington, D.C. For the first few months, Special Agent Charles Appel was the only employee in the lab. He was a document specialist who testified in the famous Lindbergh kidnapping case concerning the handwriting of Bruno Richard Hauptmann. The facilities and equipment have expanded greatly from this sparse beginning to the present site in the J. Edgar Hoover Building at 10th Street and Pennsylvania Avenue. An even larger facility is being built in Quantico, Virginia. The 463,000-square-foot building is scheduled to be completed in June 2002.

The FBI Crime Laboratory has five major subdivisions: Forensic Science Research and

Training, Special Projects, Investigative Operations and Support, Scientific Analysis, and Latent Fingerprints. Each subdivision maintains many programs. Some of these programs include the Foreign Language Program, the Bomb Data Center, the Polygraph Program, and the Photographic Equipment Program. In 1991, the FBI Crime Laboratory added a Computer Analysis and Response Team (CART) to provide computer support to crime investigations.

Among other services the FBI Crime Laboratory can provide is the analysis of blood, body fluids other than blood, hair, and fibers. It also analyzes weapons, explosives, handwriting, and the authenticity of documents. The Forensic Science and Research Training division provides research, training, and procedural support for all FBI staff and any other law-enforcement agency or crime laboratory that requests their services. DNA analysis is a recent scientific breakthrough in the fight against crime. In the 1980s, the FBI Crime Laboratory began doing DNA analysis, and in October 1997, the first positive individual identification from a DNA sample was made in court.

Also in 1997, the FBI went through some major changes. The FBI Crime Laboratory had an excellent reputation; however, its image was tarnished following allegations by a chemist in the explosive unit that were corroborated by an 18-month Justice Department investigation. The chemist, Frederic Whitehurst, had accused the laboratory of mishandling evidence, mismanagement of the lab by nonscientists, and altering of lab findings to support the prosecution in the Oklahoma City bombing case. After the corroborating Justice Department report, some laboratory employees were reassigned and nonagent professional scientists replaced them. It was at that time that the first nonagent, Donald M. Kerr, a nuclear physicist, was appointed director of the FBI Crime Laboratory. The Justice Department report noted evidence of sloppy lab work but concluded that none of the lab's findings had been compromised.

The FBI Crime Laboratory has been called

on to help with international crises such as the Kosovo war crimes in which Serbian forces captured, tortured, and murdered Albanian villagers. A 65-member team processed more than 30 crime scenes and exhumed bodies of the victims. The team documented and photographed crime scenes, collected evidence at massacre sites, and performed autopsies on victims to determine identity and cause of death. Most deaths were found to be the result of multiple gunshot wounds.

The FBI Crime Laboratory was also responsible for stopping a massive international ring of computer hackers, called Phonemasters, that had gained access to major telephone networks, air-traffic-control systems, and a number of databases. Phonemasters sold calling-card numbers, credit reports, and criminal records to individuals in the United States, Canada, Italy, and Switzerland. Calvin Cantrell and Cory Lindsay, key members of Phonemasters, were convicted of theft and possession of unauthorized devices in 1999.

Following the April 20, 1997, Columbine High School shootings in Colorado, the FBI Crime Laboratory created scaled crime-scene reconstructions of each segment of the incident. The lab duplicated, enhanced, and conducted imaging work on a series of tapes of the cafeteria area, and the lab's Computer Analysis and Response Team analyzed and restored data required by investigators. *See also* Crime Laboratories; DNA Evidence; Lindbergh Kidnapping and Trial.

References

Theoharis, Athan G., ed. *The FBI: A Comprehensive Reference Guide.* Phoenix, AZ: Oryx Press, 1999.

Zonderman, Jon. *Beyond the Crime Lab: The New Science of Investigation.* New York: John Wiley, 1990.

Fiber Evidence

Fiber evidence, such as hair, tiny pieces of glass, bits of skin, flakes of paint or metal, pieces of wood fiber, and other bits and pieces of matter that would not be seen in cursory examination, is an example of trace

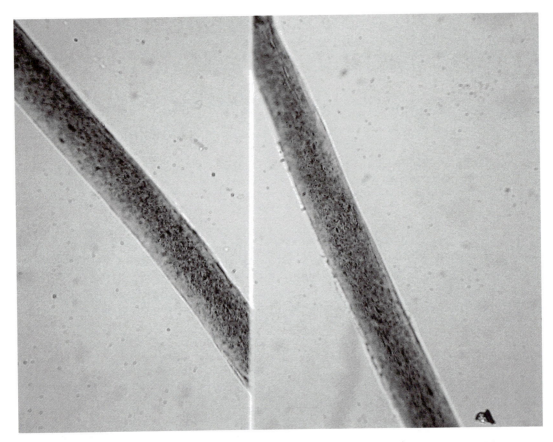

Fibers under comparison microscope. Courtesy of David Exline, ChemIcon Inc., Pittsburgh, PA.

evidence. Fibers are often overlooked by even the most careful glove-and-mask-wearing criminal. Consequently, forensic scientists will search a crime scene carefully looking for bits of fiber as well as other trace evidence. Fibers in clothes, carpets, drapes, and even AstroTurf are frequently transferred between criminal and victim or between victim and car in hit-and-run accidents. To avoid contaminating evidence, all articles showing fiber evidence should be stored and transported to the laboratory in separate paper bags. Car seats holding such evidence should be covered with polyethylene. Fibers from articles that cannot be transported should be removed with forceps, placed separately on paper, enclosed by folding, and placed in paper bags with tags identifying their source and location.

Fibers are natural or man made. Natural fibers may be from animals (wool, cashmere, silk, fur), plants (cotton, jute, sisal), or minerals (asbestos). In the United States, three-fourths of all textiles are made of synthetic (man-made) fibers such as rayon, acetate, acrylic, polyester, olefin, and others, all of which are made from polymers.

Fiber samples from a suspect, crime scene, or victim may be compared as to color, diameter, and cross-section under a comparison microscope. Dyes in the samples can be compared using microspectrophotometry and thin-layer chromatography. Comparisons may also be made using infrared spectrophotometry and pyrolysis gas chromatography. The refractive index of fibers that transmit light can be determined in the same way as that of glass.

The Roger Payne Case (1968)

The murder of Claire Josephs in Bromley, England, in 1968 was solved in large part through fiber evidence. Her husband found her body. Her throat had been cut, and police discovered that a serrated bread knife was missing from the kitchen. At the time of the murder, she had been wearing a red woolen dress.

Lacking evidence of forcible entry and given the presence of a half-empty coffee cup and cookies on the kitchen table, police surmised that the victim knew her killer. Consequently, investigators concentrated on friends and relatives. When they interviewed Roger Payne, a bank clerk who had visited the Josephses with his wife earlier that year, they noticed scratch marks on his hands. Analysis of his suit under ultraviolet light revealed 61 red wool fibers that were consistent with those on Claire Josephs's dress. Furthermore, 20 rayon fibers found on Claire Josephs's raincoat hanging inside her front door matched fibers from Payne's scarf, which he had likely hung over her raincoat upon entering the home. In addition, investigators found fibers matching the Josephs's carpet on the floor of Payne's car. The fiber evidence combined with bloodstains in Payne's car that matched Claire Josephs's blood type sealed the case. Payne was convicted and sentenced to life in prison.

Wayne Williams Case (1979–1981)

Serial killer Wayne Williams is believed to have murdered 30 young black males during a two-year period. The strangled bodies of 12 young black males were discovered. Another 18 young men were reported missing. After newspaper reports stated that police had found fibers on the victims' bodies, the killer had been stripping his victims and throwing their bodies into rivers. This attempt to conceal evidence led to his arrest.

Aware of his pattern of dropping bodies into rivers, police established surveillance crews at bridges. On May 22, 1981, police heard a splash in the Chattahoochee River. They stopped a car driven by 23-year-old photographer Wayne Williams but could find no reason to detain him. Two days later, Nathaniel Carter, the 12th known victim, was found a mile downstream. Police now had

just cause to obtain a warrant to search Williams's car and apartment.

Forensic scientists working with Du Pont chemists had determined that the fibers found on the victims were made by the Wellman Corporation and were used in carpeting manufactured for one year by the West Point Pepperell Corporation. The carpets had been sold for two years in 10 southeastern states. Based on sales, it was calculated that the probability of finding such a carpet in an Atlanta apartment was 1 in 7,792.

Fibers found on another victim were consistent with those from Williams's 1970 Chevy station wagon. On the basis of General Motors manufacturing figures and the number of vehicles of that model in the Atlanta area, police determined that the probability of finding matching fibers on a victim was 1 in 3,828.

There were also fibers from other carpets and hairs from Williams's head, from his dog, and fibers from victims on a glove; however, the crucial evidence was the probability of finding fibers from both the carpets in the accused murderer's bedroom and from the carpeting in his car on victims. Using many charts and photographs, prosecutors pointed out that the probability of finding victims with both the bedroom and car fibers was 1 in 7,792 times 3,828 or 1 in 29,827,776. With such odds against him, Wayne Williams was convicted of murder and sentenced to life in prison.

The John Serratore Case

In June 1999, a New South Wales Supreme Court jury in Sydney, Australia, found John Serratore, 27, guilty of strangling Frances Tizzone, a 21-year-old student, four years earlier. The two had planned to become engaged before Tizzone broke it off. The verdict was based to a large extent on fibers that were found on the dead woman's shoes. Forensic scientists maintained that these fibers were consistent with fibers from the carpet in Serratore's car. Furthermore, they stated that the fibers would not have remained on her footwear had she taken more than a few steps. Hence the murder must have occurred while she was either in or very near Serratore's automobile. While the evidence could not establish that Serratore was the murderer, it did show that he was at least an accomplice and led to a conviction and a sentence of 20 years in prison. *See also* Automobile Accidents; Blood; Chromatography; Comparison Microscope; Crime Scene; Glass Evidence; Spectrophotometry; Trace Evidence; Williams, Wayne.

References

Deadman, H.A. "Fiber Evidence and the Wayne Williams Trial: Conclusion." *FBI Law Enforcement Bulletin* 53(5), 10–19, 1984.

Evans, Colin. *The Casebook of Forensic Detection: How Science Solved 100 of the World's Most Baffling Crimes.* New York: Wiley, 1996.

"NSW: Obsessive Ex Boyfriend Jailed for Murdering Student." *General News* (Australia), July 20, 2000.

Saferstein, Richard. *Criminalistics: An Introduction to Forensic Science.* Englewood Cliffs, NJ: Prentice Hall, 1990.

Zonderman, Jon. *Beyond the Crime Lab: The New Science of Investigation.* New York: Wiley, 1990.

Fingerprints

Fingerprints are the patterns formed by the raised papillary ridges on fingertips, which contain rows of pores that connect to sweat glands. Fingerprints have become the twentieth-century criminalist's Rosetta Stone. Clearly distinguishable fingerprints at the scene of a crime are a forensic scientist's dream come true. Because fingerprints are unique (even identical twins have different fingerprints), they provide the best possible means of identifying a suspect.

Strangely, the fingerprints of children often disappear before they can be detected. In 1995, police detective Art Bohanan of the Knoxville, Tennessee, police department studied the problem. Bohanan, together with chemists at Oak Ridge National Laboratory, found that the fatty acids in children's prints were more volatile than those in fingerprints of adults; consequently, they evaporated more rapidly. While they were investigating children's fingerprints, these chemists discov-

ered ways to detect cholesterol and, in the prints of smokers, nicotine.

Background History

As early as the 1820s, physiologist Johann Purkinje noticed the unique nature of fingerprints, but could find no means by which he could classify them. Three decades later, William Herschel, a British magistrate stationed in India, used the illiterate natives' fingers to make inked imprints on legal documents. Herschel recognized that the imprints were unique and used them to prevent fraud. It seems that the natives would sometimes come back twice to collect their government pensions. After he began using their fingerprints to identify them, the fraud stopped.

In 1880, Henry Faulds, a Scottish physician working in a Tokyo hospital, secured the release of a man suspected of thievery when he showed police that fingerprints found at the crime scene did not match those of the man they were holding. Later, he found that they did match those of another suspect who, upon seeing the evidence, confessed to the crime. Fauld's work was publicized in *Nature*, a scientific journal, where he suggested that fingerprints might be very useful as a means of solving crimes.

In 1892, Francis Galton, an English anthropologist and eugenicist who was the first cousin of Charles Darwin, published *Finger Prints* after writing about his observations of fingerprints in the 1880s. Galton recognized the unique and unchanging nature of fingerprints and developed a means of classifying them using the three basic patterns found in such prints—loops, whorls, and arches. His system of classification was improved later in that decade by another Englishman, Edward R. Henry, who was the inspector general of the Nepal Police. In 1901, Henry was hired by Scotland Yard to establish a fingerprint division. Today, most police departments in English-speaking countries use a system based on Henry's original work to file fingerprints. In Argentina, in 1891, Juan Vucetich, a Buenos Aires police official fascinated by Galton's work, had devised his own system for classifying fingerprints. In 1904, he published his findings in book form as *Dactiloscopia comparada* (Comparative finger-printing). His system is still used in most Spanish-speaking countries.

The New York City Civil Service Commission, in 1901, was the first organization in the United States to officially use fingerprints as a means of identification. It was about this time that the Bertillon system, which relied on multiple body measurements for identification, was beginning to lose favor. Its demise was clearly foreseen in 1903 when convict Will West was sent to Fort Leavenworth Prison. Authorities there found that according to the Bertillon system West was an identical copy of another Leavenworth inmate paradoxically named William West. Although they were seen as identical by Bertillonage, their fingerprints were clearly different.

At the St. Louis World's Fair in 1904, representatives from Scotland Yard trained a number of U.S. police officials to take and analyze fingerprints. With such knowledge in hand, police departments in a number of major U.S. cities began using fingerprints in crime detection. In 1924, the fingerprint records from Fort Leavenworth Prison and the Bureau of Investigation were merged when the Federal Bureau of Investigation (FBI) was established. Today, the FBI has more than 200 million fingerprints on file—the world's largest collection.

Because fingerprints can be so incriminating, a number of criminals have tried to remove their fingerprints. In 1934, Freddy Barker and "Creepy" Karpis, members of the Ma Barker gang, had a doctor surgically remove the skin from their fingertips. When the skin healed, the fingerprints were again evident, but the doctor disappeared forever. The notorious John Dillinger had the skin on his fingertips removed with acid, but when he was shot to death a few months later, his new fingerprints quickly identified him. They were the same. Some attempts to remove fingerprints have resulted in scar tissue that partially hides the pattern of ridges; however, the scar patterns interrupting the original fin-

gerprints are as useful in identification as pure fingerprints.

Fingerprint Fundamentals

The use of fingerprints in forensic science is based on several fundamental principles. The first is that the probability of finding two people with identical fingerprints is very small. In fact, no two identical fingerprints have ever been found. Galton calculated that the probability of finding identical prints was 1 in 64 billion.

A second principle is that an individual's fingerprints do not change with time. The pattern of ridges on a person's fingertips, palms, and soles at birth remains unchanged until death. Consequently, a detective can be certain that a criminal's fingerprints will remain unchanged even if it takes years to find a suspect whose prints match those found at a crime scene.

Finally, there are enough similarities in the patterns of ridges on people's fingers that they can be classified. The photo shows the basic patterns of loops, whorls, and arches that can be found in fingerprints. About 60 to 65 percent of the population have loop patterns, 30 to 35 percent have whorls, and only about 5 percent have arches. The arches can be either plain or tented, and the whorls can be classified as plain, central pocket, double, or accidental.

Classifying Fingerprints

The detailed system used by experts to classify fingerprints is very complex; however, the initial classification is similar to that used by Henry. It consists of five fractions in which R stands for right, L for left, i for index finger, t for thumb, m for middle finger, r for ring finger, and p (pinky) for little finger. The fractions, in order, are as follows:

$$Ri/Rt + Rr/Rm + Lt/Rp + Lm/Li + Lp/Lr.$$

The numbers assigned are based on whorls. A whorl on either finger listed in the first fraction is assigned a value of 16; if the whorl is on a finger in the second fraction, it is assigned a value of 8. Similarly, on the

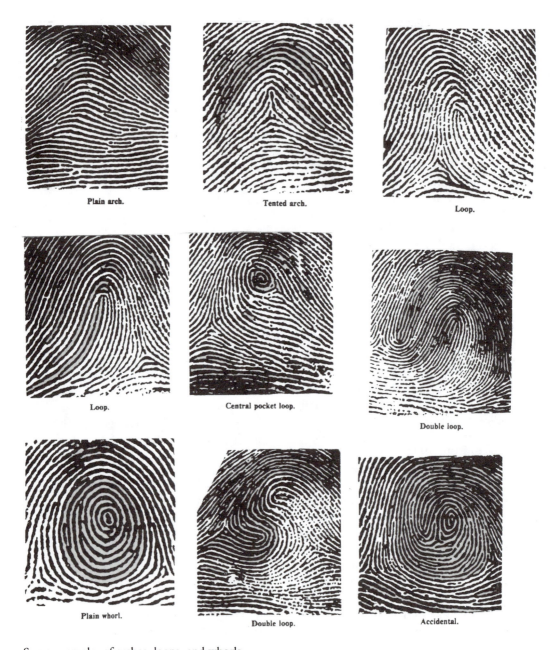

Plain arch.   Tented arch.   Loop.

Loop.   Central pocket loop.   Double loop.

Plain whorl.   Double loop.   Accidental.

Some examples of arches, loops, and whorls.

third fraction it receives a value of 4, a value of 2 on the fourth fraction, and a value of 0 on the last fraction. (Arches and loops are both assigned a value of 0.) Finally, the numbers in the numerators and denominators are added and a 1 is added to both. Suppose the right ring finger and the left index finger have whorls. Then the values assigned to the fractions would be

$$0/0 + 8/0 + 0/0 + 0/2 + 0/0 + 1/1 = 9/3 \text{ or } 3.$$

Using such a system reduces the need to search every fingerprint on file. Only those patterns with a value of 3 need be examined. Unfortunately, a complete set of prints is seldom found at a crime scene. Fortunately, today's automated fingerprints identification

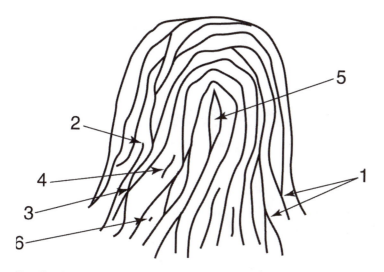

(1) Bifurcations
(2) Ridge Ending
(3) Ridge Crossing
(4) Short Ridge
(5) Enclosure
(6) Island

Details of a fingerprint.

Comparison of two prints. FBI photo.

system (AFIS) provides rapid matching of fingerprints against millions stored in databases. At a rate of 1,200/second, AFIS can search a million fingerprints for matches in less than 15 minutes. A national database of more than 30 million fingerprints developed by the FBI can now be accessed by AFIS.

The computer operator establishes a set of criteria based on ridge endings and bifurcations found in the print of a suspect. The computer then scans prints on file and converts the images to digital signals that match the criteria established by the operator. In minutes, a computer can do what would require years if it were done manually. The prints that match the criteria are then ex-

Latent prints may be developed on evidence exposed to cyanoacrylates such as super glue. FBI photo.

amined visually by an expert who decides if any selected by the computer match the print used as evidence. Ultimately, it is the eye of an expert that determines whether or not a match has been found.

Detecting and Preserving Fingerprints

Fingerprints are of three varieties: visible, plastic, and latent. Visible fingerprints are those in which fingers covered, for example, with blood or paint touch a wall, door frame, or similar surface. Plastic fingerprints are those left in soft substances such as soap or putty. The most common are latent (invisible) prints, which can be detected by using powders or chemicals. Powders are used on hard nonabsorbent surfaces such as glass, painted wood, or tiles. Chemicals are applied to porous surfaces such as paper, cloth, or unpainted wood.

Powders, such as aluminum dust (for dark surfaces) and carbon (for light surfaces), dusted onto a hard surface will adhere to perspiration or body oils left with a fingerprint. Fluorescent powders will adhere to prints and fluoresce when they are viewed under ultraviolet light.

Chemical techniques include iodine fuming, silver nitrate, ninhydrin, and cyanoacrylate ester, also known as Super Glue® fuming. For reasons as yet unknown, the violet fumes of iodine, which form when solid iodine crystals sublimate upon heating, will adhere to latent fingerprints. The surface suspected of holding prints is placed in a closed hood where it is exposed to iodine vapors. The de-

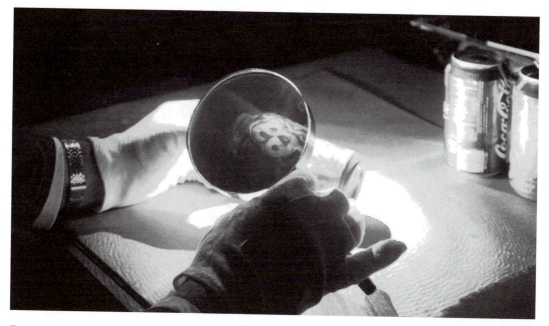

Expert viewing fingerprints with laser light under magnifier. FBI photo.

veloped prints, which fade quickly with time, can be preserved by fixing with a starch solution or by photographing them. A similar approach can be used by exposing suspected surfaces to the fumes of Super Glue® (cyanoacrylate ester). The ester adheres to substances left by fingerprints on nonporous surfaces to produce a grayish white print.

Spraying suspected surfaces with ninhydrin is the preferred approach among forensic scientists. The ninhydrin combines with the amino acids in perspiration to produce a purplish blue color. If fingerprints are present, they will appear in from 1 to 48 hours. Recently, forensic scientists have been spraying a vapor of a cyanoacrylate fluorescent dye over surfaces where fingerprints are suspected. The vapor reacts with moisture and oil in the print, revealing its characteristic pattern of friction ridges. There are now substitutes for ninhydrin, such as diazaflorin (DFO), that are more effective in making latent prints visible.

A silver nitrate ($AgNO_3$) solution will react with the salt (NaCl) left by the sweat in fingerprints to form silver chloride (AgCl). If the print is then exposed to ultraviolet light, silver will form, leaving a print that is black or reddish brown in color.

The development of portable lasers has provided investigators with yet another means of bringing latent prints into view. A variety of substances left in the sweaty remains of a fingerprint fluoresce when they are exposed to a laser's beam. Recently it has been found that certain chemicals can induce fingerprints to fluoresce when exposed to light. Furthermore, high-intensity lights, which are much less expensive than lasers, can be used with these chemicals.

To preserve fingerprints, they are first photographed at the site where they are found. Some police departments have cameras with a built-in light source and close-up lens that are specifically designed to photograph fingerprints. If the print is on a portable object, it will be carried to the laboratory and stored for possible use in court. Prints found on immovable surfaces that have been dusted can be removed by "lifting." Lifting involves applying a clear strip of plastic tape, to the powdered print. The powder sticks to the tape, providing a clear record of the pattern of friction ridges. The

Fingerprints on computer screen. Courtesy of the City of Hartford Police Crime Scene Unit.

tape is then placed on a labeled card and stored for future reference.

Often, fingerprints lifted at a crime scene are not clear. By using computers and digital-imaging software, the clarity of prints may be enhanced, making identification possible.

Cases

In 1892, Juan Vucetich, the Argentine police official who had invented his own fingerprint-classification system, was able to demonstrate the use of fingerprints in identifying criminals. Vucetich's department was investigating the murder of two children. Their mother accused a neighbor of the crime, but part of a door frame that revealed bloody fingerprints was brought to Vucetich's office. The prints were identical to those of the mother, Francesca Rojas, who, faced with the evidence, confessed.

The first admission of fingerprint testimony into a court in the United States occurred in 1910 during the trial of Thomas Jennings. Jennings was accused of murdering Clarence Hiller, a Chicago resident who encountered Jennings as he was burglarizing Hiller's apartment. Several hours before the murder, Hiller had painted his porch railings. During their investigation, police discovered four clearly defined left-hand fingerprints on the railing. The prints matched those of Jennings, who had been caught by police shortly after the crime. When fingerprint experts testified that only Jennings could have left the prints found on the railing, the jury found him guilty, and he was sentenced to be hanged.

Defense lawyers appealed the case, arguing that fingerprinting was not scientific and had never been recognized as evidence in an Illinois courtroom. A year later, the Supreme Court of Illinois upheld the use of fingerprints as evidence and refused the appeal for a new trial. The Illinois court's decision to accept fingerprints as evidence led police departments throughout the land to abandon the Bertillon system and turn to fingerprinting as a means of identifying criminals.

In 1984, a serial killer and rapist known as the Night stalker was terrorizing the Los Angeles area. He would usually enter a home through an open window, shoot any man in the building, and then rape his wife. One night in August 1985, the woman he raped after killing her husband had enough sagacity to record the license plate number of the Night stalker's car as he drove away. Police found the car and set up a surveillance. The killer never returned to the car, but he did leave a fingerprint that police were able to lift. Fortunately, the Los Angeles Police Department had computerized its fingerprint database earlier that year. Within minutes, the computer had identified possible matches, and an expert determined that the Night Stalker was Richard Ramirez, who had been arrested several years earlier for a traffic violation. His picture was circulated by the media, and he was quickly identified and captured. Four years later, he was convicted and sentenced to death. He is currently awaiting execution at California's San Quentin Prison.

At about 8:30 P.M. on October 3, 1995, Jose Rivera and a young woman entered Wayne Coonan's Boston apartment. There the three, as well as another male, smoked crack, drank alcohol, and engaged in several sexual acts. At 7:50 the next morning, Coonan called police to report a dead woman in his apartment. Police found the victim naked on Coonan's bed. Her wrists were bound, and her mouth was covered with duct tape. An autopsy showed trauma to the head and vagina. The cause of death was strangulation. Forensic evidence included stains of the victim's blood on Coonan's T-shirt, fingerprints on the duct tape that had a looped-type pattern consistent with Coonan's prints, and hair on the fly area of Coonan's boxer shorts that was similar to the victim's pubic hair. Furthermore, the victim had numerous skin particles that were consistent with the loose skin associated with a skin disorder that plagued Coonan. This evidence, together with testimony from others who had been in Coonan's apartment that night and several days before, led a jury to convict the defendant of first-degree mur-

der. *See also* Bertillon, Alphonse; Crime Scene; DNA Evidence; Galton, Sir Francis; Rojas, Francesca; Trace Evidence.

References

Asbaugh, D.R. *Quantitative-Qualitative Friction Ridge Analysis: An Introduction to Basic and Advanced Ridgeology.* Boca Raton, FL: CRC Press, 1999.

Raymond, Joan. "Forget the Pipe, Sherlock: Gear for Tomorrow's Detectives." *Newsweek*, June 22, 1998, p. 12.

Saferstein, Richard. *Criminalistics: An Introduction to Forensic Science.* Upper Saddle River, NJ: Prentice Hall, 1998.

Smith, W.C., R. Kinney, and D. Departee. "Latent Fingerprints—a Forensic Approach." *Journal of Forensic Identification* 43(6), 563–570, 1993.

Within, George and Katie Helter. "High-Tech Crime Solving." *U.S. News and World Report*, July 11, 1994, p. 30.

Footprints

Footprints, the evidence left by bare feet, are rare in industrialized nations where almost everyone wears shoes. On the other hand, they are common in India and African nations where many move on bare feet. The pattern of ridges on feet is as unique as it is on fingers; however, there are no databases by which investigators can make comparisons. On rare occasions, foot patterns can be used for purposes of identification. For example, in 1946, a female body lacking a head and hands was discovered in the mountains of California. The body was eventually identified as that of Dorothy Eggars by the podiatrist who had treated her bunions. In England, a naked rapist left footprints that a fingerprint expert was able to match with those of a suspect.

More commonly than footprints, forensic investigators find prints left by shoes, tires, or fabrics. Such prints are carefully photographed from above and at various angles to capture details of the pattern. The next step, if possible, is to bring the evidence to the laboratory, where it can be examined more closely for trace evidence and compared with

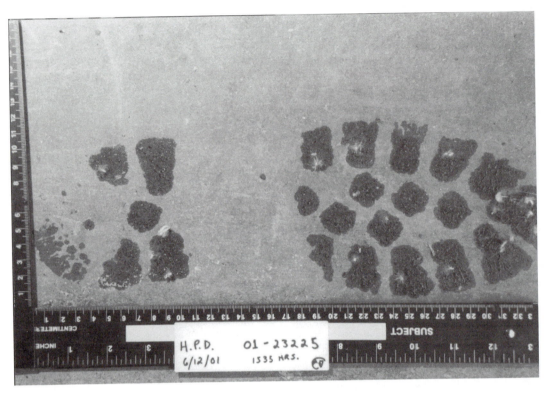

Footprint. Courtesy of the City of Hartford Police Crime Scene Unit.

known makes of shoes, tires, or fabrics. A footprint left on a piece of paper or a tiled floor can be easily transferred to a laboratory. On the other hand, a footprint left on a dusty floor presents a more formidable problem. Often these prints can be lifted in much the same way that fingerprints are lifted from glass or other surfaces. A lifting material large enough to cover the print is placed over the impression. A fingerprint roller is then used to remove any air bubbles before the print is lifted.

Some police departments have electrostatic lifting instruments that can be carried to the crime scene. A plastic film is placed over the print and connected to a high-voltage power supply. Charges on the plastic attract dust from the surface below, creating a pattern on the plastic sheet that can be photographed. Such a photograph provides a much clearer pattern than the print itself on surfaces such as carpets, chairs, and various colored materials.

Footprints and tire marks left in soft soil can be transported by pouring plaster of paris or dental stone into the impression to make a cast. After the cast dries, the impression can easily be lifted and taken to a laboratory.

At winter crime scenes, footprints and tire tracks are often detected in the snow. Surprisingly, casts can be made of such prints. People who study animal tracks have for years poured small quantities of melted paraffin into prints left in snow. The liquid quickly cools, and the process can be repeated many times until a cast is made that is thick enough to be carried away. Forensic investigators use a waxy material in an aerosol can that can be sprayed onto the impression. Three or four light sprayings are applied, with time between sprayings to allow cooling to take place. Plaster of paris or dental stone is then poured onto the waxed print. After it dries, the positive cast of the print can be brought inside. At the laboratory, the positive print in the cast can be changed to the original negative cast seen in the snow by making a cast of the cast.

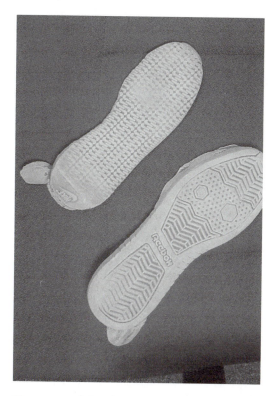

Plaster cast of shoe impressions. Courtesy of the Florida Department of Law Enforcement.

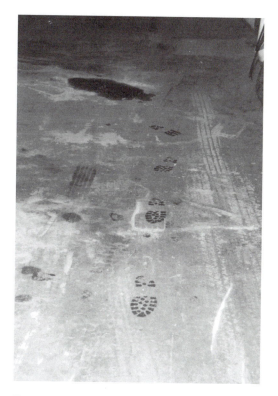

Footprints at crime scene. Courtesy of the City of Hartford Police Crime Scene Unit.

Once a print has been preserved, it can be compared with shoes worn or vehicles driven by suspects. The evidence, however, may not be very convincing unless there are correspondingly unique features between the suspect's test print and the print recovered at the scene of the crime.

There are chemicals and staining solutions that will enhance faint footwear impressions made with blood. In addition, computer software can be used to compare shoeprints. *Shoeprint Image Capture and Retrieval* (SICAR), developed in England, has databases that allow forensic scientists to compare impressions obtained at a crime scene with known footware. It can also be used to compare crime-scene samples with shoeprints left by someone in custody or with prints obtained at another crime scene.

When 18-year-old Nicole Laitner of Sheffield, England, was raped early in the morning of October 24, 1983, the killer-rapist left a bloody footprint on the stairs. When police later arrested a suspect, Arthur Hutchinson, who had been found guilty of two previous sex crimes, they found that the footprint was consistent with the pattern on the sole of a bloodstained shoe owned by the suspect. Other evidence included blood type, bite marks on cheese found at the Laitner home that matched his dentition, a palmprint on a bottle, and Laitner's identification of him as the man who had attacked her and killed her parents. He was found guilty of psychopathic murder and rape and sentenced to three consecutive life terms.

That footprints are less useful than fingerprints as evidence was made clear at the trial of 49-year-old Leonard John Fraser in Brisbane on April 22, 1999. Fraser was accused of murdering 9-year-old Keyra Steinhardt, whose partially buried body was discovered two weeks later. Dr. Paul Bennett, a forensic expert in podiatry, testified that there were a large number of correlations between the footprints found at the scene and ink imprints and plaster casts he made of Fraser's feet and a high degree of consistency between the defendant's feet and the impressions submitted as evidence. He also told prosecutor Paul Rutledge that in his opinion the footprints were made by Fraser. However, under cross-examination by defense lawyer Greg McGuire, he admitted that the footprints were not an exact match because the soil was soft, and he acknowledged that he could not be certain that the footprints were left by the defendant.

In the case of Canadian Allan Legere, a bloody footprint was found on a church bulletin at the rectory where Father James Smith, an elderly Roman Catholic priest, had been killed in Chatham Head near Newcastle, Canada, in 1989. Forensic experts from both the Royal Canadian Mounted Police and the FBI stated that it was highly probable that the footprint left at the rectory was made by boots worn by Legere. They also demonstrated that plaster molds of Legere's feet and the insoles of the boots that likely made the bloody print fit together perfectly. Furthermore, the tip of a nail that protruded through the heel of the left boot met precisely with a dark mark on Legere's left heel. This evidence, together with DNA obtained from semen found on three female murder victims attributed to Legere, led to his conviction. He was sentenced to life in prison. *See also* Bite Marks; DNA Evidence; Fingerprints; Tool Marks.

References

Bodziak, W.J. *Footwear Impression Evidence.* New York: Elsevier, 1990.

Saferstein, Richard. *Criminalistics: An Introduction to Forensic Science.* Upper Saddle River, NJ: Prentice Hall, 1998.

Zonderman, Jon. *Beyond the Crime Lab: The New Science of Investigation.* New York: Wiley, 1990.

Forensic Anthropology

Forensic anthropology is that part of forensic science that relies on the expertise of physical anthropologists. It has been popularized by novelists such as Kathy Reichs, who has written *Déjà Dead, Deadly Decisions,* and *Death du Jour,* and by the movie *The Bone Collector.*

Fossil bones constitute the raw material and central focus of physical anthropology, but until recently, anthropologists were diffident about criminal investigations. They regarded them as less professional than pure research, and police, unfortunately, did not realize that bones held clues undetected by conventional methods of forensic science. Until the 1940s, anthropologists were seldom associated with law-enforcement agencies in solving crimes. Cooperation between anthropologists and the FBI began fortuitously when FBI headquarters was moved close to the Smithsonian Institution in the 1930s. In 1942, T. Dale Stewart, who was the director of the Smithsonian's physical anthropology division, began to serve as a consultant for the FBI. By the 1970s, Smithsonian anthropologist Larry Angel was involved in so many cases that the press started calling him "Sherlock Bones" (Royte, p. 83).

Cooperation between the FBI and the anthropology department at the Smithsonian continues. The FBI often brings the bones of one of the 5,000 sets of skeletal remains found in the United States each year to the curator with the question: Are they human? The anthropologists also aid the FBI in reconstructions of faces based on recovered skulls of persons who might be murder victims. The discovery of skeletal remains always raises questions such as, how old was the person at the time of death? Was the person male or female? Young or old? Caucasian, Afro-American, Asian, or Native American? Tall or short? Was he or she muscular? Did the person have any diseases or distinct features?

Anthropologists can often provide answers. The skeleton of a pelvis is useful in determining the sex. A human female's pelvis is wider and more shallow than the male's. Height is easy to determine if the skeleton is intact, but anthropologists have developed mathematical relationships that enable them to make accurate estimates of the deceased person's height from partial remains. For ex-

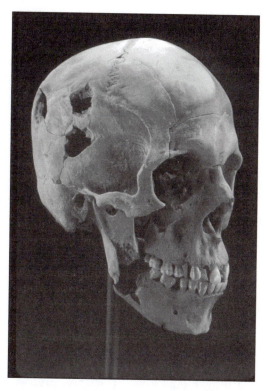

From the Smithsonian Institution Department of Anthropology, a skull with bullet holes, right frontal oblique view, showing exit wounds. © Chip Clark: National Museum of Natural History.

ample, a 45.7-cm-long femur (upper leg bone) from a Caucasian male would indicate that he was 170.2 cm (67 inches) in height. One can obtain this height by using the formula developed to determine height from length of femur:

length of femur × 2.38 + 61.4 cm = height (Evans, p. 123).

The basis for making height estimates from bone lengths was established during World War II. Mildred Totter, an anatomy professor, was in charge of a U.S. Army center in Hawaii to repatriate unidentified personnel killed in battle. Together with other anatomists and anthropologists, she established identity through dental and health records. The army also granted her permission to collect detailed data about the bones of the war dead. From that data came the correlations

that now enable anthropologists to establish height on the basis of bone length. The role of physical anthropologists in solving crimes is perhaps best illustrated by some examples.

The Josef Mengele Case

Following World War II, Dr. Josef Mengele, the Nazi "Angel of Death," who had carried out numerous atrocities and was responsible for the deaths of 400,000 people within the gates of the concentration camp at Auschwitz, escaped Germany and the war trials that followed the conflict. It was believed that he had fled to South America. The long search to find Mengele and bring him to justice was unsuccessful until 1985, when Mr. and Mrs. Wolfram, a German couple living in Brazil, led investigators to the grave of Wolfgang Gerhard. They claimed that the grave held the remains of Josef Mengele, a 67-year-old victim of accidental drowning six years earlier.

The remains were exhumed and studied by a team of forensic scientists. The narrow, steep pelvis indicated a male skeleton. Characteristic of a right-handed person, the bones on the right side of the body were longer than those on the left. The length of the skeleton's femur and other leg bones indicated a height of 173.5 cm (68.3 inches). This agreed very closely with the 174-cm height found on Mengele's military records. The skeletal teeth agreed with Mengele's dental records. Richard Helmer, a German anthropologist, superimposed images from photographs of Mengele onto photographs of the exhumed skull. The perfect match led the scientists to believe that the remains were truly Mengele's. Further confirmation came in 1992. DNA recovered from the bones in the grave of Wolfgang Gerhard was compared with the DNA of members of Mengele's family still living in Germany. The results revealed that they had found the missing "Angel of Death," but it was too late to bring him to justice.

The John List Case

In 1971, the bodies of Helen List, her three children, and her mother-in-law were found in the Westfield, New Jersey, home that the family had shared with John List. They had all been shot. John List was missing, but he had left letters explaining that he had killed his family because, threatened by bankruptcy, he could no longer support them and did not want them to suffer disgrace and ridicule. He had also left his car at John F. Kennedy Airport after withdrawing $2,100 from his mother's bank account. Police believed that it would not be long before List would be found and charged with murder. However, years passed with no further clues. As far as the police were concerned, he had vanished. Perhaps, overcome with guilt, he had committed suicide in a desolate area where his remains had yet to be discovered.

Westfield detective Bernard Tracey continued, without success, looking for clues for years after the murders. Finally, in 1989, he convinced the producers of television's *America's Most Wanted* to show a bust of John List as he would look nearly 20 years after committing the crime. Using a photograph of List, techniques developed by physical anthropologists, and a computer program designed to simulate the effects of aging, forensic sculptor Frank Bender constructed the head of John List as he would look 20 years after the photographs were taken. The television program that displayed Bender's bust of John List was followed by nearly 300 phone calls from people who claimed to know where List could be found. One tip led FBI agents to the home of Mr. and Mrs. Robert Clark in Richmond, Virginia. Robert Clark was not home, but Mrs. Clark was amazed by the resemblance between her husband and photographs of the bust sculpted by Bender. Robert Clark denied having any knowledge of John List. However, a scar behind his right ear corresponded to mastoid surgery that List had undergone years earlier. Much more condemning were his fingerprints. They matched those of John List. Subsequently, John List, alias Robert Clark, was sentenced to life in prison.

War Crimes

Clyde Snow is a well-known forensic scientist who helped determine that the body in the grave of Wolfgang Gerhard was really the buried remains of Josef Mengele. He was also the man who helped citizens in Argentina determine the fates of their loved ones after the bloody military reign in that country ended in 1983. During the 1990s war in the Balkans, Snow, together with other forensic scientists, was sent to investigate mass graves in the former Yugoslavia by a war-crimes tribunal established by the United Nations Security Council. They discovered mass graves and began the monumental task of identifying the bodies and determining the cause of death.

One of the sites was at a farm complex called Ovarca outside Vukovar where Serb troops had taken 200 hospital patients in November 1991. During an interview with Snow, a survivor of that event, who had escaped by leaping from a truck the Serbs had used to transport the patients, reported hearing gunfire nearby shortly after his escape. Three days later, near the point where the survivor had escaped, Snow spotted a skull lying in mud. Excavation revealed a mass grave with as many as 200 bodies. At one edge of the site, Snow detected the empty cartridges of bullets that matched those fired by weapons commonly carried by the Serb troops.

Following a peace treaty signed by the presidents of Bosnia-Herzegovina, Croatia, and Serbia on December 14, 1995, Snow and other forensic scientists returned to the grave site near Vukovar. There they uncovered bodies draped in white smocks and clogs, the dress and footwear worn by employees at the Croatian hospital in Vukovar. Parts of other bodies were covered with bandages, casts, slings, and even crutches. Despite Serbian denials, the evidence clearly indicated that Serbian troops had murdered the people they had taken from the hospital in 1991.

Then came the tedious task of identifying the victims. Examining a skeleton, Snow determined that the victim's sex was male by examining the pelvis, which was deep and narrow, and the bones behind the ear (mastoids) and at the nape of the neck (nuchal crest), which are larger in males. A longer right arm indicated that the person had been right-handed. Height was readily estimated from the length of the leg bones. An age estimate of 22–28 was determined by examining the pubic bones, the collarbone (clavicle), and the ends of the ribs that join the sternum (breastbone) through cartilage. Just as the bones of the skull fuse during childhood, so the pubic bones fuse with the ilium and ischium during adulthood, and a similar growth pattern takes place at the ends of the clavicle and ribs.

The cause of death was clearly visible from examination of the skull. An inwardly directed hole at the base of the skull indicated the point where a bullet had entered. Two holes that beveled outward near the upper front of the skull indicated that the bullet had split before leaving the body.

Many tentative identifications of victims were made from the skeletons and associated articles of clothing and artifacts such as rings, necklaces, watches, and the like. In an attempt to establish more definite identifications, DNA samples were extracted from the skeletons and sent to molecular geneticists who tried to establish identities by comparing the victims' DNA to the DNA obtained from blood samples provided by relatives who believed that their sons or daughters, husbands or wives, might have been buried at a particular mass-grave site. In some cases, definite identification was established and people learned the fate of their loved ones. As for justice, to date, none of those believed to be guilty of the war crimes committed at Ovarca and at other mass-grave sites have been arrested and brought to trial. *See also* Luetgert, Adolphe L.; Odontology; Webster, John.

References

Evans, Colin. *The Casebook of Forensic Detection: How Science Solved 100 of the World's Most Baffling Crimes.* New York: Wiley, 1996.

Gardner, Robert. *Crime Lab 101: Experimenting with Crime Detection.* New York: Walker, 1992.

Royte, Elizabeth. " 'Let the Bones Talk' Is the Watchword for Scientist-Sleuths." *Smithsonian,* May 1996.

Stover, Eric. "The Grave at Vukovar." *Smithsonian,* March 1997, pp. 40ff.

Thomas, Peggy. *Talking Bones: The Science of Forensic Anthropology.* New York: Facts on File, 1995.

Forensic Archaeology. *See* Forensic Anthropology.

Forensic Dentistry. *See* Odontology.

Forensic Engineering

Forensic engineering is that part of forensic science that relies on the expertise of engineers. The large number of lawsuits facing courts has made forensic engineering a profitable and growing profession. A number of firms specialize in this aspect of forensic science. They investigate and provide evidence about industrial failures, explosions, fires, and other disasters. They evaluate product liability and offer expert-witness testimony for any legal case in which an understanding of physical science can help a judge or jury understand the circumstances involved in the litigation. For example, in cases involving highway accidents, forensic engineers may be able to determine the speeds of vehicles before a collision and assess causes related to mechanical failure, highway construction, weather conditions, failure to wear seat belts, and so on.

A father and son, working as carpenters on a project to make improvements in a drainage canal, installed a temporary electrical system to provide power for their tools, as well as two submersible pumps to transfer water to a diversion channel. Two weeks later, the son was electrocuted when he picked up one of the pumps. Occupational Safety and Health Administration (OSHA) investigators were called to the scene and found a number of flaws in the way the electrical system was installed that could have contributed to the son's death. Following its investigation, OSHA fined the construction company that had hired the carpenter team $52,000. A lawyer representing the company called Technical Consultants Group in Denver to conduct an independent investigation. The forensic engineer assigned to the accident confirmed the deficiencies noted by OSHA's investigators, but he also discovered that the carpenters had buried the electrical cable beneath crushed rock over which heavy trucks traveled to and from the site. He concluded that the crushed rocks wore away the cable's insulation, which then energized the pump casing that had not been properly grounded.

Forensic engineers are not always successful in determining cause or guilt. A forensic engineering firm was asked to determine the cause of a 10th-floor fire in a 60-year-old San Francisco high-rise. The owners of the building sent claims to both the boiler/machinery and the fire-insurance companies. The two companies each hoped that the other would have to cover the entire financial loss or that the fire could be attributed to the elevator manufacturer, the roofer, or the elevator-maintenance firm. As the investigation proceeded, the engineers began to find that they had many questions and no answers. A variety of theories were proposed, but none could be substantiated by irrefutable evidence. In the end, the engineers concluded that the cause of the fire could not be determined with certainty. Rather than risk further losses, the boiler/machinery and fire-insurance companies agreed to split the costs 50–50. *See also* Automobile Accidents; Crime Scene; Glass Evidence; Paint Evidence; Trace Evidence.

References

Crawford, Ralph. "Forensic Casebook: 'No Conclusion' Results in Agreement." *Direct*, April 1, 1999.

Foley, Michael. "Forensic Casebook: Electricity and Water: A Volatile Mix." *Electrical Construction and Maintenance*, May 1, 2000.

Saferstein, Richard. *Criminalistics: An Introduction to Forensic Science.* Upper Saddle River, NJ: Prentice Hall, 1998.

Forensic Entomology

Forensic entomology is that part of forensic science that relies on the expertise of entomologists who use insects and other arthropods to obtain evidence about the time and place of death of a corpse. Examining a corpse shortly after a murder is not a pleasant task for a criminologist, but examining one that has been dead for some time can be a lot worse. In addition to the odor of decaying flesh, there are usually maggots (fly larvae) inside the body as well as on it. Twenty years ago, an investigator would probably have regarded a corpse infested with insects in any stage of their life cycle as an unpleasant nuisance. Today, the same investigator would probably collect samples of as many of the four stages (egg, larva, pupa, adult) of the insects as he or she could find on and in the body. Some would be put in alcohol to preserve them, and larvae would be placed in vials together with their favorite food.

A forensic entomologist (there are about 20 in the United States.) would be called in to examine the evidence. Often the entomologist can accurately determine the time of death by the fauna found on a dead body. Such information is very useful in determining whether or not a suspect's alibi is valid.

Entomologists know that within minutes of an animal's death, blowflies will be laying eggs on and in its body. From more than a mile away, these flies can detect the subtle odors released by blood or the gases released by a corpse shortly after death. Blowfly eggs hatch within 24 hours. The exact time after laying depends on the species. The larvae feed on flesh for 8 to 12 days and then look for a dry place nearby to pupate. Adults emerge after 18 to 24 days.

The blowfly larvae are followed by flesh flies, whose eggs hatch internally so the larvae are alive as they emerge from the females. Sometimes ants that feed on the fly larvae invade the corpse. After a day or two, beetles and wasps arrive to feast on fly eggs and larvae. They are followed by carrion beetles who feed on rotting flesh. After about a month, mites and skin beetles attack the dried hair and skin.

This sequence of fauna can be used to help determine time of death. However, it has limits. After a few weeks, investigators may be looking at second and third generations of insects. Furthermore, the sequences in an insect's life cycle are temperature dependent. Eggs hatch sooner and larvae mature faster in warm weather than in cool. Hence the season and weather conditions must be taken into account.

Forensic entomologists can often determine that a body was moved after the murder. For example, a species of green bottlefly lays eggs in sunlight, while a species of black blowfly prefers shade. When detectives showed an entomologist the black blowflies they had discovered on the body of a woman in a sunny Maryland landfill, the entomologist knew that the body had been moved from a shady area. This, together with other evidence, led police to her killer, a truck driver who had kept her body concealed in his trailer for several days before dumping the body in the landfill.

Recent research indicates that it is possible to extract human DNA from the guts of insects that have sucked blood from humans. In the future, it may be possible to use this method to identify suspects when bloodsucking insects such as mosquitoes remain at the scene of a homicide.

The first time that insects were used forensically was in China in the year 1235. A victim was found slashed to death near a village. The mayor of the town gathered all the local men together and told them to lay their rice sickles in the sun. Very soon, flies, attracted by the blood, were found swarming around one sickle. The owner, confronted by the evidence, confessed to the crime.

On November 20, 1984, Anne Swanke, a student at the University of California at San Diego, was seen carrying a can of gasoline toward her car. Four days later, her body was found in a wooded area. Her throat had been cut, and some of her clothes had been torn away. Detectives found insect eggs on Swanke's body. James Webb, a forensic en-

tomologist, determined that the eggs were from a species of blowflies that reproduce only when temperatures exceed 68° F. From weather reports it was discovered that the only day temperatures were above 68 was the day Swanke disappeared. It was also the only day for which a suspected serial killer, David Lucas, had no alibi. The information provided by Webb was instrumental in convicting Lucas, who eventually received a death sentence.

Near Winnipeg, Canada, during July 1995, two black-bear cubs were found dead, shot by summer poachers who had taken only their gall bladders. Gall bladders bring a good price in Asia, where they are used for medicinal purposes. No one had seen the killings, but several people had heard the shots and had seen the suspects leaving in a vehicle they could identify. Police found the suspects but needed evidence to convict them. A detective found blowflies crawling on the animals and collected some of the eggs they had laid. He recorded the time the eggs hatched, put some of the larvae in a vial with pieces of liver on which they could feed, and sent all the evidence to Gail Anderson, a forensic entomologist who told police that blowflies lay eggs on a carcass almost immediately after it has died. After the larvae pupated and emerged as adults, Anderson identified the species. (Insects are usually more easily identified in their adult stage.) She knew that eggs from the species of blowfly found on the bears hatch in 22 hours in summer temperatures. She subtracted 22 hours from the time the detective recorded as hatching time to establish the approximate time, within minutes, that the bears were shot. The time she determined matched the time witnesses heard shots and saw the suspects leaving the area. On the basis of the evidence, the suspects were convicted and spent six months in jail. *See also* Forensic Anthropology; Mass Spectrometry; Time of Death.

References

Chang, Maria L. "Fly Witness." *Science World*, Oct. 1997, p. 8.

Gannon, Robert. "The Body Farm." *Popular Science*, Sept. 1997, p. 178ff.

Introduction to Forensic Entomology. http://www.uio.no/~mostarke/forens_ent/introduction.shtml.

Mirsky, Steve. "Fright of the Bumblebee." *Scientific American*, July 1995, pp. 17–18.

Thomas, Peggy. *Talking Bones: The Science of Forensic Anthropology.* New York: Facts on File, 1995.

Forensic Photography

Forensic photography is that part of forensic science that utilizes photography and the expertise of photographers. Shortly after the emergence of photography, police were photographing criminals and people accused of crimes. In 1848, the Birmingham City Police in England began keeping photographic records. This police department is believed to be the first to establish such a procedure. Over the next two decades, police departments in both Europe and the United States developed "rogues' galleries." A format showing a criminal's profile, full face, and full figure soon became standard practice. As a result of the acknowledged value of photographic evidence, there is a basic principle in police procedures today. A crime scene must be photographed as completely as possible before anything at the site, other than an injured person, is moved or disturbed.

Although people commonly think of forensic photography in connection with murder cases, police make use of photography at the scene of many crimes, including those involving robbery, arson, public violence, traffic accidents, and so on. Television cameras are used as well. A verbal dialogue by an investigator working with the photographer can provide useful information while a videotape of the crime scene is being made. In any death where foul play is suspected, the coroner will see to it that photographs are taken of the body as well as the crime scene and any articles of evidence found there. If the size of a piece of evidence is significant, a ruler may be placed next to the item and included in any photography for comparison.

After the scene has been thoroughly photographed, an investigator will make detailed notes describing the scene, any evidence collected, who collected the evidence, and where and when it was obtained. One investigator will draw a rough sketch that will include careful measurements of dimensions and the positions of objects that might be related to the crime. In addition to spatial measurements, the sketch should include an arrow showing which direction is north. Later a draftsman will prepare a finished sketch using drafting tools.

Often fingerprints are found on soap, bottles, glass, leather, or other surfaces from which it is difficult to "lift" the prints. In such cases, photographs of the prints can provide the evidence needed. Prints left on striped surfaces are often photographed through a filter that cancels one or more of the colors. Anthracene, a chemical that fluoresces bright yellow light when it is exposed to ultraviolet radiation, is often sprinkled on the print. The photograph is then taken through a yellow filter to remove the ultraviolet and visible violet light to provide a clear image of the print.

Infrared film and light make it possible to photograph people or articles in dark places. The light is invisible to the human eye, but can be registered on film sensitive to such long wavelengths. Charred materials such as bank checks will often reveal fingerprints and writing when they are viewed in infrared light. Such light can also be used to reveal numbers filed from metallic surfaces such as engine blocks and guns. Ultraviolet and infrared light are often used to illuminate clothing, wood, and other surfaces because such substances as blood, gunpowder, semen, saliva, some inks, and the body oils in fingerprints fluoresce under such wavelengths and can be photographed with the appropriate film.

Macro- and Microphotography

Small pieces of evidence that will be presented in a courtroom are often photographed through a lens with a short focal length to provide images up to 10 times the size of the object. Very tiny bits of evidence such as fibers, hair, cut marks on wires, and other very small items are often photographed through the eyepiece of a microscope.

Photofit

In 1971, Jacques Penry developed what is known as the photofit system. It uses hundreds of photographs of small segments of the faces of people with different racial and physical traits to build in jigsaw fashion the face of a suspect based on the description of eyewitnesses. With the patience and cooperation of witnesses, the process often results in a composite picture that enables police to identify the suspect.

Missing Persons and Criminals

Missing children are sometimes identified from photographs on milk cartons or on post-office bulletin boards. "Wanted" criminals are often recognized from photographs posted by the police or displayed on television. John List, whom police had sought for nearly 20 years, was identified in 1989 on television's *America's Most Wanted*. Using a photograph of List taken in 1970, techniques developed by physical anthropologists, and a computer program designed to simulate the effects of aging, forensic sculptor Frank Bender constructed the head of John List as he would look 20 years later. When the bust was shown on television, a phone-call tip from a viewer eventually led to the arrest of List, who had changed his name to Robert Clark. *See also* Coroner; Crime Scene; Fingerprints; Forensic Anthropology; Trace Evidence.

References

Gardner, Robert. *Crime Lab 101: Experimenting with Crime Detection.* New York: Walker, 1992.

Newman, Andy. "Those Dimples May Be Digits." *New York Times,* May 3, 2001, p. 61ff.

Saferstein, Richard. *Criminalistics: An Introduction to Forensic Science.* Upper Saddle River, NJ: Prentice Hall, 1998.

Forensic Psychiatry

Forensic psychiatry is a branch of medicine that deals with mental health and the legal system. According to the American Academy of Psychiatry and the Law (AAPL), forensic psychiatry is "a medical subspecialty that includes research and clinical practice in the many areas in which psychiatry is applied to legal issues. While some forensic psychiatrists may specialize exclusively in legal issues, almost all psychiatrists may, at some point, have to work within one of the many areas in which the mental health and legal system overlap" (American Academy of Psychiatry and the Law). The AAPL also states that forensic psychiatry deals with the following legal areas: violence, criminal responsibility, competence, civil and criminal law, child custody and visitation, psychic injury, mental disability, malpractice, confidentiality, involuntary treatment, correctional psychiatry, juvenile justice, ethics, and human rights.

A forensic psychiatrist, like a forensic psychologist, performs evaluations and provides therapeutic counseling to people, both offenders and victims, police and lawyers, who are involved in the various aspects of legal issues. He or she can also serve as an expert witness in court concerning any of the legal areas listed earlier. For example, a forensic psychiatrist could be asked to determine if a person is competent to stand trial. A person cannot be put on trial if he or she cannot understand the proceedings and help in his or her own defense. The psychiatrist can conduct an evaluation to determine if there are any emotional, intellectual, or physical factors that would cause the person to be incompetent to stand trial. A forensic psychiatrist could also be called upon to determine criminal responsibility if a person is legally insane and should be committed to a mental facility.

Forensic psychiatrists and psychologists are sometimes called upon to help the police put together a profile of a criminal who is committing serious offenses such as serial killings and eluding traditional means of detection. In psychological profiling, a psychiatrist develops a personality sketch of a criminal based on careful analysis of evidence as well as intuition. Criminals frequently leave psychological and behavioral clues at a crime scene. These clues can reveal such characteristics as race, age, gender, physical size, marital status, sexual preferences, occupation, and socioeconomic status. Investigators can be helped in developing a list of suspects based on the characteristics revealed by the profile. *See also* Forensic Psychology; The Insanity Defense; Mad Bomber; Psychological Profiling; Unabomber.

References

American Academy of Psychiatry and the Law. http://www.emory.edu/APPL/org.htm.

Forensic Psychology & Forensic Psychiatry: An Overview. http://flash.lakeheadu.ca~pals/forensics/forensic.htm.

Lane, Brian. *The Encyclopedia of Forensic Science.* London: Headline, 1992.

Forensic Psychology

Forensic psychology deals with the emotional and behavioral issues that relate to the legal system, such as deciding on the mental stability of a person accused of a crime. Psychiatry has been involved in legal issues for many years, but it has only been since the 1960s that psychology has played a role. The turning point was the case of *Jenkins v. United States* (307 F.2d, 637, D.C. Cir. 1962). In the original case, the judge had told the jury to disregard the testimony of a psychologist regarding mental illness. He felt that a psychologist was not qualified to give a medical opinion. The court of appeals reversed that decision and stated, "Some psychologists are qualified to render expert testimony in the field of mental disorder." Since that decision, the courts have broadened the issues included within the psychologists' legally defined areas of expertise. Standards vary in state and federal law, but psychologists are given expert status in almost every area of criminal, civil, family, and administrative law.

The work of a forensic psychologist and that of a forensic psychiatrist overlap in many areas, but there are also differences between the two. A psychiatrist has medical training and can prescribe drugs. A psychologist cannot prescribe drugs and does not have a medical degree. However, both can perform comprehensive evaluations and provide therapeutic services. Both deal with the behavior of individuals and how personality and motivation shape behavior. Both can also look at the biological basis of behavior and recommend pharmacological help, but only the psychiatrist can prescribe the drugs.

Some of the questions a psychologist might be called upon to answer in forensic cases include diagnostic questions about personality characteristics and the possibility of the presence of a psychosis, a defendant's competence to stand trial or make legal decisions on his or her own behalf, the relationship of a psychological disorder to an accident or trauma, and assessments related to the best interests of a child in a custody case. Other questions might include the need for treatment and prognosis and the potential for future dangerous behavior by a criminal or parent. To answer these questions, the forensic psychologist must possess evaluation skills and be knowledgeable in relevant case law and special screening tools and tests. He or she must also be aware of confidentiality issues.

An example of the work done by a forensic psychologist occured in the case of the so-called Unabomber, Theodore Kaczynski. Kaczynski wanted to represent himself in court but was deemed mentally unable to do so in an evaluation done by a forensic psychologist. A forensic psychologist or psychiatrist could be called upon to perform an evaluation regarding a defendant's mental state at the time of a crime. The defense of "not guilty by reason of insanity" relies on the psychological or psychiatric evaluation of a defendant's inability to form criminal intent. *See also* Forensic Psychiatry; Psychological Profiling.

References

Corsini, Raymond J., ed. *Encyclopedia of Psychology.* 2nd ed. New York: Wiley, 1994.

Cyriax, Oliver. *Crime: An Encyclopedia.* North Pomfret, VT: Trafalgar Square, 1996.

Forensic-Science Errors

While forensic science is an essential part of law enforcement, errors in evaluating evidence and in determining guilt based on evidence are not uncommon. Innocent people are often imprisoned. Overworked forensic scientists make mistakes in analyzing evidence; mistaken eyewitness accounts are accepted as proof of guilt; unquestioned "expert" testimony is provided by people who are unqualified or inept; suicides are mistaken for homicides; accidental fires are regarded as arson; defendants are poorly represented by public defenders; zealous and persuasive prosecutors implore juries to find defendants guilty; judges are swayed by political factors; and "junk science" (bad science) is accepted by gullible juries.

In some cases, unqualified experts are detected by alert defense attorneys, but in many cases the testimony of "experts" obtained by the prosecution is crucial in convincing a jury that an accused person is indeed guilty. In 1997, following claims by FBI worker Frederic W. Whitehurst that forensic scientists sometimes compromised their analyses to fit the needs of prosecutors, the Justice Department's inspector general found evidence of faulty work at crime laboratories in 18 high-profile cases, including the Oklahoma City bombing and the O.J. Simpson murder case.

Junk Science

In recent years, as indicated in other entries, DNA and blood evidence have been used to prove the innocence of people wrongly charged with or convicted of crimes they did not commit, but there are other types of evidence that have been used to free or prevent the conviction of people wrongly accused of crimes. For example, in 1997, David Kunze was convicted of the murder of James McCann in Vancouver, Washington, on the basis of an earprint that police had taken from around the keyhole of the door leading to the room where McCann had been killed. Prosecutors claimed that Kunze's earprint matched the one on the door. They argued that earprints, like fingerprints, were unique and proved that the defendant was responsible for McCann's death. The jury accepted the evidence and found Kunze guilty. The conviction was overturned following an appeal in 1999. Kunze's lawyer learned that earprints were regarded as junk science by competent forensic scientists and were not accepted as valid evidence by the FBI.

According to polygraph expert Doug Williams ("Good Cop, Bad Cop"), polygraphs are another example of junk science. He maintains that polygraphs detect nervousness, not lying. A sociopath can lie without anxiety and go undetected by a polygraph, while an innocent person who is anxious and fearful by nature will be seen as lying according to a polygraph recording. Williams, who claims that people can learn how to lie and not be detected by polygraph records, advises anyone who agrees to a polygraph test to have his or her attorney present because, he claims, police interrogators can intimidate a suspect, causing confusion and the production of a record that indicates lying and guilt. According to Williams, any competent criminal lawyer will have a client take a trial polygraph before agreeing to undergo one conducted by law-enforcement personnel.

An Error by an Expert

On an early October morning in 1996, in Tabor City, North Carolina, Terri Hinkley discovered that her house was on fire. When she opened the door to her infant son's room, the flames prevented her from reaching him, and he died in the fire. Fortunately, her young daughter, though burned, escaped death.

On the basis of an investigation by an arson expert, Hinkley was accused of setting the fire and was charged with murder as well as arson. The shape of the burn patterns in the closet of the 17-month-old son's room indicated that the fire had started there. It was presumed that Terri Hinkley had set the fire.

To represent Hinkley, the court appointed a lawyer who had defended only two people facing the death penalty and had lost both cases. Fortunately, Hinkley was able to convince Ken Gibson and Gerald Hurst, two Texas forensic scientists who were arson experts, that she was innocent. When they investigated the fire, they discovered that it had actually started in the attic as a result of cloth-wrapped wires, installed when the house was built in 1935, that had overheated. In the yard, they tested the attic insulation, which was made of ground-up newspaper coated with fire retardant, and found that it readily ignited. Further examination showed that the attic fire had come down the wall and entered the child's closet at floor level, which produced an upward burn pattern on the closet wall. Following a lengthy meeting with prosecution experts and further consideration by the prosecution of its chances of obtaining a conviction given the evidence now available to the defense, charges against Hinkley were dropped. Had it not been for outside help, Terri Hinkley would probably have been falsely convicted of arson and murder on the basis of "expert" testimony. *See also* Crime Laboratories; Expert Testimony; Eyewitness Account; FBI Crime Laboratory; Forensic-Science Issues; Lineup; Lindbergh Kidnapping and Trial; Polygraph Test.

References

Frank, Laura, and John Hanchette. "Convicted on False Evidence? False Science Often Sways Juries, Judges." *USA Today,* July 19, 1994.

Freedberg, Sydney P. "Good Cop, Bad Cop." *St. Petersburg Times,* March 4, 2001.

"Good Cop, Bad Cop." www.truthinjustice.org/goodcop.htm. Accessed April 6, 2001.

Kelly, John F., and Phillip K. Wearne. *Tainting Evidence.* New York: Free Press, 1998.

Saker, Anne. "Terri's Fire." *Raleigh News and Observer,* Nov. 15–22, 1998.

"Truth in Justice." http://www.truthinjustice.org.

Forensic-Science Issues

The romanticism of Sherlock Holmes and the success of fictional scientists like Quincy and Kay Scarpetta, as well as the reliability of DNA and fingerprinting analysis, have created an air of infallibility around forensic science. Despite the public perception of certitude attached to the discipline, it has not been immune from criticism that has raised questions about the objectivity, methods, and dependability of forensic investigations. Some topics are inherently controversial. Not all accept the validity of results from voiceprints, hair analysis, or blood spatters. Even the usually reliable field of ballistics has caused difficulties for litigators as laboratories reminiscent of those established for high-school chemistry have given way to sophisticated techniques involving microbiology, nuclear physics, and detailed computer analysis. Consequently, the historical determinant of scientific evidence being established in the courtroom through the testimony of experts in the field, with the court and jury trusting in the skill and integrity of the witnesses, is even more vital to modern jurisprudence.

As reliance on forensic science and the use of expert witnesses to provide exculpatory evidence increases, there has been growing concern in legal circles that crime labs lack the objectivity they claim and create a pro-prosecution bias. Legal authorities cite as common the use of "backward reasoning"; that is, the labs presume guilt and then base their findings on that assumption. The tendency to support prosecutorial goals and theories is not surprising given that crime labs are usually affiliated with state and federal law-enforcement agencies. Even the renowned FBI facility customarily limits its services to law-enforcement agencies. As a result, access to crime labs for the defense is limited in many instances unless a defendant can afford independent testing. The potential for a conflict of interest is further complicated by the fact that large numbers of lab technicians are police officers themselves or are supervised by, or responsible to, police officials. It was not until the mid-1990s that the FBI began replacing its agents in the lab with civilian scientists. The very nature of forensic science as an ancillary discipline dedicated not to pure scientific knowledge but to providing supporting evidence for litigating a crime lends itself to the charge that crime labs are an extension of the prosecution.

In 1987, Glen Woodall was convicted of multiple felonies, including two counts of sexual assault. Subsequent DNA testing proved that Woodall was not the assailant, and it became evident that serology tests had never been done. Similarly, Henry Skinner was convicted of murdering three people in 1995. An investigation led by David Protess of Northwestern University uncovered the fact that much of the physical evidence had not been tested. Two bloody knives and an ax handle, flesh under the victim's fingernails, and DNA testing were ignored because police regarded Skinner as the perpetrator. His conviction was later overturned when that evidence was examined. In 1997, a major investigation into charges of incompetence and bias at the FBI lab led to significant changes of personnel and procedures at what many had perceived as the world's best forensic facility.

Aside from the perceived prosecutorial slant of crime labs, the image and reality of forensic science sometimes meet in bad science. In the O.J. Simpson murder trial, the defense attacked the DNA evidence as flawed because of poor collection techniques and faulty analysis. In 1993, a San Francisco lab technician certified as illegal drugs he had never tested, thus compromising a number of convictions. The 1995 Nelson Galbraith case offers an insight into the vagaries of forensic investigations. Despite signs of suicide, the medical examiner certified strangulation, or at the very least assisted suicide, as the cause of death of Galbraith's ex-wife. The medical examiner continued that tack despite the lack of defensive wounds or signs of a struggle to indicate murder. The examiner assumed that the victim could not have strangled herself due to her advanced age and arthritic physical condition. The 78-year-old Galbraith was arrested and brought to trial as the killer. The defense team brought in the chief medical examiner of Utah as a consultant. The body of the victim was exhumed, and a new autopsy was performed. The conclusion was that the first autopsy had been badly mismanaged. There were no photos of the first procedure, dissection of the neck, normal procedure for strangulation, was incomplete, and slides and tissue samples were not preserved. Defense lawyers were able to mount a definitive case of forensic bungling that led to the release of their client.

The credibility of expert forensic witnesses has increasingly come under scrutiny as the importance of scientific evidence has grown. Some well-publicized incidents involving forensic testimony have eroded the respect once accorded the field and have led to calls for some criteria for certification or licensure of both scientists and labs. West Virginia state trooper Fred Zain worked for and headed the state's crime lab while also testifying across the country as a serology expert. Persistent complaints about procedures and disturbing discrepancies in some of his cases led to a 1993 inquiry. The investigation cited Zain for fraud and falsification of reports, and the state supreme court discredited the entire body of his work. The convictions of five rapists and murderers were overturned due to Zain's lab failures. The FBI has also had to deal with the fraudulent expert. In 1974, Special Agent Thomas Curran, billing himself as a serology expert, testified he had advanced degrees in science and stated that both the victim and defendant in the trial had B blood types. Further investigation revealed that Curran had no degree and had never performed any tests. Only recently have forensic professional organizations like the American Academy of Forensic Science or the Association of Crime Lab Directors begun to call for professional standards for scientists and labs.

References

Frank, Laura, and John Hanchette. "Convicted on False Evidence? False Science Often Sways Juries, Judges." *USA Today*, July 19, 1994.

Kelly, John F., and Phillip K. Wearne. *Tainting Evidence*. New York: Free Press, 1998.

———. "Whose Body of Evidence?" *Economist*, July 11, 1998.

Forensic Taphonomy

Forensic taphonomy is evidence provided by examining buried remains. The literal meaning of taphonomy is "laws of burial"; however, the word has come to connote the portion of paleontology that investigates the processes that act on organic remains following death.

For most paleontologists, taphonomy involves thousands, if not millions, of years as they attempt to trace the paths of human evolution by locating and studying fossils. For those in the profession who are involved in forensic science, the time frame is less extensive, and the bones with which they are concerned have seldom been fossilized. A few cases, such as that of Captain Charles Hall, may involve a century, but most are usually limited to a few years at most.

Charles Hall was captain of the *Polaris*, which anchored off the northwest coast of Greenland in the autumn of 1871 on its way to find the North Pole. He died at that position and was buried. Reports indicated that Hall, after returning from a scouting trip in late October, became ill after drinking a cup of coffee. Reports of animosity between Hall and the ship's physician, Dr. Emil Bessels, who served the coffee and treated Hall after he became ill, led some to believe that Hall had been poisoned. Hall's symptoms, as reported by shipmates, were characteristic of arsenic poisoning. Nevertheless, a hearing held after the crew had returned home concluded that Hall had died of natural causes.

In the summer of 1968, pathologists Franklin Paddock and professor Chauncy Loomis found Hall's grave and his reasonably well preserved corpse. Paddock removed samples of tissue and sent them to two separate laboratories for evaluation, including neutron-activation analysis. The results showed that the tip of Hall's fingernail held arsenic at a concentration of about 25 parts per million (ppm), which was comparable to the concentration of arsenic in the soil around Hall's grave. At the base of the nail, however, the concentration was more than 75 ppm. The conclusion was that Hall had received massive doses of arsenic during the last two weeks of his life.

For bones that have been found exposed after surrounding body tissues have decomposed, the degree of weathering may be used to establish an approximate time of death. Marks on bones may reveal that death was related to the scars left by a bullet or knife. Often, some bones will be missing or scattered about, having been moved by animals.

Human corpses decompose under various conditions. Some are buried in earth or water, some are left on the ground where they fell, some are placed in the trunks of cars, and so on. Each corpse's rate of decay is related to temperature, weather, soil conditions, and insects and other fauna that may feed on the tissue. Recent studies by scientists under the direction of William Bass at the Tennessee Anthropological Research Facility (TARF) have provided valuable data regarding the time for corpses to decompose under different conditions (See Time of Death entry.) *See also* Cause of Death; Forensic Anthropology; Time of Death.

References

Lewin, Roger. *Principles of Human Evolution: A Core Textbook*. Malden, MA: Blackwell Science, 1998.

Nawrocki, Stephen P. *Ten Basic Points Concerning Human Remains Scenes*. University of Indianapolis Archeology and Forensics Laboratory. http://www.uindy.edu~archlab.

Forgery

Forgery involves the fraudulent alteration or falsification of documents. One of the most common types of forgery is a check signed with someone else's signature. Forging money or bonds is called counterfeiting. Forgery often involves erasing letters, words, or numbers using a rubber eraser, sandpaper, a knife, or a razor blade. Such mechanical erasures produce changes in the fibers of the document that can be detected with a microscope. Where changes are brought about by chemical change—oxidation of the ink—the paper may be discolored. Such discoloration, if it is not evident in visible light, may become evident under ultraviolet or infrared radiation.

Forgers will sometimes add value to a check by, for example, adding a "0" to a 9 and a "ty" to nine. If such a document is changed by substitutions or additions using a different ink, they may be made visible by using infrared luminescence. Inks differ in their capacity to absorb infrared light. If different inks are on a document such as a check, photographing that document with infrared-sensitive film in infrared light will reveal the presence of these different inks. Some inks absorb blue-green light and then emit infrared radiation. Documents with such ink can be illuminated with blue-green light and then photographed using infrared-sensitive film. Differences in the inks will be evident. Such photography may also expose erasures if traces of blue-green absorbing ink remain in the paper.

In the 1990s an ink was developed that can be infused with an individual's own DNA. Such personalized ink provides a signature that can be identified with certainty by a bank or any delegated receiver who has a scanner capable of reading the biochemical markers.

Forgery in the Arts and Sciences

Scientists have occasionally forged experimental data in order to gain professional standing or provide support for a hypothesis. Such attempts are almost always detected because other scientists are unable to obtain similar results when they repeat the experiments.

Art forgery is frequent because art collectors are willing to spend large sums of money for valuable paintings. The value of a piece of art is related to the artist who created it as well as the workmanship of the art. A painting by Picasso is worth far more than a copy by a current artist.

Scientific analysis can sometimes determine whether art is authentic or forged. Ancient as opposed to recent porcelain can be detected using thermoluminescence. When porcelain is made by heating clay, electrons trapped in the material are released, which resets the thermoluminescent clock to zero. Background radiation and radioactive materials within the porcelain cause the release of electrons, which are trapped. The porcelain can be dated by measuring the light energy released when it is heated. Heating releases the trapped electrons. By measuring the energy of the light emitted and knowing the trapping rate, the date that the material was heated during its production can be determined closely enough to reveal whether it is ancient or recent. Forged sculptures are more difficult to detect because all rock is old.

Samples of the paint in a work of art may establish the era in which the painting was done. The crystals in pre-nineteenth-century paints are rougher than the machine-ground crystals in more recent paints; analysis of pigments and binders can distinguish ancient from more modern materials; and X rays may be used to ascertain what is under the paint. However, science cannot prove authenticity. Experts familiar with a particular artist may be able to determine whether or not forgery has been committed. Their analyses are, of course, subjective, which is why expert witnesses called by the prosecution and defense often arrive at opposing conclusions.

A Case of a Lying Expert

In *Trapp v. American Trading and Production Corporation* in Kings County, NY, on March 5, 1998, a seaman was trying to recover damages for personal injuries. The trial was set aside and a new trial ordered when the defense discovered that the plaintiff's expert witness was not as qualified as he had claimed. He held no academic degrees and had completed only one year of college, although he had indicated both a B.S. and an M.S. He had served in the U.S. Navy, but as a chief warrant officer, not a captain. Furthermore, he held no license for unlimited tonnage in the merchant marine. *See also* Computer Crime; Counterfeiting; Disputed Documents.

References

Babitsky, Steven, James J. Mangraviti Jr. and Christopher J. Todd. *Comprehensive Forensic Services Manual: The Essential Resources for All Experts*. Falmouth, MA: S.E.A.K., Inc., Legal and Medical Information Systems, 2000.

Crimes and Punishment: The Illustrated Crime Encyclopedia. Westport, CT: H.S. Stuttman, 1994.

Cyriax, Oliver. *Crime: An Encyclopedia*. North Pomfret, VT: Trafalgar Square, 1996.

Gardner, Robert, and Dennis Shortelle. *From Talking Drums to the Internet: An Encyclopedia of Communications Technology*. Santa Barbara, CA: ABC-CLIO, 1997.

"Signing with Your Genes." *Popular Mechanics*, Sept. 1995.

"Work and Lies in the Promised Land." *World Press Review*, June 1995.

Fox, Sidney (1899–1930)

Sidney Fox was an Englishman convicted of killing his mother, 63-year-old Rosaline Fox, in an attempt to claim her insurance benefits. Fox attempted to conceal the murder by means of fire. The ensuing trial pitted two forensic giants against each other and emphasized the difficulty in interpreting marks on a dead body.

Fox, the youngest of four children, had a juvenile police record before he was 12 and served his first term in prison at 19. Uninterested in any kind of legitimate work, he served intermittent jail time for forgery, fraud, and theft. In 1929, the penniless mother-and-son combination reestablished their practice of drifting from hotel to hotel, leaving behind a record of bad checks, unpaid bills, and angry hoteliers. On October 16, the two checked into the Metropole Hotel in Margate, England. With his mother safely ensconced at the Metropole, Fox used a bad check to get money for a ticket to London. In the capital, he secured payment extensions on his mother's life insurance policies, which were worth £3,000. Fox then returned to the Metropole, where his mother had remained. The evening after his return, the two enjoyed dinner in the hotel dining room and retired for the night with a bottle of port.

At about 11:35 P.M., fire broke out in the Foxs' adjoining rooms. Naked except for his shirt, Fox burst into the hallway shouting a warning. A heroic traveling salesman, at first repulsed by the dense smoke, managed to drag the unconscious Rosaline Fox from the fire, and the hotel manager removed a charred and smoking armchair from her room. Oddly, Fox made no attempt to assist anyone, but watched the tragedy from a safe distance. Efforts to revive Rosaline Fox were futile. A doctor at the scene initially reported the cause of death to be shock and suffocation. This conclusion was reinforced when an inquest determined that the death was accidental.

On October 25, the grief-stricken son, without paying his bill, left the hotel to bury his mother. Rosaline Fox was laid to rest on October 29. By the afternoon of that day, her son was at the insurance company filing his claim. A suspicious agent, aroused by the survivor's lack of sadness, asked the main office for a more detailed, professional investigation. The case was ultimately referred to Scotland Yard. Fox was arrested on November 2 and held on charges of defrauding a number of hotels and filing a fraudulent insurance claim.

Rosaline Fox's body was exhumed, and Scotland Yard called in the renowned Sir Bernard Spilsbury to perform a postmortem. Finding no carbon monoxide in the blood or sooty deposits in the lungs, Spilsbury concluded that Rosaline Fox had been killed prior to the fire. He noted bruising at the back of the larynx, which indicated manual strangulation as the cause of death. This was a very controversial conclusion since the body presented none of the usual physical signs of strangulation, most notably a fracture of the hyoid bone in the neck. Spilsbury was adamant, and, given his forensic reputation, the decision took on nearly infallible proportions. Despite his denials, Fox was charged with murder, and the trial began on March 3, 1930.

From the beginning, the trial did not go well for the defendant. The defrauded hotel managers testified about Fox's character, reputation, and desperate financial state. More damaging was the analysis of the fire. A strip of carpeting between the stove, originally thought to be the fire's point of origin, and the rescued armchair was curiously unburned. How could the flame jump the gap? The Margate fire department's attempts to replicate the fire failed. The investigation turned to the armchair. From the position and burn patterns of a pile of newspapers, investigators ultimately deduced that the fire had begun beneath the armchair. Furthermore, they determined that some sort of accelerant would have been required to start the fire. Testimony revealed that Fox had gasoline in his room, which he said that he used to clean his only suit. The insurance policies and Fox's undue diligence in seeking his claims provided a neat motive for the prosecution.

Despite the prosecution's workmanlike presentation, the medical testimony during the trial attracted the most public scrutiny and threatened a conviction. Sir Bernard Spilsbury, recalling the after-dinner port the Foxes took to their room, testified that the alcohol content in the body indicated that Rosaline Fox had probably been drinking heavily prior to her death. He continued that she had most likely been asleep when her son strangled her. The highlight of Spilsbury's testimony was his demonstration, using a model of the human mouth, indicating where he had found a recent, large bruise that was symptomatic of strangulation. The Crown's pathologist asserted that a broken hyoid was not necessary for a verdict of strangulation as a cause of death. Finally, Spilsbury emphasized that he had found nothing that would indicate that death had been caused by heart problems, shock, or fire.

The defense countered with its own expert, Professor Sidney Smith. Smith, whose reputation nearly rivaled that of Spilsbury, had a different view. Working at a deficit because he never actually examined the body, Smith used Spilsbury's slides to refute his testimony. Smith showed that the slides did not reveal the neck bruise Spilsbury referred to in his testimony. Smith attributed Spilsbury's observation to postmortem discoloration, not bruising. Smith believed that the cause of death was heart failure. On the stand a second time, Spilsbury refused to recognize any other scenario than what he had proposed earlier. He insisted that the bruise was originally there and must have faded with time.

Momentum began to swing toward the defense, but the change was short-lived. Perhaps aware that his reputation and previous record were working against him, Fox took the stand in his own defense. It was a crucial mistake because he proceeded to alienate everyone in the courtroom with his cool, conceited, and very evasive testimony. When asked why, when he discovered the fire in his mother's room and was forced back due to smoke, he closed the door, Fox lamely stated that he did so to prevent smoke from spreading into the rest of the hotel. It took the jury one hour to find Sidney Fox guilty. He never appealed the conviction and was hanged on April 8, 1930. *See also* Arson; Carbon Monoxide; Smith, Sir. Sidney; Spilsbury, Sir Bernard H.; Strangulation.

References

Browne, Douglas G., and E.V. Tullett. *Bernard Spilsbury: His Life and Cases.* London: Penguin Books, 1955.

Evans, Colin. *The Casebook of Forensic Detection: How Science Solved 100 of the World's Most Baffling Crimes.* New York: Wiley, 1996.

Shew, E. Spencer. *A Companion to Murder.* New York: Knopf, 1961.

Freeman, R. Austin (1862–1943)

R. Austin Freeman was an English doctor, detective novelist, and short-story and non-fiction writer best known as the creator of Dr. John Evelyn Thorndyke, considered by critics and historians to be the first "scientific" detective. Freeman cemented his position in the detective genre by using "inverted" stories. These innovative plots, though they revealed the criminal and how the crime was committed early in the novel, focused on scientific methods of investigation as a means of establishing the guilt of the culprit.

Freeman, a surgeon and member of the Royal Colleges of Surgeons and Physicians, went to the Gold Coast (New Ghana) in 1887 as an assistant colonial surgeon. He then joined a medical expedition to the Ashanti as physician, natural scientist, and navigator; wrote two books; was appointed to the Anglo-German Boundary Commission to mediate disputed territorial claims; contracted blackwater fever, a malarial infection; and had to leave the colonial service two months short of qualifying for a government pension. Plagued by chronic illness upon his return to England, Freeman struggled to earn a living. Increasingly, he began to devote time to writing articles and short stories for magazines. In 1902, he published a novel in conjunction with a colleague, *The Adventures of Romney Pringle*, using the pseudonym Clifford Ashdown. The book was neither a financial nor a critical success.

In 1907, the 45-year-old physician finally emerged on the London literary scene with the publication of his first detective novel, *The Red Thumb Mark*. The book introduced Freeman's most enduring contribution to detective fiction, Dr. John Evelyn Thorndyke. Based upon Freeman's experience as a medical officer at Holloway Prison, the story explored the theme of fingerprint forgery. Freeman was not convinced of the reliability of the method for criminal identification and was concerned that fingerprints could be forged. In the novel, Thorndyke conducts a dramatic courtroom experiment exposing the fallibility of the fingerprinting technique.

The Red Thumb Mark gave Freeman credibility as an author, and buoyed by his modest financial success, he began to experiment with plots. These experiments were always informed by Freeman's philosophy as a writer. First, he strived for realism, not only in the story but also in the investigative methods. To this end, Freeman conducted his own tests and research, such as the one in *The Red Thumb Mark*, before including them in his books. Similarly, his knowledge of London, the setting for his novels, was a valuable tool in creating realistic situations. Second, and most significantly, Freeman theorized that readers would be attracted to an intellectual challenge. He believed that the process of detection, the gradual exposure of significant details in tracking down a criminal, was of more interest to his readers than merely discovering who committed a crime. This idea led to the "inverted" plots that contributed to the greatest of Freeman's literary creations, Thorndyke. Other Freeman works of note included *The Eye of Osiris* (1911), *A Silent Witness* (1914), and *As A Thief in the Night* (1928).

Freeman knew that he would face an inevitable comparison with his contemporary Arthur Conan Doyle and his sleuth, Sherlock Holmes. There are parallels between the two literary creations. Both are omniscient, well read, and versed in a number of different areas. The detectives are aided in their efforts by an admiring doctor friend; Thorndyke's Watson is his old medical-school friend, Dr. Christopher Jervis. The combination of lab, museum, and lodging is another commonality shared by Holmes and Thorndyke. Nonetheless, literary critics see the two investigators as different. Few doubt that Holmes is the better literary figure. Freeman has been criticized for his difficult language and old-fashioned style. However, these same commentators acknowledge that Thorndyke was the better scientific detective and the more accurate thinker. *See also* Thorndyke, Dr. John Evelyn.

References

Donaldson, Norman. *In Search of Dr. Thorndyke: The Story of R. Austin Freeman's Great Scientific Investigator and His Creator*. Bowling Green, OH: Bowling Green University Popular Press, 1971.

McAleer, John. "R. Austin Freeman." *Dictionary of Literary Biography*. Vol. 70. *British Mystery Writers, 1860–1919*. Detroit: Michigan Gale Research, 1988.

G

Galton, Sir Francis (1822–1911)

Sir Francis Galton, a cousin of Charles Darwin, was an English gentleman scientist, explorer, and statistician who was a significant contributor to the nascent field of fingerprinting. Galton had a variety of scientific interests, including biology, anthropology, meteorology, and experimental psychology. The talented amateur also did extensive studies on human heredity and intelligence.

In scientific circles, Galton is primarily known as the father of eugenics. Eugenics, a term he coined, called for the physical and mental improvement of human beings by selective parenthood. Galton was convinced that the application of scientific breeding to human populations would lead to improvement of the species. As a result of his studies on the influence of heredity on the human race, Galton investigated the use of fingerprints as a reliable source of personal identification. He published the definitive early text on fingerprinting and, more significantly, developed a methodology for classifying and organizing fingerprints to facilitate more immediate identification.

Galton studied medicine and mathematics, traveled extensively, and became a renowned explorer. He became a fellow of the Royal Society in 1856. After his marriage in 1853, Galton settled in London, where he spent the remainder of his life devoted to scientific pursuits. Since private research was his passion, Galton never held an academic or professional scientific position. He studied anthropometry and invented a heliograph, a mirrorlike device to send messages using the sun's rays. He studied blood transfusions, twins and even attempted to investigate, statistically, the efficacy of prayer. It was the publication of Charles Darwin's book *On The Origin of Species* in 1859 that inspired Galton to begin his studies of heredity.

Galton was intrigued by the idea of using fingerprints as an alternative to the popular anthropometric Bertillon system of identification. His interest was stimulated by an article in the periodical *Nature* written by fingerprint pioneers William Herschel and Henry Faulds. Galton began a lively communication with Herschel about the potential use of fingerprints in the identification process. Galton was very statistical in his approach to projects. Believing that everything was quantifiable, he set out to prove the value of fingerprints through the laws of probability.

Having established an anthropometric lab in London, Galton requested that any visitor to the facility leave a set of fingerprints. In addition, he borrowed Herschel's collection of prints, which encompassed three decades of fingerprints from the same subjects. These sources provided Galton with much of the raw data for his analysis. With access to thousands of sets of fingerprints and the use of photographic enlargement as necessary, the statistician began his analysis and discovered no matches at all. He concluded that there was a one-in-four chance of similarity in one fingerprint. If all 10 prints were taken, the odds of ever achieving a uniform match with another subject were inconceivably small. The odds, he calculated, were about 64,000,000,000 to 1. Establishing the uniqueness of fingerprints was not the result of a scientifically controlled study but rather of Galton's statistical analysis. In the course of his inquiry, Galton also verified that fingerprints were stable over time, remaining the same from youth to adulthood. Finally, he concluded that there was no connection between race, gender, or family and fingerprints.

After establishing the uniqueness and consistency of fingerprints, Galton set about finding some way to classify the data to make identification more efficient and reliable. Once again, there was some preliminary work to guide the scientist. The treatise of Prague physiologist Johann Purkinje offered the best early suggestions and provided a base for Galton's labor. Purkinje recognized recurring patterns in fingerprints, and Galton reduced these gross patterns to three categories: arches, whorls, and loops. The gross patterns were useful for a rough identification but lacked the precision necessary for a definitive identification. Galton then uncovered the key factor in fingerprinting. He discovered what he termed minutiae, small components that were distinctive and unchanging in each person's prints. While he labeled 700 characteristics, Galton referred to the most important feature as the delta. This triangular-shaped feature occurred in some way in almost every fingerprint. The triangle might be on the right or left, there could be several deltas or, in some cases, none at all. It was the minutiae of a print that allowed for a specific identification. In 1892, Galton wrote his primary work, *Finger Prints*, which detailed his findings. It was the first of three works he authored on fingerprints between 1892 and 1895. In recognition of his scientific endeavors, Frances Galton was knighted in 1909. *See also* Fingerprints.

References

Darwin, George H. "Sir Francis Galton." *Dictionary of National Biography: Second Supplement.* New York: Oxford University Press, 1958.

Forrest, D.W. *Francis Galton: The Life and Work of a Victorian Genius.* New York: Taplinger, 1974.

Stigler, Stephen M. "Galton and Identification by Fingerprints." *Genetics.* July, 1995.

Gardner, Erle Stanley (1889–1970)

Erle Stanley Gardner was a prolific American author most famous for the creation of the fictional lawyer-detective Perry Mason, a character who is not afraid to defy the police or tamper with evidence in order to help his client. Gardner studied law, passed the bar exam at age 21, and was a successful courtroom lawyer for over 20 years. During his time as an attorney, he began writing short stories with legal backgrounds. His first Perry Mason novel, *The Case of the Velvet Claws*, was published in 1933. After that, according to Gardner, he became a fiction factory, writing an average of 12 books a year. He penned over 80 Perry Mason novels. He also wrote a succession of novels about a district attorney, Doug Selby, and under the pseudonym A.A. Fair he wrote a series about a pair of private investigators, Bertha Cool and Donald Lam.

The character Gardner was trying to develop in Perry Mason was that of a fighter who possessed infinite patience. In *The Case of the Velvet Claws*, Mason calls himself a paid gladiator because he will intervene in cases involving the authorities and people in trouble. Many observers say that the character Perry Mason has many qualities that Gardner displayed as a courtroom lawyer. Gardner was known for his dramatic and innovative defenses and his intense cross-examinations of hostile witnesses. He would boldly use legal tricks and forgotten statutes to defend his clients.

As a lawyer, Gardner had defended many poor Chinese and Mexican citizens. He established the Court of Last Resort to help the convicted who strongly maintained their innocence. It incorporated experts from various fields of criminology who reviewed cases. Gardner himself became quite knowledgeable in the areas of forensic psychology, polygraph testing, police procedures, and criminal medicine. (The convicted murderer Sam Sheppard is one person who was helped by this organization and acquitted.)

In addition to legal issues, Gardner also wrote western novels and science-fiction books. Having lived in the American Southwest, he was motivated by his love of the area and his interest in nature to write a number of nonfiction books, including *Hidden Heart of Baja* (1962), *The Desert Is Yours* (1963), and *Mexico's Magic Square* (1968).

The Perry Mason stories were so popular that a radio series based on the novels ran for 12 years, starting in 1943. Warner Brothers had bought motion-picture rights to some of Gardner's novels in the 1930s and produced six Perry Mason films. Gardner was not pleased with the films and made sure that he had creative control of the radio series. Later, he guaranteed his control of the Perry Mason television series by starting his own production company, Paisano Productions, in 1956. The popular television series starred Raymond Burr as Perry Mason.

Gardner died at the age of 80 on March 11, 1970. His literary career stretched over five decades, and he wrote more than 700 fictional stories. After his death, William F. Nolan wrote, "Gardner, more than any other writer, popularized the law profession for a mass-market audience, melding fact and fiction to achieve a unique blend; no one ever handled courtroom drama better than he did. . . . His books, however imaginative, were authentic." *See also* Mason, Perry.

References

Nolan, William F. *Erle Stanley Gardner (1889–1970)*. http://erlestanleygardner.com/nolan.htm.

Senate, Richard. *A Brief Biography of Erle Stanley Gardner*. http://erlestanleygardner.com/senate.htm.

Steinbrunner, Chris, and Otto Penzler, eds. *Encyclopedia of Mystery and Detection*. New York: McGraw-Hill, 1976.

Gas Chromatography. *See* Chromatography.

Glass Evidence

Glass evidence is commonly found by forensic investigators at a crime scene or on a suspect. A burglar who has broken a window to enter a building will often carry bits of the glass in his shoes or clothing. Glass from a car's broken headlight may be found at the scene of a hit-and-run accident, and fragments of the same glass may also be found on the victim's body. Investigators will check headlight filaments because if they were on when the glass was broken, they will be oxidized as a result of being exposed to air while they are hot.

Glass evidence must be carefully examined and collected because the best use of glass is to find pieces at the scene that fit those associated with a suspect or his vehicle. If the glass pieces fit together like pieces of a jigsaw puzzle, it shows that both pieces came from the same source. The likelihood of finding two such pieces due to chance is so small that a perfect fit is considered identifying evidence and will almost invariably lead to conviction. Lacking pieces that fit, forensic scientists will seek to find matching properties. The two that are most often tested are refractive index and density.

Refractive Index

The refractive index of the glass, *n*, is related to the fact that glass bends (refracts) light as it passes from air into the glass. Refractive index is defined as

sine i/sine $r = n$.

Sine i is the sine of the angle of incidence, i, the angle between a ray of light and a perpendicular to the glass surface at a point where the ray strikes the glass in air. Sine r is the sine of the angle between the same perpendicular and the ray of light after it enters the glass. The value of n for glass is about 1.5. This means, for example, that if a ray of light strikes the glass at an angle of 30 degrees, the angle that the ray makes with a perpendicular to the glass surface once it enters the glass will be about 19.5 degrees. The exact value of the angle will depend on the nature of the glass. As far as light is concerned, two substances with the same index of refraction are the same, and the light will be neither reflected nor refracted when it passes from one into the other.

Of course, it would be difficult to use angles made by light rays to measure the index of refraction for glass fragments because of their size and roughness; however, the index can be determined by comparing them with a substance whose precise index is known. The refractive index of such fragments is found by placing them in a clear oil and gently heating it. The index of refraction of the oil changes with temperature. When the glass particles disappear, their index must be identical with that of the oil, which has been determined at various temperatures to within ± 0.0002. Software has been developed that allows the comparison to be done automatically using a computer.

Density

The density of glass fragments can be found in a similar manner. Substances more dense than water will sink in that liquid, while substances less dense than water will float. An object with the same density as water will remain suspended wherever it is placed in the liquid. To find the density of glass particles, they are placed in a mixture of two liquids with densities greater and less than the density range of various glasses. The composition of the liquids is adjusted until the glass neither sinks nor floats but remains suspended. The density of the glass now matches that of the liquid, which can be easily determined to within ± 0.0003 g/ml.

If glass particles from two places, such as a suspect and a crime scene, differ significantly, they could not have come from the same source. If, on the other hand, they match in density and refractive index to within experimental error, they may have come from the same source. The FBI Crime Laboratory has a database of refractive indexes for a great variety of flat glass specimens. That database can be used to see how common a glass with an index of 1.5279 might be. If only 1 in 1,200 specimens has such a value, the data could be very valuable in court.

Bullets through Glass

When a bullet or stone passes through a sheet of glass, it leaves a hole with radial and concentric fracture lines. A high-speed bullet leaves a wider opening on its exit than on its entrance side of the glass. The shape of the opening left by a slow-moving projectile, however, is not indicative of direction, but fracture lines may be. Radial lines will be left on the exit side of the glass and concentric lines on the side where the projectile entered. As viewed from the edge of the glass, radial stress lines will end perpendicular to the side of the glass on which the projectile entered. If the glass was penetrated several times, it may be possible to determine the sequence in which the projectiles penetrated the glass. A new fracture line always terminates on a previously existing fracture line.

Cases

On April 5, 1950, a badly damaged stolen car was found on a streetcar platform in Washington, D.C. Meanwhile, a police officer saw a man, later identified as Edward W. Patalas, emerging from a nearby alley. Based on the man's staggering gait, the officer judged Patalas to be intoxicated. He was arrested and taken to a hospital for treatment of a cut on his forehead.

At the hospital, police found a pair of spec-

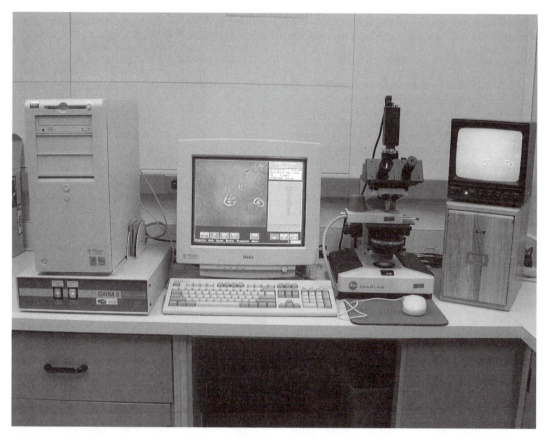

GRIM2 (Glass Refractive Index Measurement, by Foster & Freeman, U.K.) being used for the measuring of the refractive index of a glass fragment. Courtesy of the Biology Section, Centre of Forensic Sciences, Toronto, Canada.

tacles with a broken lens in Patalas's coat pocket. When police examined the damaged car, they found several pieces of broken glass beneath the steering wheel. The pieces of glass found in the car fit the pieces missing from Patalas's glasses perfectly. This evidence, coupled with his proximity to the accident, led to his conviction for automobile theft. A later request for an appeal was denied.

During a 1966 bank robbery in England, the robber fired his sawed-off shotgun into the glass separating him from the young teller, who died from the wound. Later, police found the stolen car they believed the robber had used to make his getaway. In the car, they found and collected glass fragments. A forensic scientist measured the refractive index and density of the glass particles found in the car as well as those at the bank and found that they matched. Eventually, through questioning of witnesses, police were able to arrest a suspect who was driving a stolen car and carrying a stolen gun. The man, Michael George Hart, tried to commit suicide and later confessed to the crime.

In 1988, Glenn Carroll, a forensic scientist at the Royal Canadian Mounted Police's forensic laboratory in Ottawa, was asked to do a glass analysis for a case involving two policemen who fired shots at a stolen car. One of the bullets struck Wade Lawson, the 17-year-old driver, in the back of the head. He died the day after being shot. The policemen said that they had fired at the rear tires. Nevertheless, their final shot went through the rear window, which was made of tinted tempered safety glass. Carroll set out to determine exactly where the bullet had struck the glass, which would enable him to determine the direction in which the gun had been pointed. He had the window removed, encased in foam and plywood, and moved to his lab. He also removed the lined shelf beneath the window, which held thousands of tiny glass fragments, and moved it to his lab as well. He also collected all the tiny glass fragments from the vicinity. By projecting straight lines back along the fractures, he could see that they converged about 40 centimeters from the frame on the passenger's side of the car and approximately the same distance from the bottom of the frame. The results indicated that the gun had not been aimed at the tires. Nevertheless, the police officers' lawyer raised so many questions about the validity of the evidence that the policemen were acquitted. *See also* Ballistics; Crime Scene; Trace Evidence.

References

Barrett, Sylvia. *The Arsenic Milkshake and other Mysteries Solved by Forensic Science.* Toronto: Doubleday Canada, 1994.

Gardner, Robert. *Crime Lab 101: Experimenting with Crime Detection.* New York: Walker, 1992.

Patalas v. United States. No. 10687, U.S. Court of Appeals, District of Columbia Circuit, 1950.

Saferstein, Richard. *Criminalistics: An Introduction to Forensic Science.* Upper Saddle River, NJ: Prentice Hall, 1998.

Wilson, Colin. *Clues! A History of Forensic Detection.* New York: Warner Books, 1991.

Goddard, Calvin Hooker (1891–1955)

American criminologist and military historian Calvin Hooker Goddard was the father of forensic ballistics, the study of bullets and firearms. Although he received a medical degree from Johns Hopkins University in 1915, Goddard chose not to go into private practice and, instead enlisted in the Army Medical Corps. In 1920, he resigned from the military with the rank of major. Goddard then became the assistant director of Johns Hopkins Hospital and also taught in the medical school. In 1924, he moved again, accepting a position as assistant professor of

C.H. Goddard. Courtesy of the Northwestern University Archives.

chemical medicine at Cornell Medical School.

By 1925, Goddard had determined to pursue his firearms hobby full-time and joined the Bureau of Forensic Ballistics in New York City. Here he pioneered the techniques for identifying a weapon that fired a specific bullet. Goddard was convinced that each weapon left specific signature marks on a bullet, not unlike fingerprints, that no other gun could reproduce. He devoted himself to proving that ballistics was an exact science, traveling extensively, demonstrating his techniques, and instructing law enforcement officials. His efforts and work won Goddard recognition as the foremost firearms expert in the nation.

Beyond professional circles, Goddard achieved international recognition for his role as a consultant in several high-profile criminal investigations. Goddard's participation and influence were most prominent following the 1921 Sacco-Vanzetti murder trial. A committe investigating the trial asked Goddard to test the gun that killed a policeman. Insisting it was not within his charge to ascertain if Nicole had Sacco pulled the trigger, Goddard concluded that the bullet that had killed a payroll guard in the Braintree, Massachusetts, holdup could only have come from Sacco's gun. Goddard's controversial decision was upheld by additional ballistics tests in 1961 and 1983.

In 1929, the ballistics expert was called to Chicago to aid in the investigation of a gangland massacre that had left seven mobsters dead. Goddard's work in easing public suspicion about police involvement was so impressive that two wealthy men financed a new Scientific Crime Detection Laboratory as a part of Northwestern University. Goddard organized and led the lab from its inception until the depression ended funding in 1933. Goddard's dream was to establish a central crime lab, accessible to police forces nationwide, that would handle forensic issues. The lab used the most recent scientific methods in ballistics, fingerprinting, and handwriting analysis to prevent and detect crimes. Under Goddard's guidance, the facility operated a training program for police and experimented with truth serum and the polygraph. The director also served on the Northwestern Law School faculty, and his police-science classes were among the first offered in the United States. In 1930, Goddard founded the *American Journal of Police Science*, the first scientific periodical in the nation to assist in solving crimes.

The closure of his lab did not diminish Goddard's reputation. In 1935, he won a Guggenheim Fellowship to write the definitive book on firearms identification. He did his part, but the book was never published. In 1940, he was named military editor of the *Encyclopaedia Britannica*, a position he held until his death.

Goddard returned to service in World War II. As a lieutenant colonel, he was assigned to the history section of the Army War College and in 1942 was appointed chief historian of the Ordnance Department. Goddard did not forget his crime-fighting roots. In 1948, he was in Tokyo, where he served as the founder and chief of the Far East Command's Criminal Investigation Lab. Illness forced Goddard to resign from the service in 1954. *See also* Ballistics.

Reference

Bodayla, Stephen D. "Goddard, Calvin Hooker." *Dictionary of American Biography: Supplement Five*. New York: Charles Scribner's Sons, 1977.

Gonzales, Thomas A. (1892–1964)

Thomas A. Gonzales was New York City's medical examiner when botanical evidence was first used to nullify a seemingly good alibi. Also, in 1938, during Gonzales's term as medical examiner, the first serological laboratory was established in New York. This lab's task was to study blood and bloodstains often collected as evidence at a crime scene. Alexander Weiner was appointed by Gonzales to head this lab.

The case involving botanical evidence was the 1942 murder of a young woman named Louise Almodovar. She had been found strangled in Central Park. At first she appeared to be the victim of a mugging; however, a gold necklace had been left around her neck. After talking to her parents, investigators suspected her estranged husband, Anibal Almodovar, but he had an alibi. He had been at a dance club with another woman and had many witnesses to confirm his presence.

Alexander Gettler, who was in charge of New York City's Chemical and Toxicological Lab, noticed that the crime-scene photographs revealed that the murder had taken place on a rare type of grass. Grass seeds of the same type were found in Almodovar's clothing. He denied that he had been in Central Park, but Joseph J. Copeland, professor of botany and biology at New York's City College, testified that in New York City this rare type of grass grew only in Central Park and only on the very hill where the murder took place. Investigators found that the dance hall where Almodovar had established his alibi was only a few hundred yards away from the murder scene. He could have left the dance hall, gone to the hill where he had arranged to meet his wife, killed her, and returned to the dance hall without anyone even noticing that he was gone. The evidence was convincing. Almovodar was convicted and was executed in 1943. *See also* Medical Examiner; Weiner, Alexander.

References

Evans, Colin. *The Casebook of Forensic Detection: How Science Solved 100 of the World's Most Baffling Crimes.* New York: Wiley, 1996.

Thorwald, Jürgen. *Crime and Science.* New York: Harcourt, Brace & World, 1967.

Gross, Hans (1847–1915)

Hans Gross was an Austrian examining judge considered by many to be the father of criminalistics, the application of science to criminal and civil laws enforced by a criminal justice system. Gross established an international reputation with the publication of his authoritative and seminal work *Criminal Investigation* (1893). He believed that science, if it was used correctly, could solve any crime, and he spent his career advocating a revision of outdated investigative procedures.

Gross's attraction to science as a means of criminal identification and apprehension was borne of necessity. Trained as a lawyer, Gross became an examining judge at a time when the police were primarily concerned with peacekeeping activities. It was left to the judges to investigate the facts of a criminal case. Gross found himself totally unprepared for the investigative facet of his post and turned to science to strengthen his professional decision making. *Criminal Investigation* was the product of his nearly 20 years of scientific investigative experiences.

Gross intended *Criminal Investigation* to be a handbook of scientific procedures to be used in criminal inquiries. The near-encyclopedic volume covered topics ranging from blood spatters to poisoning to criminal psychology. Gross detailed, from experience, the assistance investigators could expect from disciplines like botany, zoology, chemistry, and anthropometry. He tried to impress on detectives how important securing a crime scene was and wrote about successful interrogation techniques as well. A second edition was published the following year, and the book was translated into several languages.

Gross spent a good portion of his career as a popular lecturer on criminology and penal law at the Universities of Prague and Graz. He also continued his writing, focusing on criminal psychology. His career influenced Arthur Conan Doyle, who incorporated much of Gross's research into the popular detective stories he wrote.

References

Block, Eugene B. *Science vs. Crime: The Evolution of the Police Lab.* San Francisco: Cragmont, 1979.

Lane, Brian. *The Encyclopedia of Forensic Science.* London: Headline, 1992.

Gunshot Wounds

Gunshot wounds can be the result of an accident, a suicide, or a homicide. It is often the task of forensic scientists to determine which of these possibilities was responsible for the gunshot wound found on a victim and to determine what gun was used and from what distance.

According to *The World Almanac and Book of Facts, 2000,* of the 33,750 deaths due to guns in the United States in 1996, 14,037 were homicides, 18,166 were suicides, 1,134 were accidents, and 413 were of undetermined cause. Accidental shootings are rare, but occur most frequently among children and young adults. Data from the *Statistical Abstract of the United States 1999* revealed that in 1996, more than two-thirds (67.5 percent) of all homicides in the United States were due to firearms, and 54.6 percent were caused by handguns.

If a gunshot wound is believed to be the result of suicide, investigators must decide if the entrance wound is one that could have

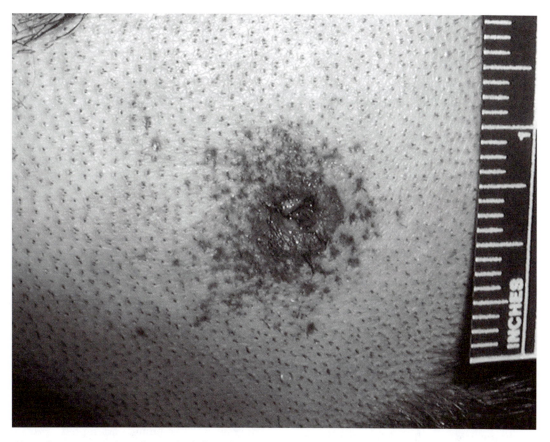

Close shot at approximately one inch from barrel to head. Smaller marks show stippling from gunpowder. Courtesy of the Office of the Medical Examiner, Suffolk County, NY.

been self-inflicted. Obviously a shot in the back of the head or a wound indicating that the bullet or shell traveled more than a few inches will suggest that suicide was not the cause of death. Absence of a suicide note, however, is not indicative of murder or accident. Only about 25 percent of suicide victims leave a note.

To establish the cause of a gunshot wound, investigators consider the location of the wound, the number of wounds, the presence or absence of a gun, the range from which the shot was fired as estimated by the nature of the wound (discussed later), the angle of the bullet's entry, and circumstances such as evidence of robbery or rape. The point of the bullet's entry will be circular or oval (if entry was at an angle), surrounded by an inflamed ring. If the bullet passed through the body, the exit wound will be larger and

more irregular than the point of entry. If the powder burns—the minute spots around the wound—are brownish orange, then the bullet entered the body before the victim died. If the same spots are grayish yellow, the victim was dead before being shot. Any bullets, shells, or guns recovered at the scene will be referred to ballistics experts by investigators.

Estimating Range

The nature of the wound provides evidence about the range from which the bullet was fired. If the gun was held against the body, there will be a heavy concentration of lead deposited around the entrance wound. Skin and clothing fibers will be scorched from the heat released at the end of the barrel when the gun was fired. As the distance from the point of entry increases, the diameter of the sooty ring around the wound increases, but the concentration of the particles decreases. However, it is difficult to make accurate estimates of the distance the bullet traveled without testing the actual weapon and ammunition used. Generally, if the bullet traveled more than a yard, there will be no evidence of gunpowder on the victim, but there will be a dark ring around the entrance wound resulting from carbon, dirt, grease, primer, and lead powder that rubs off the bullet's surface (the so-called bullet wipe) as it enters the body. An infrared photograph may reveal powder stains on clothing that were not visible in ordinary light.

Testing the Hands of a Suspect

The hands of a suspect, if he or she is captured soon after the shooting, may be tested for primer powder that is exhausted when a gun is fired. The powder is removed quite readily when hands are washed or wiped on clothes or other articles. Consequently, some laboratories will not accept evidence relating to primer powder if it has been collected more than six hours after a shooting. The suspect's hands should be swabbed with dilute nitric acid and the swabs tested for barium and antimony using absorption spectrophotometry. Another approach is to use tape to remove any tiny particles of powder from the suspect's hands. The tape can then be examined with a scanning electron microscope. The primer-powder particles have a characteristic size and shape.

Shotgun Wounds

The damage resulting from a shotgun wound depends on the gauge as well as the distance, but at close range (contact or within an inch or two), which is the only way a shotgun can be used in suicide cases, there will be no scatter of pellets. The solid mass of metal and wadding will essentially destroy a victim's head or make a huge hole in the chest or abdomen. At a range of two to three feet, the wound will be ragged and about two inches in diameter with scorching and powder burns. Generally, the spread of pellets from a 12-gauge shotgun is about one

inch for every yard. Thus, at six yards, the pellets will produce a wound about 6 inches in diameter. At such distances, the wound is usually not fatal, but the pellets penetrate the skin, producing a number of small, painful wounds.

The Nathan Wells Case

Forensic scientist Peter Wilson told jurors in Nelson, New Zealand, that the barrel of the shotgun that had wounded Rebecca Hannah, who was allegedly shot by her former boyfriend, Nathan Wells, was between 3.4 and 4 meters from her body when it was fired. Wilson based his assessment on the minimal scattering of the pellets and the fact that wadding from the cartridge had penetrated the young woman's skin. In response to Wells's claim that the gun had discharged accidentally, Robert Ngamoki, another forensic expert, stated that the weapon could not have been fired accidentally. The gun did not have a sensitive trigger, and to fire it, the user had to pull back the hammer prior to pulling the trigger. Wells, who pleaded not guilty, was charged with attempted murder, aggravated wounding, attempted kidnapping, and theft of a shotgun. The gun allegedly used in the shooting was the property of Antony Robinson, whose son was a friend of Wells. Wells was found guilty in May 2000 and sentenced to eight years in prison. *See also* Ballistics; Crime Scene; Paraffin Testing; Spectrophotometry.

References

"Jury Inspects Site of Shooting." *Nelson Mail* (Nelson, New Zealand), May 18, 2000, p. 2.

Lane, Brian. *The Encyclopedia of Forensic Science.* London: Headline, 1992.

"Nelson Man Gets Eight Years after Shooting." June 22 2000. http://www.planet/fm.co.nz. news/june00/220600.htm.

Saferstein, Richard. *Criminalistics: An Introduction to Forensic Science.* Upper Saddle River, NJ: Prentice Hall, 1998.

U.S. Census Bureau. *Statistical Abstract of the United States, 2000.* 120th ed. http://www.census.gov/prod/2001pubs/statab/sec05.pdf.

The World Almanac and Book of Facts, 2000. Mahwah, NJ: World Almanac Books, 1999.

H

Hair Evidence

Hair evidence is often found at a crime scene, on a victim, or on a suspect. Before the development of DNA analysis, hair evidence was used to indicate the possibility of guilt, but it could not provide definite identification. Today, hair can be analyzed for DNA, which can be used to identify both a victim and a criminal.

Human hair grows out of follicles located in the skin. The part of a hair within a follicle is the root; the portion that is visible above the skin is called the shaft. Each hair is composed of three layers. The outer layer, known as the cuticle, is a single layer of overlapping scalelike cells. The middle layer, the cortex, consists of elongated cells that contain pigment. The cells of the medulla, the inner layer, are many sided and in rows.

When strands of hair are viewed under a microscope, it is possible to distinguish human from animal hair. The cuticle cells of animal hairs are larger, less regular in shape, and characteristic of the species. The medulla is continuous and well defined, unlike human hair, in which often the core cannot be distinguished or is interrupted or fragmented. The medulla index, which is the ratio of the diameter of the medulla to the diameter of the entire shaft, can also be used to help distinguish between animal and human hair.

For humans, the medulla index is usually less than 0.33; for animals, it is generally 0.5 or greater.

The best way to compare hair samples is side by side under a comparison microscope. The forensic scientist can then look for a match between hairs in terms of color, length, diameter, the nature of the medullas, and signs of bleaching or coloring. In fact, since hair grows at a rate of approximately 1.3 centimeters per month, it is possible to estimate the time that has passed since a hair was dyed or bleached. Often a forensic expert can differentiate Negroid hair from Caucasian, and he or she can usually determine whether the hair came from the head, armpit, pubic region, scrotum, or face. Hairs from a scalp are usually circular in cross-section, show little variation in diameter, and exhibit a uniform distribution of pigment. Hair that has been pulled out of a victim's head will usually have tissue from the follicle attached to the root. Pubic hair is short, curly, and triangular or oval in cross-section, with diameters that vary in size and with continuous medullas. Hair from a beard is usually coarse, triangular in cross-section, and with blunt tips due to shaving or cutting.

The first Australian murder in which forensic science played a major role was the Colin Ross case (1921). A 13-year-old girl's body was found wrapped in a blanket in a

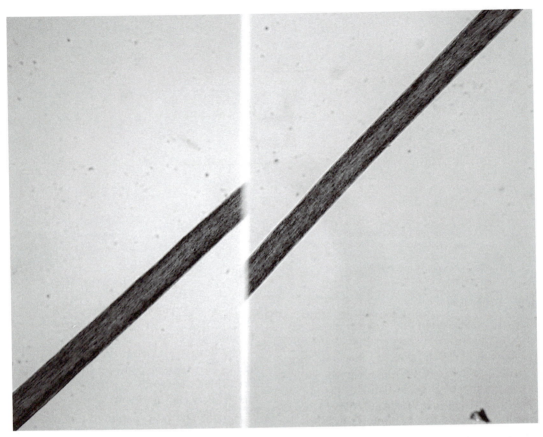

Two hairs under comparison microscope. Courtesy of David Exline, ChemIcon Inc., Pittsburgh, PA.

Melbourne alley. Lack of blood on the bruised body, which had been raped, strangled, and beaten, led police to believe that the murderer had washed and dried the body to remove evidence and then carried it to the alley. During questioning of people who had been in the area the night of the crime, the name Colin Ross, a local bar owner, was frequently mentioned. In fact, one witness said that he had seen Ross carrying a large object wrapped in a blanket.

When police searched Ross's home, they found two blankets. By carefully examining them, a chemist, Dr. Charles Taylor, found a number of strands of golden red hair that matched the color of the victim's hair. The incriminating hairs, some of which had been pulled out of the victim's head, were clearly human, with a length that indicated that they were female in origin. Under a microscope, they were identical to those of the victim. In

an effort to show that the prosecution could not identify a victim by using hair analysis, the defense lawyers challenged Taylor to distinguish between hairs taken from the victim and those of another fair-haired young woman. When he succeeded in doing so, the jury was impressed. It found Ross guilty, and he was later hanged.

René Castellani was suspected of killing his wife, Esther Castellani, by intermittent small doses of arsenic. In order to prove that the concentration of arsenic in the victim's body at different times corresponded to the dates when she had exhibited symptoms of arsenic poisoning, Norm Erickson, a Canadian forensic chemist, sliced hair samples from the victim into lengths of 1.0 cm, 0.5 cm, and, finally, 0.25 cm—the lengths that hair normally grows in a month, two weeks, and one week, respectively. Erickson then analyzed the samples using neutron-activation analysis.

His results revealed beyond reasonable doubt that the dates of Mrs. Esther Castellani's symptoms of poisoning corresponded to the times when the concentrations of arsenic in her hair and, therefore, body were peaking.

In the case of Richard Crafts (1986), the woodchipper murderer, strands of hair found on his wife's hair brush matched those found along the bank of the Housatonic River and on his chain saw. Hair evidence was particularly significant in this case because Helle Crafts's hair exhibited "shouldering." That is, there were extensions along the shaft of the hairs, a phenomenon rarely found. *See also* Arsenic Poisoning; Comparison Microscope; Crime Scene; DNA Evidence; Neutron-Activation Analysis; Trace Evidence; Woodchipper Murder.

References

Barrett, Sylvia. *The Arsenic Milkshake and Other Mysteries Solved by Forensic Science.* Toronto: Doubleday Canada, 1994.

Cyriax, Oliver. *Crime: An Encyclopedia.* North Pomfret, VT: Trafalgar Square, 1996.

Evans, Colin. *The Casebook of Forensic Detection: How Science Solved 100 of the World's Most Baffling Crimes.* New York: Wiley, 1996.

Saferstein, Richard. *Criminalistics: An Introduction to Forensic Science.* Upper Saddle River, NJ: Prentice Hall, 1998.

Wilson, Colin. *Clues! A History of Forensic Detection.* New York: Warner Books, 1991.

Zonderman, Jon. *Beyond the Crime Lab: The New Science of Investigation.* New York: Wiley, 1990.

Hall-Mills Murder (1926)

The Hall-Mills murder was never solved because the crime scene, vital to forensic evidence, was destroyed by unauthorized people who were allowed to walk about the murder scene. The murder and the trial that followed set off a national news frenzy. Americans believed that they were witnessing the trial of the decade, if not the century. The combination of illicit love, unknown killers, and courtroom dramatics enthralled the nation.

On September 16, 1922, near New Brunswick, New Jersey, two young people strolling down DeRussey Lane (a lovers' lane) discov-

ered a ghastly murder scene. The bodies of 34-year-old Mrs. Eleanor Mills, the wife of a local janitor and church sexton, and the Reverend Edward Wheeler Hall, the rector of St. John the Evangelist Episcopal Church, were carefully arranged side by side under a crabapple tree. Mills was a member of Hall's congregation, sang in the choir, and was very active in church affairs. Her husband was the sexton at St. John's. According to church gossip, the minister and former child bride had been carrying on an affair for almost two years. Neither of their spouses apparently suspected anything unseemly.

The two had been shot with a .32-caliber handgun. The 41-year-old minister had been dispatched with a single shot to the head. The woman had been shot three times in the forehead, with the wounds forming a triangular pattern. Most disturbingly, Mills's throat had been cut from ear to ear, nearly decapitating the woman, and her tongue and vocal chords had been removed. Mills's head was nestled in the crook of Hall's arms, their faces were covered, and the crime scene was littered with Mills's explicit love letters to her lover.

The ensuing investigation was marked by incompetence and bungling from the beginning. First, there was a jurisdictional controversy. The crime was committed just over the line in Somerset County, but all the principals in the case, victims and eventual suspects, were from Middlesex County. Most significantly, there was no security at the crime scene. Thousands of curiosity seekers traipsed over the scene, and by the first weekend, the crab-apple tree had been literally destroyed by souvenir hunters. Authorities did find shell casings but were never able to match them to a weapon. Finally, there were no autopsies or even a coroner's inquest.

In most homicide cases, the first suspects are usually family members. This investigation was no exception. The police quickly narrowed their attention to Mrs. Frances Stevens Hall. Police were convinced that the socially prominent heiress, seven years older than her husband, had to be involved. Law-enforcement officials theorized that with the aid of her brothers, Henry Stevens, a gentleman and crack shot, and William "Willie" Stevens, the town eccentric, Mrs. Hall had administered her own brand of justice. Though it was handicapped by mishandled evidence and police incompetence, the state pressed for an indictment. After a five-day hearing, the grand jury decided that there was insufficient evidence to charge anyone in the murders.

Four years later, the case resurfaced as a result of a tabloid circulation war. William Randolph Hearst's *New York Daily Mirror* and the more successful *New York Daily News* were fighting for readers when the *Mirror* launched a new inquiry. The tabloid's interest was sparked by a nondescript legal proceeding. In July, 1926, Art Riehl was seeking an annulment after 10 months of marriage to Louise Geist. Geist was a former maid in the Hall household and allegedly confided in her husband, after the wedding, that she had informed Frances Hall of the minister's plan to elope with his paramour. Riehl also asserted that Geist received $5,000 in hush money not to reveal that the offended wife and her brothers had left the Hall house on the night of the killings. Riehl concluded by saying that the next morning, Willie Stevens had told the maid that something terrible had happened. Due to public pressure, the governor of New Jersey ordered the case reopened. Frances Hall, her brothers, and a cousin were arrested and charged with murder.

When the 18-day trial opened on November 3, 1926, the prosecution's case rested mainly on the more complete testimony of the "Pig Woman," as she was known by the press. Fifty-six-year-old Jan Gibson was an eccentric local pig farmer who provided much of the color and drama for the carnival-like trial. She regularly rode a mule, was involved in an ongoing feud with a pig-farming neighbor, and, most poignantly, was dying from cancer. During the trial, she collapsed and ultimately testified from a hospital bed in the courtroom, attended by a nurse and a physician. Gibson testified that on the night of September 14, 1922, at about 10:30 P.M.,

she rode her mule out after a thief. While passing DeRussey Lane, she saw a couple struggling with three men and a large woman with gray hair. She noticed the hair due to the moonlight and the glare of an auto's headlights. The Pig Woman closed her dramatic testimony by saying that she heard the woman yell, "Explain these letters." A number of shots followed, and Gibson hurried away. All during the testimony, Gibson's mother sat in the front row of the courtroom muttering that her daughter was a liar.

The state's claims were further bolstered by evidence from the crime scene. One of the minister's visiting cards was found propped against his foot. A fingerprint analysis uncovered Willie Stevens's left index fingerprint on the back of the card.

The defendants were represented by the best legal talent their considerable resources could buy, and their legal team was dubbed the "million-dollar defense" by the tabloids. Lead lawyer Robert H. McCarter, the former attorney general of New Jersey, quickly pointed out discrepancies in Gibson's testimony compared to her 1922 grand-jury statements. He then attacked her credibility by getting her to declare that she could not remember if she was married, divorced, or remarried. Harry Stevens swore that he had been fishing the weekend of the murders, which was corroborated by three witnesses. "Crazy Willie" Stevens did an outstanding job on the witness stand. He calmly and confidently went over his story three times for the prosecutor, who hoped that he could show that the testimony was memorized and rehearsed. Stevens never faltered and emphatically denied that he had had anything to do with the murders. The mismanagement of the crime scene returned to haunt the state when police admitted that the visiting card had been exposed to the weather for 36 hours, which could have impaired the print. As the trial concluded, Frances Hall took the stand. She admitted that she had left her home on the night of the murder, but insisted that she had done so to look for her husband, who was late getting home. She denied all allegations and was so composed and impressive in the face of a relentless examination that the press referred to her as the "Iron Widow." A five-hour deliberation produced a not-guilty verdict for all the defendants on December 3, 1926. Frances Hall and her family subsequently sued the *Daily Mirror* for libel, asking $3 million in damages. The case was settled out of court for a reputed $50,000.

References

Kunstler, William. *The Minister and the Choir Singer: The Hall-Mills Murder Case.* New York: Morrow, 1964.

Sifakis, Carl. *The Encyclopedia of American Crime.* New York: Smithmark, 1992.

Thurber, James. "A Sort of Genius." *Unsolved: Classic True Murder Cases,* selected by Richard Glyn Jones. New York: Peter Bedrick Books, 1987.

Handwriting. *See* Disputed Documents.

Harris, Carlyle (1868–1893)

Carlyle Harris was an American man convicted of killing his wife, 19-year-old Helen Potts, with an overdose of morphine. The murder was considered by many, including the killer, to be the perfect crime, and only shrewd scientific detection into poison coverups led to a conviction.

Harris was the grandson of Benjamin McCready, a highly respected surgeon and medical professor in New York City, and was attending medical school at his grandfather's expense. The young man's mother was a minor celebrity in her own right as a lecturer and writer. On the surface, Harris was considered a model young man, with a fine personality, who seemed assured of success. Outside the family, Harris actively cultivated a reputation as a gambler, womanizer, and stylish rogue.

In the summer of 1889, the 21-year-old Harris, as a third-year medical student, met Helen Potts while he was on holiday at Ocean Grove, New Jersey. The two fell in love, but her family was opposed to a marriage so early in the courtship. On February 3, 1890, the couple was secretly married in a New York City civil ceremony, using the names Charles Harris and Helen Neilson. Neilson was Potts's middle name.

In the spring following the marriage, the new wife became pregnant. The exact situation is unclear. Either the child was stillborn or Harris botched an abortion that had to be rectified by Potts's uncle, a physician in Scranton, Pennsylvania. In any event, the frightened girl confided her secret to her mother. The older woman urged Harris to acknowledge the marriage and to sanctify the union by a church ceremony. Pleading the pressure of school and developing a career, Harris asked for time. Potts was enrolled in the Comstock School for Girls in New York City. Eventually tiring of his mother-in-law's badgering, Harris agreed to her requests. However, at some point it appears that the medical student lost interest in his bride and decided on a different course of action.

In January 1891, Harris visited his wife at school, where she complained of headaches and insomnia. Harris offered his professional assistance and went to a very reputable drugstore, Ewen McIntyre and Son. There the medical student wrote a prescription for a mild and fairly standard order of six capsules. Each capsule was to be a quarter of a grain of morphine and four and a half grains of quinine. He initialed the prescription receipt and added "student" after his acknowledgment. He gave the pills to his wife, but kept two because, as he later stated, he was concerned about the girl medicating herself. Since Harris was leaving town for a few days, he promised to check back with her upon his return. Potts took the pills as directed, one each night before bed.

Beginning at 10:30 on the night of January 31, 1891, Potts complained of numbness and difficulty breathing. When the school doctor arrived, around midnight, he immediately recognized the symptoms as profound narcotic poisoning. He treated the girl with atropine and caffeine, but to no avail. Helen Potts died on the morning of February 1, 1891.

The doctor noticed the pillbox on the

nightstand and called Harris to the school. It appeared that the death was an unfortunate and terrible accident. Either the girl had misunderstood the directions and taken too many capsules, overdosing, or the pharmacy had made a mistake. Ewen McIntyre and Son insisted that it was not its error. Nonetheless, the coroner needed no investigation to declare the death an accident, and Potts was all but forgotten. Harris was censured by the medical-school faculty, but more felt sorry for him losing a patient so early in his career. Harris was too distraught to attend the funeral.

Rumors persisted about Potts's death, and Ike White, a reporter for the *New York World*, proved tenacious in his investigation. White uncovered the secret marriage and Harris's reluctance to publicly acknowledge the union. Perhaps more devastating was the exposure of Harris as a disreputable husband and womanizer. The revelations created a national sensation, and the authorities ordered an exhumation and analyses of the remains. New York's leading toxicologist, Dr. Rudolph Witthaus, performed the tests and discovered morphine in the body but no quinine.

Carlyle Harris was indicted for the murder of his wife on March 30, 1891, and his trial began on January 1, 1892. During the three-week trial, prosecutor Francis L. Wellman hammered at the motive for murder and the fact that Harris had access to the drugs that killed his wife. The prosecutor asserted that Harris was afraid that a marriage below his social status would anger his grandfather, who would cut off the student's educational funds. The state further contended that Harris tampered with the capsules, reloading one with a lethal dose of morphine and the others with harmless quinine. The two clerks from the drugstore testified about their diligence in preparing the compound. The state's attorney even introduced one of the capsules Harris had not given Potts (Harris had lost the other), as evidence. An analysis of the medicine proved that the dosage was as prescribed and that the drugstore was evidently not to blame for the death. Suicide was quickly dismissed as a possibility since to all who knew her, Potts had taken in good spirits, cheerful, and in good health. If there was any preexisting physical condition, the prosecution wondered why the previous three capsules had not caused some type of reaction. Finally, there were Harris's reputation as a playboy and the purported abortion incident to consider.

The defense countered with several arguments. First, it tried to argue that the postmortem was inconclusive. It tried to make a case for kidney disease as the cause of death, this despite the fact that there was no mention of the ailment in Witthaus's report. Failing in this attempt, the lawyers tried to establish that Potts had been a morphine addict and had died of an overdose by her own hand. Throughout the proceedings, Harris insisted that he was innocent to anyone who would listen. However, he never testified in his own behalf.

It took the jury one hour to return a guilty verdict, and Harris was sentenced to death. There was an appeal, but the higher court upheld the initial verdict. Harris was executed at Sing Sing Prison on May 8, 1893. His bitter mother, who refused to believe that her son was guilty, had his coffin labeled "CWH Murdered, May 8, 1893." *See also* Atropine; Morphine; Buchanan-Sutherland Murder; Opium.

References

Boswell, Charles and Lewis Thompson. *The Carlyle Harris Case*. New York: Collier Books, 1961.

Pearson, Edmond. *Masterpieces of Murder*. Ed. Gerald Gross. Boston: Little, Brown, 1963.

Hauptmann, Bruno Richard: Murder Trial. *See* Lindbergh Kidnapping and Trial.

Heinrich, Edward Oscar (1881–1953)

Edward Oscar Heinrich was a pioneer American forensic scientist and criminologist dubbed the "Edison of Crime Detection" or the "Wizard of Berkeley" by the contemporary press. Largely self-trained, Heinrich became an expert in a number of scientific fields, including geology, physics, botany, and biochemistry, and was an authority on handwriting, ink, and paper. His credo was that every criminal left clues behind, and it was the detective's responsibility to examine the evidence and interpret what he found. Over a 45-year career, he is credited with solving more than 2,000 mysteries, and he figured prominently in some of the nation's major murder trials. Heinrich was also a lecturer in criminology and political science at the University of California.

Heinrich, who never finished high school, became a pharmacist at age 18, upon the death of his father. Becoming acutely aware of the need for further education, he saved money and dabbled in the commodities market to finance his dream of attending college. He was admitted to the University of California as a special student because he had no high-school diploma and graduated in 1908 with a B.S. in chemistry.

Following graduation, the young scientist moved back to the Pacific Northwest and became a city chemist in Tacoma. Nine years on the job, working closely with the coroner and police, introduced Heinrich to the developing field of forensic science. He established a solid investigative reputation, and in 1916, his success led to a position as police chief in Alameda, California, just across the bay from San Francisco. After a short stint as city manager in Boulder, Colorado, Heinrich returned to the Bay Area in 1918. He opened an investigative office in San Francisco and a lab at Berkeley and took up duties as a lecturer at the University of California. Heinrich's private practice was an immediate success. He was involved in a wide range of investigations, from industrial theft to civil controversies to cases of disputed documents, that made good use of his chemical expertise.

While the locally respected investigator was still at Alameda, he won international acclaim for his work on the 1916 Hindu conspiracy. As World War I raged on, British intelligence uncovered a plot by Hindu nationalists to overthrow British control of In-

dia. British and American agents believed that the scheme was being financed by the German government through its consul in San Francisco. The probe turned up what investigators believed were large amounts of evidence, though much of it appeared to be unusable. The problem was that the documents were in three different Indian dialects. Heinrich was called in and agreed to help. To be of any aid, he insisted on learning the dialects and translating the documents himself in order to better understand each individual author's style, diction, and syntax. For weeks he analyzed the handwriting styles and chemically evaluated the various inks authors had used. His most painstaking work was the study of the typewritten documents agents provided. He was especially careful in investigating the individual letters of documents in order to provide definitive proof about the typewriter they originated from. So conclusive were Heinrich's analyses and courtroom presentation that 31 of 32 indictees were convicted.

Further evidence of Heinrich's capabilities came with the Black Tom Island explosions. Black Tom Island, in New York harbor, was a shipping terminal for the Lehigh Railroad. On Sunday, July 30, 1916, war supplies awaiting shipment overseas exploded, killing six and wounding hundreds of others. The initial shock was felt as far away as Philadelphia. New York City and New Jersey citizens became hysterical in the face of a perceived German attack. When calm was restored, the nation's only question was, sabotage or accident? There was a strong suspicion of German espionage but no decisive proof.

Dogged by insurance claims from the Lehigh Railroad and others, the case for compensation dragged through the courts and a number of hearings. The ongoing inquiry did turn up some evidence of German involvement, though the postwar government vehemently denied any culpability. Key in the case against Germany was a magazine containing a code based upon pinholes beneath certain letters. When the pinpricks were decoded, they named German diplomats who knew of the Black Tom Island explosion.

German experts said that this was part of an American plot to implicate the Germans. In fact, they believed that the evidence had been manufactured in the postwar period. Heinrich was called in to verify the authenticity of the evidence. He studied the migration, spreading, and penetration of the ink, as well as the fibers of the paper, and determined that both were of World War I vintage. Consequently, the courts found that the Black Tom Island disaster was the work of German agents under orders from the kaiser's government.

The case that secured Heinrich's place in forensic history was the 1923 robbery of a Southern Pacific Railroad train in the mountains of southern Oregon. On October 11, 1923, three men stopped the San Francisco–bound train in search of money, securities, and other valuables in the mail car. During the failed robbery, four railroaders were murdered. Local deputies and railroad detectives rushed to the scene, and an extensive search turned up little. What they did uncover, a pair of greasy blue-jean overalls, a revolver, detonation batteries, and shoe covers soaked in creosote to throw off tracking dogs, did not seem very promising. On the basis of some slim circumstantial evidence, the authorities arrested a local auto mechanic. The best police efforts to connect the mechanic to the crime failed.

Stymied, reluctant authorities called on Heinrich, though no one believed that he could be successful. Heinrich had the overalls sent to his lab and began a methodical examination. After a two-day investigation, Heinrich offered a description of the perpetrator. He was most likely a left-handed lumberjack, probably from the Pacific Northwest. Further, the culprit was Caucasian, was about 5'10" tall, weighed about 165 pounds, and had light brown hair. How had Heinrich come up with such a complete description based upon some old coveralls?

His analysis revealed, first, that the grease on the coveralls was really pine tar. Second, he discovered some material in the right pocket of the coveralls that he microscopically determined to be fir needles, native to the Pacific Northwest. The killer was probably a left-handed lumberjack because while he was chopping, his right side would be closest to the tree, and that would explain the needles in the pocket. Heinrich also noticed that the left-side pockets were more worn. The position of the suspender buckles determined the height of the wearer. That the murderer was Caucasian with light brown hair was deduced from one strand of hair found on a button. This was Heinrich's initial report. Further examination of the overalls revealed a washed-out, illegible piece of paper in the pencil pocket. Treating the paper with iodine vapor, Heinrich identified it as a registered-mail receipt and partially established the number.

Postal inspectors traced the receipt to the d'Autremont family of Eugene, Oregon. The father told police that he was very concerned about his sons, Ray, Roy, and Hugh, who had been missing since early October. His description of Roy d'Autremont was a perfect fit, even to the left handed lumberjack, of Heinrich's suspect. A hair sample from Roy d'Autremont's hair brush proved identical to the evidence from the overalls. Despite a major manhunt, the three brothers eluded capture for three years. When they were finally arrested, they confessed and were sentenced to life in prison. *See also* Disputed Documents; Fiber Evidence; Hair Evidence; Trace Evidence.

References

Block, Eugene B. *Famous Detectives.* Garden City, NY: Doubleday, 1967.

———. *The Wizard of Berkeley.* New York: Coward-McCann, 1958.

Nash, Jay Robert. *Almanac of World Crime.* New York: Bonanza Books, 1986.

Helpern, Milton (1902–1977)

Milton Helpern was the chief medical examiner of New York City for many years. He was much respected, and New York University named its Milton Helpern Institute of Forensic Medicine after him. He was one of the cofounders of the National Association of

Medical Examiners in 1966 and its first president, from 1966 to 1970.

In the Coppolino case, his experience, carefully conducted autopsy, and scientific curiosity led to the conviction of a murderer who almost succeeded in hiding his guilt. Carl Coppolino, a retired anesthesiologist, and his wife had moved to Florida from New Jersey. In 1965, Coppolino found his wife Carmela dead. She was supposedly a victim of a heart attack, though she was only in her 30s. There was no reason at the time to suspect Coppolino, so an autopsy was not performed. However, when a former friend and probably a scorned lover of Coppolino, Marjorie Farber, heard that he had remarried quickly after Carmela Cappolino's death, she confided that Coppolino had killed her husband, William Farber, back in New Jersey. The authorities now looked more closely at Coppolino and discovered that he had increased Carmela Coppolino's life insurance shortly before her death.

Helpern performed autopsies on both Carmela Coppolino and William Farber. William Farber's death had been ruled a heart attack also. Helpern found no signs of heart disease, but there was evidence of strangulation. The cricoid cartilage in the neck was fractured. Marjorie Farber said that Coppolino had strangled her husband when she did not have the nerve to give him an injection of something Coppolino had given her. Helpern found no signs of heart disease in Carmela Coppolino either. He did find a very small hypodermic puncture mark on one of her buttocks, but forensic testing found no signs of drugs in her system.

Helpern recalled that Coppolino was a former anesthesiologist who, therefore, would have known that the drug succinylcholine chloride could kill without leaving traces in the victim's system. However, chemists did know that succinylcholine chloride would degrade in the body to products that had yet to be determined. To see if these chemicals could be identified, Helpern recruited the help of Dr. Charles Umberger, who was in charge of the New York City toxicology department. Umberger did an experiment in which he injected succinylcholine chloride into rabbits and frogs and then buried them to see if any chemicals related to the degradation of succinylcholine chloride could be found. He discovered a large amount of succinic acid in the brains of both the animals. The same substance was identified in Carmela Coppolino's brain.

The well-known lawyer F. Lee Bailey defended Coppolino in New Jersey for the murder of William Farber. Bailey did not believe that there had been a murder. He argued that Marjorie Farber had made up the story to hurt Coppolino because he had rejected her. He had Richard Ford and Joseph Spelman, both well-established medical examiners, testify that Farber had shown signs of heart disease and that the cricoid fracture had occurred after death, probably when the body had been buried or exhumed. With the conflicting testimony by the medical examiners, the jury could not find Coppolino "guilty beyond a reasonable doubt" of murdering William Farber. However, when Coppolino was tried for the murder of his wife, Helpern's and Umberger's experiments were convincing evidence to the jury, and Coppolino was convicted for the murder of his wife.

In 1949, Mrs. Abbie Borroto of New Hampshire was terminally ill with cancer. There was nothing more her doctor could do for her except ease the pain. One night, her physician, Dr. Hermann Sander, prescribed a large dose of her painkilling drugs, an amount that could be considered an overdose. The next morning, with a nurse present, Dr Sander injected 40 cc of air into Abbie Borroto's arm. She was pronounced dead. Sander recorded what he had done in Abbie Borroto's medical record. Somehow it reached a newspaper, and eventually Sander was put on trial for Abbie Borroto's death. Borroto's body had been embalmed, so when an autopsy was performed, as requested by the defense, it was impossible to tell if the injection of air (an air embolism) was the cause of death. At the autopsy, Helpern was an observer for the attorney general prosecuting Sander. Helpern admitted that there was no proof of embolism as the cause of death, but there was no proof against it either.

When the case went to trial in February 1950, Sander stated that Mrs Borroto was already dead when he injected her with air. He had done it to make sure she was dead. In court, Dr. Richard Ford, who had actually performed the autopsy, stated that for death to occur from air embolism, the amount of air injected into a vein would have to be 200–300 cc. Helpern, testifying for the prosecution, stated that the amount needed to kill a person could be as low as 40 cc, the amount injected into Borroto. The jury found Sander not guilty. Today, the amount of air needed to kill a person by injection is still debated but is believed to be between 50 cc and 300 cc, depending on the health and size of the individual. *See also* Camps, Francis; Cause of Death; Medical Examiner; Norris, Charles; Sheppard, Sam; Bailey Toxicologists.

References

Cyriax, Oliver. *Crime: An Encyclopedia*. North Pomfret, VT: Trafalgar Square, 1996.

Evans, Colin. *The Casebook of Forensic Detection: How Science Solved 100 of the World's Most Baffling Crimes*. New York: Wiley, 1996.

Henry, Edward R. *See* Fingerprints.

Heroin. *See* Drugs; Opium.

Herschel, William. *See* Fingerprints.

Hitler Diaries. *See* Disputed Documents.

Holmes, Sherlock

Sherlock Holmes was the world's first fictional detective capable of solving the most difficult cases through his formidable deductive powers, intellect, and scientific approach. Sir Arthur Conan Doyle's Holmes is the main character in a series of novels and short stories and is probably the most famous de-

Basil Rathbone as Sherlock Holmes in the 1939 film *The Hound of the Baskervilles*. © Bettmann/CORBIS.

tective in all of fiction. Holmes's friend, Dr. John H. Watson, a retired military doctor, serves as the narrator of the stories and aids him in solving the crimes.

The first Sherlock Holmes story, *A Study in Scarlet*, was published in 1887. In this story, Holmes discovers a simple test for identifying blood that allows him to determine that stains found on a suspect's clothes were from blood, not rust or food. When he is introduced to Watson, Watson is warned that Holmes performs strange experiments such as beating corpses to find out how long after death blows can result in bruises. His adventures continued in other novels and a series of stories published in the *Strand*. In these stories, Holmes worked with fingerprinting, ballistics, disputed documents, and soil analysis. In *The Adventure of the Devil's Foot*, Holmes thinks that a woman's death and brain damage to her two brothers may have resulted from poisonous fumes. In testing his theory, Holmes almost kills both Watson and himself with the same fumes.

In many stories, Watson and Holmes help the unfortunate and downtrodden as they engage in a constant quest for criminals such as Professor Moriarty. In *The Final Problem*, it is implied that during a struggle both Moriarty and Holmes fall to their deaths above Reichenbach Falls in Switzerland. However, after three years, Holmes returns. He had gone to Tibet, conducted research, and studied with the Dalai Lama.

Other characters in the stories include Mrs. Hudson, the housekeeper; Lestrade, the not-so-talented Scotland Yard inspector; Hopkins and Gregson, more gifted Scotland Yarders; Mycroft Holmes, Sherlock's older and wiser brother; and the Baker Street Irregulars, a pack of street ragamuffins whom Holmes uses to track down information or evidence.

It is believed that the character Sherlock Holmes was based on one of Doyle's medical professors, Dr. Joseph Bell, at Edinburgh University. Dr. Bell was known for his ability to diagnose diseases by using deductive reasoning, just as Holmes does in solving crimes. Holmes, like Bell, is a man of science. He analyzes and solves crimes using the same scientific and deductive skills he would use in solving a problem in science. Even Doyle's physical description of Holmes emphasizes his powers of observation. He has sharp piercing eyes and combs crime scenes with a magnifying glass. His powers of observation hold the key to finding solutions to the crimes he investigates.

Watson also shares with the reading audience that Holmes is remarkably knowledgeable in the areas of anatomy, chemistry, mathematics, sensational literature, and British law, but completely ignorant in others such as philosophy, astronomy, and politics. Holmes admits to Watson that he might not be the best roommate since he can be aloof, smokes strong tobacco, possesses chemicals to do experiments, and plays the violin. Holmes is also addicted to cocaine, which was legal at the time. Later on in his life, Holmes breaks the addiction with Watson's support.

Holmes conducted his private consultation practice for 23 years. In 1903, he retired to keep bees on the slopes of the Sussex Downs. He came out of retirement for a short time after World War I but has lived a quiet life since that time. In Tibet, Holmes learned the secret of long life from the Dalai Lama and is believed to be still alive.

The Sherlock Holmes stories were very popular when they were first published and continue to be popular today. A group of devoted fans started a club named Baker Street Irregulars after the group of characters in the stories. The group is active today with chapters all over Europe and the United States. Sherlock Holmes is known worldwide for his astute sense of observation, his deerstalker hat, his cape, his magnifying glass, and his curved meerschaum pipe. His stories have been translated into 50 languages, adapted into plays, films, and comic books, and used in advertisements. In 1993, England honored Holmes (and, thereby, Doyle) by placing him on a series of postage stamps. *See also* Doyle, Sir Arthur Conan.

References

Arthur Conan Doyle: Doyle vs. Holmes. http://www.letsfindout.com/subjects/art/arthur-conan-doyle.html.

Arthur Conan Doyle (1859–1930). http://www.kirjasto.sci.fi/acdoyle.htm.

Huntington, Tom. "Sherlock Holmes." *Historic Traveler*, Feb. 1997.

Redmond, Chris. Sherlockian.Net. http://www.sherlockian.net.

Sherlock Holmes: History of the Mystery. http://www.mysterynet.com/history/holmes/main.shtml.

Sherlock Holmes Museum. http://www.sherlock-holmes.co.uk.

Homicide

Homicide occurs when one person kills another. However, the killer may not be guilty of committing a crime. If the killing was done in self-defense or was the result of an accident, it may not even be prosecuted. Often, forensic science is required to determine whether a death was murder, self-defense, suicide, or the result of an accident.

In May 1957, Kenneth Barlow, a nurse

from Yorkshire, England, called a doctor after (he claimed) trying to resuscitate his wife Elizabeth using artificial respiration. He said that he had found his wife, who had been ill and vomiting that evening, submerged in her bathtub. The doctor, noticing the victim's dilated pupils, called the police. Suspicion of homicide increased when police observed Barlow wearing dry pajamas after claiming to have tried to pull his wife's body from the tub before administering artificial respiration as she lay in the tub. The medical examiner found water on the inside of Elizabeth Barlow's elbows, an unlikely circumstance if she had received artificial respiration. The examiner also found hypodermic syringes in the kitchen, which Barlow, a nurse, claimed he had used to administer penicillin to himself.

David Price, the suspicious medical examiner, discovered three tiny marks indicating injections in the victim's buttocks. A careful analysis of Mrs. Barlow's symptoms, as described by her husband, indicated low blood sugar, which could have been caused by a lethal dose of insulin; however, her blood revealed a sugar concentration greater than normal. Research by forensic experts led to the discovery that in violent deaths the victim's liver often responds by sending large amounts of stored sugar into the bloodstream. Forensic scientists then injected mice with insulin. The mice died very quickly. When mice were injected with tissue extracts from the victim's injection sites, they too died. Injections of extracts from other tissues taken from the body had no such effect. The evidence developed by the forensic team, together with testimony that Barlow had told others that insulin was a way to commit the perfect murder, led to a conviction of homicide. He was sentenced to life in prison.

In South Africa, in 1925, Petrus Hauptfleisch tried to make it appear that his mother had died from a fall down a flight of stairs. After killing his mother, Hauptfleisch had left her face down for several hours, but had turned her on her back as he staged the fall. The coroner, who found evidence of lividity on the front of her body, realized that the body had been turned. By denying that he had turned the body, Hauptfleisch sealed his fate.

English common law, which is the basis for the American legal system, has long recognized two types of homicide: murder and manslaughter. A killing that is deliberate or premeditated—committed with "malice aforethought"—is classified as murder. However, murders are either of the first or second degree. First-degree murders are killings that were planned or were committed during another crime such as rape or robbery. Second-degree murders, for which punishment is less severe, are those performed as a result of momentary anger or impulse.

Killings classified as manslaughter are wrongful but unplanned and committed without malice. Again, there are two types of manslaughter, voluntary and involuntary. Voluntary manslaughter is associated with intentional killings performed in a heat of passion provoked by the victim, as, for example, in a fight. Involuntary manslaughter is unintentional killing such as that resulting from an automobile accident or a riot. *See also* Crime Scene.

References

Gerberth, V.J., ed., *Practical Homicide Investigation: Tactics, Procedures, and Forensic Techniques.* 3rd ed. Boca Raton, FL: CRC Press, 1996.

Renstrom, Peter G. *The American Law Dictionary.* Santa Barbara, CA: ABC-CLIO, 1991.

Hyoscyamine. *See* Atropine.

I

Immunoassay

Immunoassay involves testing substances for antigen-antibody reactions. Forensic scientists have used the technique to detect drugs in the blood and urine of people suspected of taking illegal substances. Antigens, such as bacteria, viruses, pollen, dust, drugs, and blood, produce an immune response upon entering the human body. Of course, the antigen must be foreign to the body. A patient can, without fear of an immune response, accept his own transfused blood. Immunoassay techniques have also been widely used to test athletes who are required to be "clean" before competing. Police use the technique to test apprehended people who are suspected of drug use and often, as a case described later reveals, to ensure that people engaged in occupations where public safety is involved are not using drugs.

Although there are a number of different ways to detect drugs in an individual's body, one widely used approach to drug detection is the enzyme-multiplied immunoassay technique (EMIT), in which blood or urine is tested by adding antibodies specific for the drug in question. Laboratories that test athletes and other people suspected of drug use analyze in two stages. In the first part of the test, a sample of a body fluid such as urine is mixed with a substance that contains anti-bodies for a number of types of drugs as well as compounds made from the drugs themselves that are bonded to chemical enzymes. In a drug-free sample, the added antibodies and compounds combine, leaving the sample essentially free of enzyme-tagged drug compounds. In samples that contain metabolized drugs, most of the antibodies combine with the metabolized drugs, leaving many of the enzyme-tagged compounds in the fluid. By using spectrophotometry, the amount of light of a particular wavelength absorbed by the enzyme-tagged drugs in the sample can be determined. High absorption indicates that the antibodies have bonded with metabolized drugs that were in the urine, leaving a high concentration of the tagged drugs.

For example, the urine of those smoking marijuana contains THC-9-carboxylic acid (THC is tetrahydrocannabinol). Antibodies specific for this chemical are prepared by combining THC-9-carboxylic acid with a protein and injecting the substance into an animal such as a rabbit. The animal responds by producing antibodies that will react with the injected drug-protein compound. Serum recovered from the animal's clotted blood will contain the antibodies because these antibodies will react with quantities of THC-9-carboxylic acid as small as a millionth of a gram.

If the first test is positive, a second test can

be performed using gas chromatography and mass spectrometry. The mass spectrometer is used to confirm the identification of substances detected by chromatography.

It was this testing in 1988 that cost an Olympic athlete his gold medal. In September of that year, Canadian sprinter Ben Johnson, Olympic winner of the 100-meter dash the previous summer, was stripped of his title when he tested positive for anabolic steroids. Johnson is but one of many athletes who have been disciplined for using drugs believed to enhance their athletic prowess.

In a sodomy case, *State v. Smith* (Mo. Ct. App. W.D. 1982), a forensic chemist who had used neutron-activation analysis to analyze hair taken from a comb at the crime scene testified that, within reasonable scientific certainty, hairs on the comb matched those from the suspect. The probability of their being identical was 4,250,000:1 for seven comparable concentrations of elements. If nine comparable concentrations of elements were considered, the probability of their being identical was 33,300,000:1.

In *Amalgamated Transit Union v. Cambria County Transit Authority* (1988), division 1279 of the Amalgamated Transit Union sought a preliminary injunction against mandatory drug and alcohol testing of bus drivers and mechanics during annual physical examinations. The union argued that taking employees' urine and blood samples without any suspicion of illegal conduct constitutes unreasonable searches that intrude upon the integrity of a person's body, exposing private facts to the government and public. Since the Fourth Amendment of the U.S. Constitution forbids unreasonable searches, the union maintained that the requirement established by the Cambria County Transit Authority of Pennsylvania was in violation of the employees' constitutional rights.

The proposed testing program required that drivers and mechanics submit blood and urine samples to a local hospital, where an EMIT screening test would determine the presence or absence of 10 substances, including alcohol, opiates, cocaine, and cannabinoids. Should the test be positive, the employee had the right to challenge the result by having the sample tested by a laboratory of his or her own choice. If that test were also positive, the employee was required to undergo a rehabilitation program before returning to work. Refusal would be cause for termination of employment.

The court concluded that the testing program proposed by the transit authority did not violate the Fourth Amendment. The decision maintained that such testing, which was uniform and nondiscretionary, was a reasonable means of enhancing both employee and public safety by helping to ensure that bus drivers and mechanics were free of any effects due to alcohol or controlled substances. *See also* Blood; Chromatography; Mass Spectrometry; Spectrophotometry.

References

Amalgamated Transit Union, Division 1279 v. Cambria County Transit Authority. Civ. A. No. 88–796, U.S. District Court W.D. Pennsylvania, July 19, 1988.

Chemistry and Crime: From Sherlock Holmes to Today's Courtroom. Ed. Samuel M. Gerber. Washington, DC: American Chemical Society, 1983.

Saferstein, Richard. *Criminalistics: An Introduction to Forensic Science.* Upper Saddle River, NJ: Prentice Hall, 1998.

Infanticide

In criminal law, infanticide is the killing of a baby by its parent or with a parent's consent. In the United States, England, and most European countries, it is considered a form of murder.

Most commonly, infanticide is the result of smothering or suffocating the child, which poses a problem for forensic scientists because an autopsy cannot distinguish death by smothering from sudden infant death syndrome (SIDS). Most states now require an examination of the scene of death and/or a review of the child's medical history to determine probable cause of death. Investigators look for any suspicious circumstances that could indicate foul play. If, for example, there has been another SIDS death in the family or the child has been close to death before at home with no similar difficulty while under observation at a hospital, it would raise a red flag for investigators and indicate possible child abuse or a homicide.

This was not always true. Earlier it was thought that SIDS might have a hereditary basis and be more likely to occur in some families than in others. As a result, serial homicides in some families may have gone undetected. In 1972, Dr. Alfred Steinschneider of Syracuse, New York, published a study on SIDS that featured the children of Waneta and Tim Hoyt of Newark Valley, New York. The Hoyt family's apparently healthy babies suffered spells of prolonged apnea. The children would suddenly stop breathing and turn blue due to lack of oxygen. Three of five children in this family had died of SIDS. The other two children were fine in a hospital, with only brief periods of apnea, but died at home while alone with their mother. Steinschneider felt that SIDS could run in families and that bouts of apnea could be a SIDS indicator. SIDS and apnea remain challenging medical problems, but no more evidence has been found to link them since Steinschneider's research, which has since largely been discounted. In 1976, Dr. Linda Norton, a Dallas forensic pathologist specializing in violent death, read the report of Steinschneider's research and believed that the babies in the study had been murdered, not that they were victims of SIDS. Her concerns and those of other medical and law-enforcement authorities called into question Steinschneider's 1972 study. However, it was not until 1994 that police were able to make an arrest in the case of the Hoyt babies. In that year, Waneta Hoyt admitted that she had smothered all five of her children. She later recanted but was found guilty of five counts of murder and sentenced to 75 years to life. She died in prison in 1998.

In another study of baby deaths, Dr. David Southhall, an English physician, installed surveillance cameras in hospital rooms in the United Kingdom to observe parents with babies who had all experienced life-threatening events. In the 39 cases he studied over the course of several years (1986–1994), some mothers were caught choking their babies on videotape. Some of the babies also had siblings who had died under suspicious circumstances.

Some of these mothers may suffer from Munchausen's syndrome, a mental illness in which a person feels compelled to fake or induce illness to obtain care and attention from doctors. In Munchausen's syndrome by proxy, a person injures her child instead of herself to gain attention. Besides smothering their children, such people may poison or drug their child or mix blood in their urine to induce doctors to test the child.

Another common form of infanticide involves teenage mothers. They may conceal their pregnancy, give birth while alone, and then kill and dispose of the child.

It is believed that there are many cases of infanticide that are never discovered. Relatives may hide instances of infanticide to prevent criminal charges and family pain and stress. Pediatricians may also overlook symptoms, not wanting to believe that the families for whom they care are capable of such acts.

Mothers commit most infanticides. One reason is that mothers usually have much more contact with their babies than the fathers. Mothers can also suffer from emotional disturbances brought on by postpartum depression. In recognition of this fact, England passed the Infanticide Act in 1938. It follows the principle that a woman may experience an abnormal state of mind after childbirth and, therefore, should not be put to death for her wrongdoing. The United States and much of Europe recognize this principle also.

In some primitive cultures, infanticide is still an accepted practice. In these cultures, where survival is a struggle, the reason for killing a baby is a shortage of food. Infanticide is also accepted in other primitive cultures where the parents must pay to procure marriage partners for their children. The lack of material wealth in a family will mean that the parents can afford to marry off only a limited number of children.

Any type of abnormal birth, such as twins,

Jay Hoyt wipes away tears as his mother looks at the jury in disbelief when a guilty verdict is given on five counts of second degree murder. To the far right is Timothy Hoyt, Waneta's husband. October 24, 1997. © *The Post Standard*/Dennis Nett, Syracuse, NY.

often led to infanticide in superstitious societies of the past. In ancient Greece and Rome, deformed infants were killed. It is believed that it was the custom of ancient Hebrews to kill their firstborn to appease the gods. In India, many families continued this practice until the nineteenth century. *See also* Cause of Death; Homicide; Sudden Infant Death Syndrome (SIDS).

References

Burton, William C. *Burton's Legal Thesaurus.* 2nd ed. New York: Macmillan, 1992.

Evans, Colin. *The Casebook of Forensic Detection: How Science Solved 100 of the World's Most Baffling Crimes.* New York: John Wiley, Inc., 1996.

"False Alarm: The Failed Promise of Apnea Monitors." Syracuse.com. May 5–7, 1996. http://www.syracuse.com/features/apnea.

Renstrom, Peter G. *The American Law Dictionary.* Santa Barbara, CA: ABC-CLIO, 1991.

"Sudden Death: The Search for the Truth about Cot Death." BBC Online. Feb. 25, 1999. http://www.bbc.co.uk/science/horizon/sudden death.shtml.

Ink Evidence. *See* Disputed Documents.

Inquest. *See* Coroner.

Insanity Defense

The use of the insanity defense in court depends on a psychological assessment performed by a psychiatrist or a psychologist to determine whether a person should be held responsible for a crime he or she committed. The assessment is then used as evidence by lawyers in the case. If a person is determined to be insane on the basis of the psychological assessment, that person's legal responsibility for the crime can be annuled.

Insanity, according to law, refers to a person's mental condition at the time he or she committed a crime. Consequently, the person may not be insane at the time of the assessment but can still be determined to have been temporarily insane at the time the criminal act was committed.

Insanity is not easily determined. In the United States, there is no real consensus on how mentally ill a person must be to be excused from criminal responsibility. There are several different standards used among the 50 states. Some states use the old English standard developed in 1843, the M'Naughten rule. This standard states that a person is legally mentally ill (insane) if he or she does not understand the nature of the criminal act or know that the act is wrong. Other states use a modified form of the M'Naughten rule that includes the idea of the irresistible impulse. That is, while the person may have known that an action was wrong, the mental illness prevented the person from controlling his or her behavior. Other states use the standard set by the American Law Institute 1962 Model Penal Code. In this standard, a person is not criminally responsible if the mental illness prevents him or her from possessing "substantial capacity" to "appreciate the criminality" of an action or to control behavior to the "requirements of the law."

The insanity defense has always been controversial because of the lack of a set standard, and the states of Montana, Idaho, and Utah have actually abolished the insanity plea. An example of a controversial case is that of John Hinckley, Jr., who, by using the insanity defense, was acquitted of charges stemming from his attempt on President Ronald Reagan's life in 1981. Many people worried that he would be released too quickly from the mental hospital and try to harm someone again.

Another high-profile case where the insanity defense was of concern involved the Un-abomber, Theodore Kaczynski, who was responsible for 16 bombings causing 3 deaths and 23 major injuries over a period of 17 years. His lawyers wanted to use the insanity defense, but Kaczynski refused and decided to admit his complicity.

Many states have established an option for juries in insanity-defense cases to find a defendant guilty but mentally ill. This verdict requires the jury, the judge, or a board of mental health professionals to determine whether the defendant will be sent to prison or to a mental hospital. This option has developed because of the difficulty juries have with the issue of factual guilt and the defendants' ability to judge the morality of their actions. The verdict of guilty but mentally ill is seen as one way to allow a jury to sidestep difficult questions by shuttling those who might otherwise "escape" by an insanity plea into a new category where they can be judged guilty.

In civil law, insanity or mental illness can be a factor also. A person can be committed to a mental institution against his or her will if he or she is deemed mentally ill. A person can be declared "incompetent" to manage his or her own affairs. In this instance, care of all property may be transferred to some form of guardianship. A court can decide if a person lacks the ability to make a will, get married or divorced, be sued, or be a witness in a trial based on his or her mental state. *See also* M'Naughten Rule; Unabomber.

References

American Psychiatric Association. Public Information Insanity Defense Fact Sheet. http://www.psych.org/public_info/insanity.cfm.

Burton, William C. *Burton's Legal Thesaurus.* 2nd ed. New York: Macmillan, 1992.

Microsoft Encarta. *Insanity.* Funk & Wagnall. 1993. http://encarta.msn.com/.

Renstrom, Peter G. *The American Law Dictionary.* Santa Barbara, CA: ABC-CLIO, 1991.

J

Jack the Ripper

Jack the Ripper was an unidentified nineteenth-century English serial killer who, between August and November 1888, killed at least five prostitutes in East London. Lacking the tools of modern-day forensic science, police never discovered his true identity, although an entire enterprise known as Ripperology continues to probe the case searching for an identity.

All his victims were murdered in the early morning hours on either the first or last weekend of the month within a 38-acre area of Whitechapel, London, a region that police would never enter alone. All the victims had had their throats cut, and all had been mutilated, but none of them had been sexually assaulted. After each murder, the killer taunted the police with letters pertaining to the crime. There were many suspects, but evidence was minimal; however, some believe that the killer was a surgeon because the eviscerations he performed were skillfully done. Known only as Jack the Ripper, he was popularized by a number of books and, later, films. There was even a musical based on his exploits, and he was introduced as a character in Alban Berg's opera *LuLu*. *See also* Serial Killers.

References

Cyriax, Oliver. *Crime: An Encyclopedia*. North Pomfret, VT: Trafalgar Square, 1996.
Rumbelow, Donald. *Jack the Ripper: The Complete Casebook*. London: Penguin, 1988.
Zonderman, Jon. *Beyond the Crime Lab: The New Science of Investigation*. New York: Wiley, 1990.

K

Kennedy, Craig

Craig Kennedy, a fictional chemistry professor at Columbia University, has been called the "American Sherlock Holmes" because Kennedy, like Holmes, used scientific techniques to solve mysteries. Kennedy was created by author Arthur B. Reeve early in the twentieth century.

Kennedy used scientific equipment such as a gyroscope, X rays, a lie detector, and a portable seismograph that can contrast footstep sizes. Again, as with Holmes, the use of these investigative methods was predictive of the future since the stories were written in the early 1900s when forensic science was in its infancy in the United States.

Kennedy had a roommate, Walter Jameson, a newspaper reporter who recorded Kennedy's escapades in a manner similar to Dr. Watson's chronicling of Sherlock Holmes's cases. Both Holmes and Kennedy had a passion for music. Holmes demonstrated his fondness by playing the violin, while Kennedy enjoyed opera. Kennedy was one of the first detectives ever to use psychoanalytic techniques to help solve mysteries and was also a master of disguise. But Kennedy was a man of action as well, and if the need arose, he used a gun.

Kennedy was portrayed in a series of very popular silent films in the 1910s and 1920s.

In many of these short films, a young pretty heroine, Elaine, was in the clutches of a very evil villain, and Kennedy used scientific methods to save her. In one episode, he brought a dead girl back to life with an electric resuscitation machine; in another, he escaped death rays. Kennedy is not popular today because many of his scientific methods are now outdated and some are too bizarre for the more technically sophisticated public. *See also* Holmes, Sherlock; Reeve, Arthur Benjamin.

References

Penzler, Otto, Chris Steinbrunner, and Marvin Lachman, eds. *Detectionary.* Woodstock, NY: Overlook Press, 1977.

Steinbrunner, Chris, and Otto Penzler, eds. *Encyclopedia of Mystery and Detection.* New York: McGraw-Hill, 1976.

Kidnapping

Kidnapping, in the realm of criminal law, is the taking and carrying away of a human being by force, fraud, threats, or intimidation against the victim's will and without lawful authority. Kidnapping cases are difficult for investigators because even if evidence leads them to the perpetrator, they must be careful not to jeopardize the safety of the victim. The purposes of kidnapping include holding someone for ransom or as a hostage to expedite the commission of a felony or flight thereafter, inflicting bodily injury or terrorizing the victim, or interfering with the performance of any governmental function, as in an act of terrorism carried out for political reasons.

A high-profile case of kidnapping for governmental reasons was the 1979 Iran hostage kidnapping. Iranian students protesting U.S. involvement in Iranian affairs and its connection with the deposed shah stormed the U.S. Embassy in Tehran and took Americans hostage. For 444 days, 52 people remained captive through a failed rescue attempt and intense negotiations. They were not released until Iran became engaged in a war with Iraq and needed assets seized in American and European banks and held as part of the sanctions leveled against Iran.

Kidnapping in the United States is a first-degree felony unless the kidnapper releases the victim unharmed. In that case, it is a second-degree felony. In most states, it is also a crime to attempt or plan to commit a kidnapping. This severe penalty is a result of the ill-fated kidnapping of Charles Lindbergh's young son in 1932. Charles Lindbergh was the famous American aviator who was the first person to fly across the Atlantic Ocean solo. Lindbergh's kidnapped son was found murdered. As a result of this crime, the Federal Kidnapping Act, known as the Lindbergh Law, was passed in 1932. This act makes kidnapping a federal crime when the victim is transported from one state to another or to a foreign country. The Lindbergh Law was amended in 1934 to make conspiracy to commit kidnapping a crime. In 1968, the Supreme Court nullified the section of the Lindbergh Law that gave the jury the power to recommend the death penalty for kidnapping.

In the Lindbergh kidnapping case, the gold notes used to pay the ransom money helped investigators locate the kidnapper, Bruno Richard Hauptmann. The $20 gold notes were no longer in circulation, which alerted a gas-station attendant when Hauptman paid for his gas with one of the gold notes. (Hauptmann was given the death penalty for the murder of Lindbergh's son.)

Another child kidnapping case where the ransom money helped catch the kidnapper was the kidnapping of George Weyerhauser. The nine-year-old boy was taken from his home on May 24, 1935. He was returned safely on June 1, 1935, after his parents paid the $200,000 ransom. The bills had been marked, and an observant store clerk identified one and notified authorities. The kidnapper, William Mahan, was caught and sentenced to 60 years in jail.

The kidnappers in a 1953 case were convicted through forensic evidence. In Kansas City, Missouri, on September 28, 1953, six-year-old Bobby Greenlease, the son of a wealthy automobile dealer, was kidnapped and held for $600,000 ransom. The ransom was paid, but the kidnappers, Carl Hall and Bonnie Heady, killed the boy and buried his body in the yard of Heady's house in St. Joseph, Missouri. The boy's body was found on October 7, 1953, buried, wrapped in a plastic bag, and covered with lime. A dentist identified the body as that of Bobby Greenlease. Bloodstains were found on a floor and steps of the Heady residence and on a nylon blouse and fiber rug. Thirty-eight-caliber shell casings found in the house and a lead bullet found in Heady's car were examined by the FBI Crimes Laboratory. The forensic scientists ascertained that the bullets had been fired from a .38-caliber snub-nose Smith and Wesson in Carl Hall's possession at the time of the arrest. On November 19, 1953, Carl Hall and Bonnie Heady were convicted and sentenced to death. They were executed together on December 18, 1953. *See also* Lindbergh Kidnapping and Trial.

References

Black's Law Dictionary. 7th ed. Bryan A. Garner, editor-in-chief. St. Paul, MN: West Group Publishing, 1999.

Burton, William C. *Burton's Legal Thesaurus.* 2nd ed. New York: Macmillan, 1992.

Federal Bureau of Investigation Web Site. "The Greenlease Kidnapping." http://www.fbi.gov /fbinbrief/historic/famcases/greenlease/ greenleasenew.htm

Oran, Daniel. *Oran's Dictionary of the Law.* 2nd ed. New York: West Publishing Company, 1991.

Statsky, William P., Bruce L. Hussey, Michael R. Diamond, and Richard H. Nakamura. *West's Legal Desk Reference.* New York: West Publishing, 1991.

King, Dot: Murder (1923)

Dot King was an American playgirl and social butterfly. She was murdered in 1923 in a crime that remains unsolved. Forensic science was used to eliminate suspects, but there was not enough evidence for police to reach any conclusions with regard to guilt. Born in 1894, Anne Marie Keenan was the fourth child of a poor Irish family. In 1912, she married a chauffeur, but the ill-fated union ended in divorce after only a short time. The blonde, blue-eyed beauty moved to New York City, where she secured a modeling position at a Fifth Avenue dress shop. To be more fashionable in the trendy city, as well as to further her show-business aspirations, Keenan changed her name to Dot King. Christened the "Broadway Butterfly" in death, King was a fixture on the New York party scene. For a time, the small, perky playgirl was a hostess at a speakeasy. At the speakeasy, she met and established a relationship with a man she only knew as "Mr. Marshall." A sugar daddy of the first magnitude, Marshall, claiming to be a Boston banker, created a love nest for the couple on West 57th Street. He lavished jewels on King (estimates of their value range as high as $30,000), paid her expenses, and gave her an allowance. In return, Marshall visited twice a week on a regular basis.

Marshall did not seem to mind that King shared the apartment with another man, Alberto Santos Guitmars. Guitmars was a small-time thug, but King was convinced that their relationship was true love and marriage was in the future. She showered her "boy toy" with gifts and gave him money. In return, Guitmars beat her.

On March 15, 1923, King's maid discovered Dot King dead in bed. Recognizing the victim's questionable character, the initial police response was to report it as a suicide. There was no sign of forced entry, the telephone was moved a distance from the bed, and a chloroform bottle lay nearby. Closer examination of the scene totally subverted the initial judgment. The apartment was ransacked, and it was obvious that some of King's jewels, furs, and gowns were missing. A number of her personal letters were also gone, though police did find one note from the mysterious Marshall. The body itself was mute testimony of murder. There were scratches and burns around King's eyes and nose indicating that there had been a struggle in the application of a sufficient amount of chloroform to blister her face. Finally, King's arm was in a hammerlock position behind her body. This was clearly no suicide, and with few concrete clues, the police began an investigation.

The most immediate suspect was Marshall. Police interrogated the maid and elevator boys to get a description, but the inquiry seemed at a dead end when they discovered that there was no Boston banker named Marshall. On March 24, in the face of a continuing investigation, Marshall stepped forward. To the tabloids' delight, he was prominent Philadelphia socialite J. Kearsley Mitchell, son-in-law of millionaire E.T. Stotesbury. The contrite Mitchell admitted that he had spent the day prior to her death with King. The couple went out to dinner and returned to the apartment about midnight. Mitchell insisted that he had left King alive at 2:30 A.M. Fired by tabloid sensationalism, a secret identity, and articles about blackmail, few believed Mitchell. Nonetheless, he did convince the authorities, who absolved him of any involvement in the crime. Mitchell and his family immediately left the country for Europe.

A second, more reasonable suspect was the hot tempered Guitmars. Though police suspected him of the killing, he had an apparently air tight alibi. The petty hoodlum was involved with another woman, socialite Aurelia Dreyfus, who vouched for her man. Left with no suspects, the police could only speculate what might have happened. They hypothesized that the murder was the result of a robbery gone bad. The felon had probably entered the apartment as Mitchell left and had used too much chloroform on King, killing her. Then it had become a snatch-and-run affair: the killer had taken what he could carry and run off. About a year later, another Broadway showgirl, also a kept woman, suffered a strikingly similar death.

Although the crime was never solved, the police were convinced that Guitmars was the perpetrator. In 1924, he was with Dreyfus when she plunged to her death from a Washington, D.C. high-rise suite. Among the woman's belongings was an affidavit, probably to protect herself from Guitmars, declaring that she had perjured herself in supplying Guitmars with an alibi. After a lengthy investigation, police once again felt that there was not enough evidence for an indictment in the Dreyfus death or to reopen the King inquiry. Guitmars died in 1953 but never revealed any new information in either case.

References

Churchill, Allen. *A Pictorial History of American Crime, 1849–1929*. New York: Holt, Rinehart & Winston, 1964.

Lane, Brian. *Chronicle of 20th Century Murder*. Vol. 1, *1900–1938*. New York: Berkley Books, 1995.

Sifakis, Carl. *The Encyclopedia of American Crime*. New York: Smithmark, 1992.

L

Lacassagne, Alexandre (1844–1921)

French-born Alexandre Lacassagne has been called the father of forensic science. During his long career, he made a number of significant contributions to forensic detection.

After attending the Military Academy at Strasbourg, he served as an army physician in North Africa. His experiences in North Africa motivated his interests in bullet wounds and tattoos. In 1878, he wrote a long treatise, *Précis de médecine* (Medical abstract), on the value of tattoos for identification of dead bodies and his experiences with gunshot wounds. He was the first professor of forensic medicine at the University of Lyon, where his extensive knowledge of medicine, biology, and philosophy helped him develop and discover many new techniques in forensic medicine. In 1889, he used new techniques of bone study and dental comparison to identify the corpse of the 49-year-old Parisian bailiff August Gouffe, who had been murdered and sent in a trunk to Lyon. He picked out a deformity on the skeleton's right knee that would affect the attached muscles. The wasting away of these muscles would have resulted in a limp, an impediment that was confirmed by Gouffe's family. This helped positively identify the corpse.

In the same year, Lacassagne was the first person to recognize the significance of the grooves on the surface of a bullet that had been extracted from a murder victim and suggested that it might be identified with the gun from which it had been fired. This study of guns and bullets became the area of forensic science known as ballistics. Lacassagne was also one of the first to see the relationship between an attack on a victim and the shape and design of resulting bloodstains.

In 1898, Lacassagne was asked to do an evaluation of the French Ripper, a butcher named Joseph Vacher who had committed a string of sex murders on children that included mutilation and dismemberment. Vacher pleaded insanity. After a comprehensive five-month examination, Lacassagne concluded that Vacher was not insane. His testimony led to a guilty verdict, and Vacher was guillotined on December 31, 1898.

Lacassagne went on to investigate better methods for determining whether or not a subject was dead. Some morgues actually had pull cords near the corpses in case they woke up. One doctor was said to have made an incision into the body so he could feel the heart to see if it was still beating. Lacassagne made some very interesting discoveries during his investigation. From prolonged observation, he determined that the purple blotches that appeared on dead bodies resulted from blood falling to the lowest levels of the body after circulation had ceased. The accumulated blood produced purple discoloration of the skin. The falling of the blood could be timed predictably, which helped determine the time of the victim's death.

Lacassagne also studied rigor mortis and discovered that it did not start in the jaw, as had been believed, but in the heart. Cadaveric spasm is a form of rigor mortis that occurs in violent death. The fingers grip whatever they hold at the time of death with such strength that they cannot be pried open. Lacassagne contrasted this grip with standard rigor mortis. In one case of apparent suicide, a body was found in a locked room holding a gun. Lacassagne established experimentally that the fingers of a newly dead body could be made to hold objects like a gun loosely, and the commencement of regular rigor mortis would temporarily imitate cadaveric spasm as the hand tightened around the gun.

Lacassagne also studied the rate at which the body cools after death. He realized that a great deal depended on the temperature of the environment. The rule of thumb up to this point had been that body temperature dropped one degree centigrade per hour during the first four hours after death. Lacassagne discovered that there were too many inconsistencies. The unreliability of this index prompted Lacassagne to coin the saying: "One must know how to doubt." *See also* Cause of Death; Time of Death.

References

Cyriax, Oliver. *Crime: An Encyclopedia*. North Pomfret, VT: Trafalgar Square, 1996.

Evans, Colin. *The Casebook of Forensic Detection: How Science Solved 100 of the World's Most Baffling Crimes*. New York: Wiley, 1996.

Lane, Brian. *The Encyclopedia of Forensic Science*. London: Headline, 1992.

Thorwald, Jürgen. *The Century of the Detective*. New York: Harcourt, Brace & World, 1965.

Larson, John (1892–1965)

John Larson, born in Canada, devised the precursor of today's polygraph, a device widely used by forensic scientists to determine whether or not a person is telling the truth. Larson received his bachelor's degree from Boston University in 1914. In 1921, Larson was studying for his doctorate in criminology at the University of California and working as a sergeant in the Berkeley, California, Police Department, when he was approached by his supervisor, August Vollmer, and asked to build a device that would be able to detect when a subject was lying. After a few weeks of research and experimentation, Larson presented his first model. It was basically a sphygmomanometer (the inflatable cuff physicians use to measure blood pressure) connected to a needle that would record changes in blood pressure by pressing against a soot-covered roll of paper that turned beneath it.

Larson tested his device, which he called a cardio-pneumo-psychogram, on Vollmer. (His pedantic name for his machine was soon replaced by "lie detector.") Vollmer was very impressed. Each time he deliberately lied, he could see the needle rise dramatically. Further tests with real criminals revealed the value of Larson's lie detector and led to the development of the polygraph, which enables a subject's heart rate, breathing rate, and galvanic skin response as well as blood pressure to be recorded. *See also* Polygraph Test; Vollmer, August.

References

Cyriax, Oliver. *Crime: An Encyclopedia*. North Pomfret, VT: Trafalgar Square, 1996.

Wilson, Colin. *Clues! A History of Forensic Detection*. New York: Warner Books, 1991.

Lattes, Leone (1887–1954)

Leone Lattes was an Italian physician who studied blood, bloodstains, and body fluids and believed that they could play an important role in crime detection. His discoveries made it possible to determine a person's blood type from either blood or secretory stains. His interest in this area began after he had completed his medical studies and spent some time at German universities. There he met Max Richter, who had done experiments with blood. Lattes found this area interesting and began his own study of blood and serology. He was fluent in many languages and read everything that was published on blood groups.

It was not a crime that helped Lattes make some significant discoveries but rather a marital altercation. In Turin, Italy, in 1915, Lattes became involved in an amusing situation. A man named Girardi had a very jealous wife named Andrea who thought that Girardi had been cheating on her. This belief was based on a bloodstain she found on Girardi's shirt after a late-night outing. She was sure that the blood came from another woman. The incident in question had occurred three months earlier, and Girardi could not convince his wife that the bloodstain was not from another woman. Girardi begged Lattes to help him convince his wife of his fidelity. After talking to Girardi, Lattes found that there had been three people in the house on the day the bloodstain appeared—Girardi, his wife, Andrea, and a family friend, Teresa Einaudi.

Lattes said that he would try to help but would need to have a blood sample from all three of them. Luckily for Girardi, the two women agreed to let Lattes take a sample of their blood. Lattes also learned that at the time of the incident, Einaudi had been menstruating and, though she denied it, could have wiped her hands with Girardi's shirt after using the bathroom. Lattes determined that Girardi had type-A blood, Andrea Girardi had type-O, and Einaudi had type-A.

The problem was to determine the type of blood in the bloodstain, which was three months old. There was no method at the time for determining the blood type of dried bloodstains. Lattes went through a lengthy, complicated procedure to remove the blood from the material. He actually weighed the material with the bloodstain and another small piece of material with exactly the same thread count to determine the weight of the blood. Then he dissolved the blood with the proper amount of liquid and was able to determine that the blood was type A, so it was definitely not Andrea Girardi's. From cytology, the science of body cells, Lattes knew that the cells of the vagina are different from any other body cells and that these cells are found in menstrual blood. Lattes found no vaginal cells in the bloodstains, so now he could rule out vaginal blood from Einaudi, but if it was Girardi's own blood, where did it come from? Lattes did a complete physical examination of Girardi and discovered that he had prostate trouble that led to some bleeding from the urethra. The mystery of the old bloodstain was solved, and marital bliss was brought back to the Girardis' home.

Lattes continued his study of bloodstains and blood groups and in 1922 published a book, *L'Individualità del sangue nella biologia, nella clinica, Nella medicina legale* (very roughly translated, The individuality of blood). For many years it was the standard textbook on blood. In the book, he covered the areas of blood transfusion, inheritance of blood groups, and the determination of blood type in bloodstains. By this time, he had simplified the method he had used in the Girardi case. He had found that if he let flakes of blood dry on wood or other smooth surfaces, he could put the small flakes on a microscope slide, add fresh blood, and place another slide on top. The serum of the fresh blood did the work of dissolving the other blood cells, and if the blood was from a different group, it would clump. This simplified method is known as the Lattes method and is still utilized today by forensic investigators. *See also* Blood.

References

Thorwald, Jürgen. *Crime and Science.* New York: Harcourt, Brace & World, 1967.

Wilson, Colin. *Clues! A History of Forensic Detection.* New York: Warner Books, 1991.

Laudanum. *See* Opium.

Lindbergh Kidnapping and Trial (1932–1935)

The kidnapping of aviator Charles Lindbergh's son has been called the crime of the (twentieth) century and is still considered by many a controversial case because of questionable evidence used to convict the man accused of the crime. Charles Lindbergh, the first man to fly solo across the Atlantic Ocean, was only 25 years old at the time of his historic flight in 1927. He was recognized as a national hero and was given the rank of colonel in the U.S. Army Air Corps. However, after March 1, 1932, he was also known for a tragic family event. On that evening, his 20-month-old son, Charles Augustus Lindbergh, Jr., was kidnapped from his nursery.

Lindbergh, his wife, Anne, and their young son lived in Englewood, New Jersey. On weekends they retreated to their newly completed house in a rural area of New Jersey near the town of Hopewell. Usually, they would return to Englewood on Monday mornings, but on the fateful day they had delayed their return because the baby had a cold. On the evening of Tuesday March 1, the family's nurse went up to check on the baby and discovered that he was missing. Lindbergh ran upstairs and found a poorly written note inside an open window, the one window in the nursery that had a warped shutter and could not be completely shut. The letter ended with a design consisting of two interlocking circles.

Both the Hopewell police and New Jersey state police were notified. The police found no fingerprints in the nursery. They did find footprints in the ground below the open window, but they did not measure the prints or make plaster casts of them. A broken ladder and chisel were found, and nearby, on a small dirt road, tire tracks were discovered. The estate was soon bustling with policemen and reporters as the news of the kidnapping spread. Any other evidence was lost because of the multitude of people that descended on the estate.

Colonel H. Norman Schwarzkopf, head of the New Jersey state police, was in command of the state investigation. (He was the father of H. Norman Schwarzkopf, Jr., commander of the U.S. forces during the Desert Storm war in 1991.) However, at the request of Colonel Lindbergh, a New York detective, Lieutenant James J. Finn, was put in charge of the more than 100 detectives who were assigned to the kidnapping case. Finn had been one of Lindbergh's bodyguards when he returned home from his historic solo flight.

On March 4, a second ransom letter was received. The kidnapper was upset that Lindbergh had notified the police and demanded more money. The same symbol of interlocking circles appeared on this note.

The kidnapping of an American hero's son outraged the nation. Many people offered help. On March 7, a Bronx newspaper received a letter from a schoolteacher, Dr. John F. Condon. He offered $1,000 of his own savings to the kidnapper if he would return the Lindbergh baby. He also offered to act as a negotiator between the Lindbergh family and the kidnapper. Both the kidnapper and Lindbergh accepted Condon's offer. Condon placed an ad in the *New York American* to inform the kidnapper that the money was ready. Condon was then instructed to go to the Woodlawn Cemetery. On April 2, Lindbergh went with Condon to the cemetery. The man at the cemetery told Condon that his name was John, but he refused to let Condon see the baby. However, he did send him the baby's sleeping suit.

The kidnapping brought out the best in some people such as John Condon; however, it brought out the worst in others. Gaston B. Means, a former FBI agent who had been fired, contacted some prominent people, telling them that he was in contact with the kidnapper. One of these individuals was Evalyn Walsh McLean, a wealthy woman from Washington, who gave Means $100,000 on his promise that the child would be returned. When the child was not returned, McLean went to the authorities, and Means was arrested and indicted. He was found guilty of larceny and given a 15-year prison sentence.

Another man, John Hughes Curtis, a shipbuilder who had lost his fortune during the depression, told Lindbergh that he had made contact with the kidnapper. Curtis said that the baby was on a boat along the mid-Atlantic seacoast. During the search that followed, Lindbergh was called home and told that the body of his son had been found in the woods near his home. Curtis was fined $1,000 and given a one-year suspended sentence for his deception.

The cause of the baby's death was determined to be a blow to the head. The baby was cremated shortly after Lindbergh identified it. This destroyed any chance of finding more evidence from the body. The only physical evidence was the ladder, the chisel, and the notes from the kidnapper.

Lindbergh had assumed that the kidnapper was a professional. Some of the detectives involved in the investigation thought that the kidnapping was an inside job or at least local and unprofessional. This belief was based on the kidnapper's knowledge that contrary to the family's usual habit, they were still at their weekend home on a Tuesday evening and the kidnapper knew which room was the nursery and which window was warped and could not be latched shut. However, all of Lindbergh's family and hired staff were cleared. Later, one member of the staff, Violet Sharpe, committed suicide, probably from the stress of the situation.

Finn thought that the ransom money could be used to trap the kidnapper. The ransom was paid in unmarked gold notes, but the serial numbers had been recorded. Finn's hypothesis was enhanced accidentally by a presidential order that took the United States off the gold standard. After May 1, 1933, the possession of gold notes would be illegal. Finn kept a map indicating where the bills with the correct serial numbers were turning up. A break came in September 1934 when a gas-station worker wrote down the license-plate number of a car whose owner had paid for gas with a gold note. The gas-station worker had feared that the note was counterfeit.

The owner of the car was Bruno Richard Hauptmann, a former carpenter who had entered the United States illegally in 1923 at the age of 23. He had been in the German army during World War I and had supposedly jumped ship to enter the United States. He married a German waitress, Anne Schoeffler, in 1925, and they had one son in 1933.

In the late spring of 1932, Hauptmann, no longer a carpenter, was an investor in stocks. Hauptmann was arrested on September 19, 1934, for passing some of the Lindbergh ransom money to the gas-station attendant. At the time of his arrest, another $20 gold certificate was found in his wallet. It was one of the bills in the ransom money.

Hauptmann's trial began on January 2, 1935, in Flemington, New Jersey, the county seat. The attorney general of the state of New Jersey, David T. Wilentz, led the prosecution. A New York newspaper provided Hauptmann with a Brooklyn defense attorney, Edward J. Reilly. The prosecution used circumstantial evidence to build a convincing case. One example of the circumstantial evidence was the fact that $14,600 of the ransom bills were found in Hauptmann's garage. Another example was the fact that a missing board from Hauptmann's attic was found to match a part of the ladder used in the kidnapping. The prosecution also had many handwriting experts, including Albert S. Osborn and FBI agent Charles Appel, testify that Hauptmann wrote the ransom notes after they examined some of his writing samples.

Along with this evidence, the prosecution produced several witnesses who reported seeing Hauptmann or his car near the Lindbergh estate before the night of the kidnapping. But one of the most crucial segments of the prosecution was the positive identification of Hauptmann as "Cemetery John" by Condon and Lindbergh. Lindbergh testified that the voice he had heard in the cemetery while he was delivering the ransom money with Condon was that of Hauptmann. The defense was unable to satisfactorily account for how Hauptmann got the ransom money without being the kidnapper, the handwriting similarities, and the connection between Hauptmann's attic board and the kidnapper's ladder.

The principal defense witness was Hauptmann, who came across as defiant. On the witness stand, Hauptmann was not given an interpreter, though he spoke English with difficulty. Hauptmann had to withstand a

Ransom note. Courtesy of the New Jersey State Police Museum.

very long cross-examination by the prosecution and could not satisfactorily prove that the money found in his possession had come from Isidor Fisch, a man who Hauptmann claimed had owed him money. His claim could not be checked because Fisch had returned to Germany where he died of tuberculosis. Furthermore, Hauptmann was unable to account for his own whereabouts on March 1, the night of the kidnapping, or April 2, the night the ransom money was exchanged. After six weeks, 1,500,000 words of testimony, and over 380 exhibits, the case went to the jury. About 12 hours later, the jury delivered a verdict of guilty. The standard number of appeals and a brief stay of execution followed, but on April 3, 1936, Hauptmann was executed for the crime.

Hauptmann, however, maintained his innocence until his death.

Since the conviction of Bruno Hauptmann, much has been written about the possibility of his innocence. Some believe that he was framed and that much of the evidence was fraudulent. Hauptmann was a skilled and meticulous carpenter. To some, it seems unlikely that he would crudely repair a ladder with a timber from his attic. It has also been said that the first time Hauptmann wrote the handwriting samples, he spelled the words correctly, unlike the original ransom notes, and that the police insisted that he write them again with the misspellings. Both Condon and Lindbergh identified Hauptmann's voice as the voice they heard on April 2, 1932, in the cemetery. Their identification

Arthur Kohler, brought in as an expert witness in 1932 and who later testified at the Lindbergh kidnapping trial, works with a duplicate of the kidnap ladder. Courtesy of the New Jersey State Police Museum.

came over two years after the brief encounter in the cemetery. Finally, some of the witnesses who stated that they had seen Hauptmann or his car in the area of the Lindbergh estate near the time of the kidnapping later said that they were interested in a share of the reward money. Other theories exist about the kidnapping case. Some researchers have even indicated that the kidnapper could have been a family member. Greg Ahlgren and Stephen Monier in their book *Crime of the Century: The Lindbergh Kidnapping Hoax* theorize that Lindbergh himself killed his son while playing a practical joke that went tragically wrong. He then covered up the accident and invented the kidnapping. Noel Behn indicates in his book *Lindbergh: The Crime* that Anne Morrow Lindbergh's sister, Elisabeth Morrow, had killed the baby be-

cause she was in love with Lindbergh and jealous that he had married her younger sister, Anne. There is enough controversy over Hauptmann's conviction that his descendants are petitioning for a posthumous pardon.

At the time (1932), kidnapping was not a federal crime. Hauptmann was accused of causing the child's death and taking something of value from the house, the child's clothing. As a result of this very public case, Congress passed a law called the Lindbergh Law that made kidnapping with intent to seek ransom or reward a federal offense. *See also* FBI Crime Laboratory; Hauptmann, Bruno Richard; Kidnapping; Osborn, Albert S.

References

Ahlgren, Gregory and Stephen R. Monier. *Crime of the Century: The Lindbergh Kidnapping Hoax*. Boston: Brandon Books, 1993.

Aiuto, Russell. *Lindbergh*. Crime Library. Dark Horse Multimedia. http://www.crimelibrary.com/lindbergh/main.htm. Accessed August 21, 2001.

Behn, Noel. *Lindbergh: The Crime*. New York: NAL-Dutton, 1994.

Horan, James D. *The Desperate Years*. New York: Bonanza Books, 1962.

Kennedy, Ludovic. *Crime of the Century*. New York: Viking Penguin, 1996.

Lineup

A lineup is a police identification process in which a single suspect in a crime is presented with other individuals, who generally resemble one another, to a witness or victim to determine if the suspect was involved in a crime. The term *lineup* derives from the fact that the "suspects" are placed in a row from which the victim or witness may select the offender. Usually one person in the lineup is a police suspect, while others are fillers or foils, police officers, nonsuspect prisoners, or other available people. The due-process clause of the U.S. Constitution requires that the identification not be made under suggestive circumstances that point to a specific suspect or in a situation that, in the court's estimate, is conducive to a mistaken identity. For example, a blatant physical difference among the people in the lineup would make the selection inadmissible in court.

According to a U.S. Supreme Court ruling in 1972, a lineup prior to filing formal charges or the issuance of an indictment does not require a lawyer's presence. After charges or an indictment, an attorney representing the person indicted must be present at any lineup. The right to counsel thus depends on when a lineup is held. The Supreme Court further found that a witness who identified a defendant in a tainted lineup may not refer to or describe the process in any way at the trial. Moreover, the witness or victim is not allowed to identify the defendant in open court unless the prosecutor can convince the judge, away from the jury, that the identification is the result of memory or some other reason than the lineup.

The Terry Leon Chalmers Case

On August 8, 1986, a young woman was raped near White Plains, New York. The police arrested Terry Leon Chalmers based on his identification by the victim after viewing mug shots. The victim later identified Chalmers in two separate lineups and in court. Because of the young woman's repeated ability to identify him and his own inability to corroborate his alibi, Chalmers was convicted in a Westchester County court on June 9, 1987, and sentenced to 12 to 24 years in prison.

Chalmers continued to claim that he was not guilty and persuaded the Innocence Project to consider his case. Lawyers were able to obtain physical evidence, which was sent to the Forensic Science Association for DNA testing. Using blood samples from the victim and from Chalmers as well as vaginal and cervical swabs from the victim that were in the original rape kit, the tests revealed that the semen on the swabs could not have come from Chalmers.

As a result of the DNA tests, Chalmers' conviction was overturned after he had served eight years in prison. Charges against him were dismissed in April 1995.

The Ronald Cotton Case

In July 1984, Jennifer Thompson was raped at knifepoint, but, retaining her rationality, she claimed she carefully observed her attacker. Shortly after the crime, she sat in a police station and helped a detective and artist produce a composite sketch of her attacker. A week later, she identified the man who had raped her from a lineup of six men. During the trial that followed, she pointed to Ronald Cotton and stated with conviction that he had been the man who raped her. Cotton was convicted and sentenced to life in prison on January 17, 1985.

A year later, a rapist named Bobby Poole

Jennifer Thompson and Ronald Cotton have managed to form a friendship over the past several years, despite the fact that she testified against him in her rape case in 1984. Cotton remained in prison for 11 years before a DNA test cleared him of the crime. The two make an effort to meet eight to ten times a year to catch up with one another's lives. © *The News and Observer*, Raleigh-Durham, NC.

arrived at the prison where Cotton was serving his sentence. Poole bragged to others that Cotton was serving some of the time he (Poole) should have received. This information, coupled with the fact that another woman had been raped in a similar manner an hour after Thompson in the same town in North Carolina, enabled Cotton's lawyer to obtain a second trial. Unfortunately for Cotton, Jennifer Thompson still identified Cotton, not Poole, as the rapist. Disconsolate, Cotton returned to prison.

In 1995, Cotton convinced Richard Rosen, a University of North Carolina law professor, that he was innocent. Rosen was concerned that Cotton had received a life sentence on the basis of a lineup identification, which he regarded as unreliable. The professor was successful in his request that DNA tests be conducted on both Cotton and Poole. The results showed that Rosen's concern was justified. The DNA tests proved that Poole, not Cotton, was Thompson's rapist.

Reference

Grano, Joseph. "Lineups." *The Guide to American Law*. St. Paul, MN: West Publishing, 1984, vol. 7, p. 194.

Lip Prints

Lip prints, the markings left by lips, are often found at a crime scene. While they are not as valuable as fingerprints, lip prints left on objects such as a drinking glass or cup, facial tissue, a handkerchief, or a cigarette butt can be used by forensic scientists to help in the identification of a victim or suspect. They are particularly evident when they are left by a woman who wore lipstick. It is possible to lift lip prints as well as fingerprints from objects. Such prints can then be compared to those of a suspect. Unfortunately, many police departments are unaware of how useful lip prints can be as a means of identifying or eliminating suspects. Consequently, this valuable piece of evidence is often ignored.

There is evidence that lip prints, like fingerprints, are unique. In a 1970 study carried out in Japan, researchers examined lip prints from 1,364 people using both photographs and a system similar to taking fingerprints. The study also discovered that inflammation can alter lip prints, but the prints return to normal when the swelling goes down. Furthermore, it was found that lip prints, like fingerprints, do not change with age.

A more limited study involving 150 subjects was conducted in the United States in 1991 by Mary Lee Schnuth. Again, the prints proved to be unique even among identical twins; however, there was evidence that heredity plays a role because similarities in lip prints exist between parents and their offspring.

As with fingerprints, lip prints have certain typical patterns that make it possible to categorize them. Some lips have grooves that run vertically across the entire lip. Others have straight grooves that disappear halfway into the lip. There are grooves that fork, grooves that intersect, netlike grooves, and grooves that do not fall into any of the types. There are two basic categories: simple and compound. The simple category is subdivided into four groups: straight, curved, angled, or sine-shaped curves. Compound prints are classified as bifurcated, trifurcated, or anomalous, such as vertical grooves, partial vertical grooves, branched grooves, crossing grooves, grooves that intersect to form rectangles, and random groove patterns.

Although American law-enforcement agencies have made little use of lip prints, Japanese forensic scientists, in particular, have found lip prints to be a useful tool. In one case, Tokyo police found that lip prints on the envelope of a threat letter sent to their department matched those of a suspect who was then arrested. *See also* Fingerprints; Footprints.

References

Lane, Brian. *The Encyclopedia of Forensic Science*. London: Headline, 1992.
Schnuth, Mary Lee. "Lip Prints." *FBI Law Enforcement Bulletin*, Nov. 1992, vol. 61, pp. 18–19.

Suzuki, K., and Y. Tsuchihashi. "Personal Identification by Means of Lip Prints." *Journal of Forensic Medicine*, 1970, pp. 52–57.

Tsuchihashi, Y. "Studies on Personal Identification by Means of Lip Prints." *Forensic Science*, 1974, pp. 233–248.

List, John. *See* Forensic Anthropology.

Lividity (Hypostasis). *See* Time of Death.

Locard, Edmond (1877–1966)

Edmond Locard was a French criminologist, lecturer, and author who articulated the theory that is the basis of forensic science. The Locard principle postulated that every criminal could be connected to the crime scene as a result of contact with particles carried from the site. This trace evidence could provide valuable clues if it was interpreted correctly.

Locard remarked that as a youth he was fascinated by the Sherlock Holmes mysteries, and this led to his interest in scientific criminology. He received his education at the University of Lyon, where he graduated with both medical and law degrees. He had the good fortune to study with two giants of forensic science, Alphonse Bertillon, whose system of body measurements provided an initial method of criminal identification, and Alexandre Lacassagne, a professor of pathology specializing in forensic medicine. Upon graduation, Locard worked with his mentor, Lacassagne, until 1910. He resigned that post to establish the Institute of Criminalistics for the Rhone Prefecture, a branch of the University of Lyon.

Locard's ideas were established by several successful and well-publicized investigations. On one occasion, Locard was confronted by widespread coin counterfeiting and a suspect who adamantly denied all knowledge of the crime. Locard obtained the clothes the man had been wearing at the time of his arrest and searched them for dust. He uncovered small particles of earth containing minute specks of metal. His Frenchman's analysis revealed that the coins were made of the same metallic substance as the traces found in the accused's garments. Confronted with this new evidence, the counterfeiter confessed, which led to the capture of his whole gang.

Another Locard inquiry not only established the international reputation of the Lyon lab, but was also humorous. The police were baffled by a series of jewel heists from wealthy homes. In each instance, the theft took place in broad daylight with no signs of forced entry. The burglar appeared to have entered via an upper-story window. Additionally, the thief took only one piece of jewelry, despite access to a large amount of other baubles. Detectives attributed the robberies to local boys but could find no proof. As a last resort, the police called in Locard.

As a proponent of fingerprints, Locard asked to see all the available pictures of the victims' windowsills. As he studied the evidence, he noticed that the recovered prints resembled human prints but were definitely not human. Locard's hypothesis was that the thief was a monkey. The already-skeptical authorities were incredulous and wondered how to arrest a monkey. Locard had a solution. He directed each local organ-grinder to appear at headquarters with their monkeys in three days. The monkeys were fingerprinted, and Locard did an on-site analysis, comparing the evidence from the robberies with the new prints. Before long, the mastermind was arrested and jailed, while his accomplice was banished to the zoo.

In 1912, Locard again achieved distinction for his work on the Gourbin inquiry. Emil Gourbin, a bank clerk in Lyon, was accused of the murder of a lady friend. It was well known that the bank clerk was a very jealous man and that his lady friend, Marie Latelle, enjoyed flirting with others to raise Gourbin's ire. When Marie Latelle was discovered dead, Gourbin became the chief suspect. Gourbin, however, appeared to have an unbreakable alibi. Forensic evidence placed Latelle's death at approximately midnight, the same time that Gourbin was playing cards and drinking with a number of friends. Locard's examination determined that the woman had died of manual strangulation. The scientist took scrapings from under the accused's fingernails and under a microscope noted a strange pink dust. The fine debris was face powder—makeup. Police, upon searching the victim's belongings, uncovered the identical makeup. Gourbin confessed and admitted that he had created his alibi by advancing the wall clocks from 11:30 to 1:00, breaking up the card-playing evening early.

Like many of the forensic pioneers, Locard became an expert in a number of fields, including handwriting, forgery, and dental comparisons. He even made his own eccentric contribution to fingerprinting through poroscopy, an examination of the sweat-pore patterns in fingerprints. *See also* Bertillon, Alphonse; Lacassagne, Alexandre.

References

Block, Eugene B. *Science vs. Crime: The Evolution of the Police Lab*. San Francisco: Cragmont, 1979.

Cyriax, Oliver. *Crime: An Encyclopedia*. North Pomfret, VT: Trafalgar Square, 1996.

Lombroso, Cesare (1835–1909)

Italian physician, pioneer criminologist, and inventor of the world's first polygraph, Cesare Lombroso caused a stir with the publication of his book *Criminal Man* (1876), which formulated his now-outmoded theory of the "atavistic" or born criminal. He is sometimes referred to as the father of criminology in recognition of his efforts to study crime scientifically.

Lombroso was convinced that criminal deviance was the result of inherited biological traits. He believed that criminals were throwbacks, less evolved members of a more dangerous, primitive segment of human society. Such "atavistic" hoodlums could be readily identified by a set of physical attributes he called stigmata. These included, but were not limited to, such apelike traits as a low forehead, bushy eyebrows, and excessively long arms. Other peculiarities sure to indicate degeneracy included prominent jaws and cheekbones, hands that were exaggeratedly small or large, fleshy lips, and ears of unusual size. Tattooing was also indicative of primitive instincts as far as Lombroso was concerned. He even went so far as to postulate that different kinds of deviance could be associated with particular patterns of stigmata.

In 1893 Lombroso wrote *The Female Offender*, in which he asserted that women were not generally criminals, and those that were were only occasional lawbreakers. There were atavistic female criminals, but they were not as easily distinguishable in the general population as men. He did add that these women could be particularly vicious. Responding to critics, Lombroso conceded that societal and psychological factors could be important in exacerbating criminal tendencies, but he never completely abandoned his original theories.

Lombroso received his medical degree from the University of Parma in 1858, followed by a surgical specialty from the University of Genoa. He held a variety of positions, but his major contributions came while he was at the University of Turin, where, at various times, he was professor of legal medicine, psychiatry, and criminal anthropology. His major work, *Criminal Man*, went through five editions and was translated into a number of European languages, but never a complete English edition.

Lombroso's view of the born criminal resulted from a combination of his extensive anatomical research and the celebrated Darwinian theory of natural selection. Beginning in 1870, the Italian academic systematically studied a number of criminal skulls and over 5,000 individual felons to identify specific similar characteristics. He identified two groups. Born criminals, which he estimated made up 40 percent of the deviant population, were biologically doomed to degeneracy. A second clique, less biologically determined and not as physically distinguishable as major felons, he called "criminaloids." The crimes of these offenders were the result of external factors. So-called crimes of passion were the result of criminaloid actions. So popular were Lombroso's theories about deviant physical characteristics that Arthur

Conan Doyle's villains are straight from the Lombroso mold. The theory of the hereditary criminal type is no longer accepted; the emphasis today is overwhelmingly on environmental factors.

The man some refer to as the father of criminology should also be remembered for his call for more humane treatment of convicts. The criminal population was immoral because they were incapable of restraining their antisocial inclinations due to biological determinism. Similarly, Lombroso did not believe in the wholesale employment of the death penalty. He felt that it should be used only in the most extreme cases. He believed that it was not enough to segregate criminals. The government should use the skills and abilities of the offenders in some way. He advocated that victims of crime be compensated from the proceeds of work done by the prison population.

Reference

Cyriax, Oliver. *Crime: An Encyclopedia*. North Pomfret, VT: Trafalgar Square, 1996.

Luetgert, Adolphe L.: Case (1897)

Adolphe Luetgert was a Chicago sausage maker who was also a wife beater and womanizer. His attempt to destroy evidence after murdering his wife was thwarted by a determined investigation involving forensic science. Analysis of a ghastly mixture discovered in Luetgert's factory and fragments of remains were all that tied the would-be "Sausage King" to the murder of his wife. The evidence was enough to convince a jury of his guilt, and the gruesome disposal of his wife's remains shocked Chicago and the nation.

Luetgert immigrated from Germany in the 1870s and built a successful sausage factory on the Windy City's North Side. The businessman was almost as well known for his immense sexual appetite as he was for his sausage. So serious was Luetgert in his pursuit of extramarital conquests that he had a bed moved into his office at the plant.

The sausage king's second wife, Louisa, was not unaware of her husband's dalliances. The marriage was strained, and the two quarreled frequently. On one occasion, neighbors had to intervene when Luetgert grabbed his wife by the throat and nearly strangled her. During another marital confrontation, the burly German reportedly chased his wife down the street brandishing a revolver.

In May 1897, after Luetgert's sons realized that their mother had been missing for several days, they began asking neighbors if they had seen her. When Louisa Luetgert's brother, Dietrich Bicknese, stopped to visit and learned that his sister was missing, he questioned Luetgert, who told him that Louisa Luetgert had run off with a lover. Pleading personal shame and embarrassment and concern that a scandal might damage his business, Luetgert reluctantly revealed that he had not contacted the police, but he assured his brother-in-law that he had two private detectives working on the case. A reluctant Dietrich left Chicago and returned home to await further word from Luetgert. When none was forthcoming, the suspicious relative returned to Chicago and reported his concern to the police.

When Leutgert was approached by the police, he told them that his wife had walked out with $18 in her handbag following an argument. After talking to neighbors and relatives, police became suspicious and questioned workers at the sausage factory. The initial investigation disclosed that Louisa Luetgert had last been seen walking in an alley behind the factory with her husband. One employee, Frank Odorowsky, reported that a few days before Louisa Leutgert disappeared, he and his boss had filled a vat with a solution containing more than 300 pounds of potash. The strange brew simmered for a week, and Luetgert told curious employees that he was making soap for a factory cleanup. About a week later, on the evening Louisa Leutgert disappeared, the night watchman was told to stoke the fire under the vat of simmering potash solution. He was then told that he could have the rest of the night off. On the following day, both Odorowsky and the watchman noticed that a red-brown slime had clogged the drain.

Comments about sludge and slime sparked the investigators' curiosity. When Luetgert was questioned about the potash by Captain Herman Schuettler, who conducted the investigation, he said that because he needed a lot of soap to clean the factory, he had prepared a vatful of it. Sure enough, the vat contained a soft brown soap, but around the vat investigators found strands of hair, a false tooth, a piece of leather, and a hairpin. Among the ashes in a smokehouse behind the factory, they found pieces of charred bone and a steel corset stay. The evidence suggested that Luetgert had tried to burn whatever remained of his wife after boiling her in the vat. When Schuettler had the vat emptied through improvised gunnysack filters, he found two gold rings, one with the letters "LL" engraved on it, and several bone fragments. Chemists told police that a concentrated solution of potash would dissolve a body within hours. Just to be sure, police threw the body of an unknown corpse into a vat filled with potash solution. After they saw a similar brown soap form several hours later, Luetgert was arrested and charged with murder.

His trial began on August 30, 1897. It was packed with spectators who had incorrectly heard that Luetgert had ground his wife into sausage. They were disappointed when the trial ended with a deadlocked jury. The defense had produced a witness who claimed to have seen Louisa Leutgert at a railroad station well outside Chicago. At the second trial, although Luetgert insisted that his wife had run off and that she would return, the prosecution presented a devastating array of circumstantial evidence. Luetgert insisted that bone fragments were common in making sausage and that those presented into evidence were from animals. The prosecution called forensic investigator Dr. George Dorsey to the stand. Jurists were impressed by the testimony of Dorsey, a physical anthropologist at Chicago's Field Museum of Nat-

ural History. It was probably the first time an anthropologist served as a witness and forensic expert in a criminal case. Dorsey said that the bone fragments were definitely human, and that one of the small bones that was similar to a sesame seed was, as its shape implied, a sesamoid bone found in the tendon of a human's big toe. He was emphatic when he stated that the few ounces of bone were from a human female. Today his bold assertion might not be accepted in court, but at the time it was a stunning refutation.

Luetgert never addressed the discovery of the rings. His wife's family testified that they were Louisa Leutgert's rings and further stipulated that due to swollen arthritic fingers she could not remove the jewelry. The defendant's claim that the potash mixture was for cleaning purposes was easily dismissed when the prosecution revealed that a large number of detergents were available in factory storage. Luetgert's compulsive womanizing also caught up with him at the trial. His mistresses delivered deadly testimony by swearing that Luetgert had said that he would crush his wife. One of them said that he had asked her to hold a bloody knife for him, though no such weapon was ever produced. Portrayed as a vicious beast who flaunted his numerous affairs and beat his wife, Luetgert could only protest his innocence. Convinced that the bones were human, the jury found Luetgert guilty. He was remanded to Joliet prison to serve a life sentence. He died without ever saying anything further about the death of his wife. *See also* Forensic Anthropology; Webster, John: Murder.

References

Cyriax, Oliver. *Crime: An Encyclopedia.* North Pomfret, VT: Trafalgar Square, 1996.

Duke, Thomas. *Celebrated Criminal Cases of America.* Reprint. Montclair, NJ: Patterson Smith, 1991.

Evans, Colin. *The Casebook of Forensic Detection: How Science Solved 100 of the World's Most Baffling Crimes.* New York: Wiley, 1996.

Thomas, Peggy. *Talking Bones: The Science of Fo-

rensic Anthropology. New York: Facts on File, 1995.

Lung Flotation

Lung flotation involves determining whether or not a dead baby's lungs will float. It can be used to answer the question, Was the baby born dead (stillborn), or did the mother smother her unwanted child? To answer this question, a medical examiner may request that the child's lungs be tested to see if they float in water.

Prior to birth, the lungs of a fetus contain no air. If a child is stillborn, it will never take any air into its lungs, and the lungs will not float. If respirations were made after birth, the lungs will never collapse completely. When the thorax is opened during an autopsy, the lungs deflate, driving out most of the air. However, before the alveoli (the tiny sacs in the lungs where gases are exchanged between blood and air) are completely emptied, the small bronchi leading to the alveoli collapse, trapping what is called the minimal air. This small amount of air is enough to cause the lungs to float, indicating that the baby did inhale air and, consequently, was alive at birth. *See also* Carbon Monoxide; Cause of Death; Coroner; Crime Scene; Drowning; Infanticide; Neutron-Activation Analysis; Pathology.

M

Mad Bomber, The (1940–1957)

For 16 years, the Mad Bomber terrorized the citizens of New York City by planting a number of homemade bombs in very public places, including Radio City Music Hall, Grand Central Station, and a variety of theaters in the metropolitan area. The conclusion of the case represented one of the first successful uses of psychological profiling. Although there were no fatalities during the course of the bomber's rampage, eight people were seriously injured.

The initial incident occurred on November 16, 1940, when a crudely constructed bomb was discovered on a windowsill at the West 64th Street Consolidated Edison plant. The accompanying handwritten note, printed neatly in capital letters, read, "CON EDISON CROOKS—THIS IS FOR YOU." Police dismissed the dud bomb as a onetime prank. The appearance of a second unexploded bomb in September 1941 and another note with similar sentiments caught the city's attention. Apparently the bomber's patriotic convictions as the country entered World War II got the best of him. The culprit announced that officials could expect no problems from him for the duration of the war, but Con Edison would pay later. The note was signed "F.P." During the war years, "F.P." periodically wrote threatening letters to newspapers, hotels, department stores, and Con Edison, but there were no bombs.

On March 29, 1950, the terror campaign reopened with the discovery of a bomb at Grand Central Station. A fourth bomb, in a phone booth outside the New York Public Library, announced the bomber's new proficiency to the public when it exploded on April 24, 1950. Authorities were distressed by the improved construction, power, and size of the new bombs. Between 1951 and 1956, 14 bombs drove the city to near hysteria. On December 2, 1956, the explosion of the biggest bomb in Brooklyn's Paramount Theater injured 15 citizens, 6 seriously. The public was panicking. Attendance at movie theaters dropped significantly as police began an all out manhunt for the Mad Bomber.

Stymied by lack of information and clues, the police were willing to try any means to end the bombing. Two major breaks in late 1956 and early 1957 paved the way for a resolution. Inspector Howard E. Finney, who held a degree in forensic psychiatry, called upon Dr. James Brussel, the New York State assistant commissioner of mental health, for help. Brussel was considered by many to be the father of psychological detection. Dubbed by the press the "Sherlock Holmes of the couch" or a "psychiatric seer," Brussel was given complete access to police evidence, including notes, pictures of unexploded devices, and letters. The resulting psychological profile developed by Brussel was astounding in its specificity and ultimate accuracy.

Brussel began with the understanding that on average, bombers are most frequently men. He believed that the bombings represented a vendetta against Con Edison and that the suspect was probably paranoid. Statistically, paranoia manifests itself in middle age. Since the bombings began in 1940, it seemed logical that the suspect was between 40 and 50 years old. Middle-aged paranoids usually lived either by themselves or with elderly relatives. Brussel further inferred that "F.P." was not a native New Yorker because he used "Con Edison" rather than the popular Con Ed. Based on the syntax of the bomber's letters and notes, Brussel predicted that he was foreign, probably Slavic, given a penchant for revenge, grudges, and the use of bombs. Most of the letters were mailed from Westchester County, so it seemed reasonable to believe that the felon mailed his notes while passing through the area because he would not be so foolish as to mail them from his hometown. This pointed to the fact that he was most likely from Southern Connecticut, perhaps Bridgeport, where there was a high concentration of Polish Americans. The letters and notes alluded to the fact that the bomber had been ill for some time, and Brussel speculated that he might suffer from chronic heart disease or a related ailment. Almost as a throwaway comment, Brussel concluded that the paranoid suspect was middle-aged, single, obsessively neat, intelligent, and well proportioned in build. He even went so far as to predict that the bomber would probably be wearing a neat double-breasted suit when he was finally apprehended. Brussel thought that the motive for the bombings was either a reprimand or a discharge from the company. His prediction of a well-proportioned build came from research at the time that had found that 85 percent of paranoiacs were of well-proportioned build. Brussel's belief that the bomber was single and lived with older female relatives came from the formation of

letters. His *W*'s were made from two rounded *U*'s that looked like a woman's breasts, which implied a sexual problem. The letters, which were very neat and meticulous, led Brussels to the prediction of neatness and the double-breasted suit.

As significant as Brussel's profile was, it was the bomber himself who offered the final clue to his identity. In December 1956, a New York newspaper, the *Journal-American*, printed a letter to the bomber asking him to give himself up and offering to print his grievances. In a series of three letters to the *Journal-American*, the bomber said that he would not give himself up, but revealed his motive for the onslaught and more about himself. Aside from railing at Con Edison, the author stated that he had been injured and ultimately debilitated for life while working for the utility. When he had sought legal redress, he had discovered that due to the two-year New York State statute of limitations, there was nothing that could be done. Obligingly for the police, "F.P." added the date of his accident, September 5, 1931. An intensive search of old files began, and an office assistant, Alice Kelly, found a letter from an unhappy former employee in an old file. The letter was from George Metesky, who had been knocked down by a boiler explosion while working at Con Edison on September 5, 1931. Metesky had complained of persistent headaches after the boiler accident. Follow-up medical exams had found nothing wrong with the complainant. He had been paid 12 months' insurance and fired. Metesky had later tried to sue Con Edison but had been told that it was too late.

Brussel's psychological profile proved to be uncanny. The disgruntled ex-worker was taken into custody at the Waterbury, Connecticut, home of his two older sisters, with whom he lived. The 54-year-old bachelor freely admitted his crimes and revealed that "F.P." stood for "Fair Play." He asked only that he be allowed to get his coat before being taken to the police station. It was a double-breasted pinstripe suit coat. Brussel's only error was his prediction that the bomber suffered from a chronic heart disease. Me-

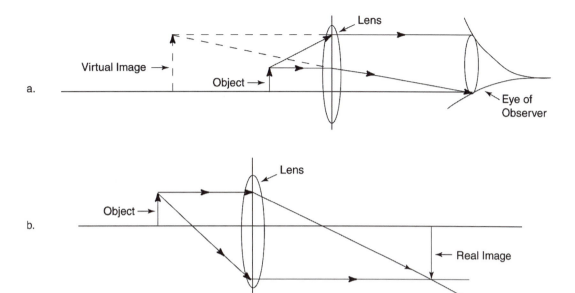

(a) An object placed less than one focal length from a convex lens produces a virtual image when viewed through a lens by an observer. Notice that the rays entering the observer's eye seem to come from distances greater than the object and cause the image to appear larger than the object. (b) An object placed outside the focal length of a convex lens produces a real image; that is, the rays really do come together to form an image that can be "captured" on a screen.

tesky actually had tuberculosis, which he believed was caused by the workplace accident.

Facing assorted charges that could have led to sentences totaling more than a thousand years, Metesky was found criminally insane. In April 1957, he was remanded to a New York State mental facility. On December 12, 1973, after sixteen years in a number of hospitals, he was considered cured and was released to the custody of his sisters. All charges were dropped. *See also* Forensic Psychiatry; Psychological Profiling.

References

Cyriax, Oliver. *Crime: An Encyclopedia*. North Pomfret, VT: Trafalgar Square, 1996.

Evans, Colin. *The Casebook of Forensic Detection: How Science Solved 100 of the World's Most Baffling Crimes*. New York: Wiley, 1996.

Lane, Brian. *The Encyclopedia of Forensic Science*. London: Headline, 1992.

Nash, Jay Robert. *Encyclopedia of World Crime*. Wilmette, IL: Crime Books, Inc. 1990, vol. 3, p. 2172.

Wilson, Colin. *Clues! A History of Forensic Detection*. New York: Warner Books, 1991.

Magnifiers

Magnifiers—microscopes and magnifying glasses—remain the forensic scientist's most commonly used tools. Microscopes are used to provide greater magnification in examining trace evidence. With a compound microscope—two lenses mounted at either end of an adjustable tube—forensic scientists can examine hair, fibers, wood fibers in paper, and other small objects that may provide valuable clues to a crime. The magnifying glass, which is simply a convex lens in a handle, is used to search for small bits of evidence at a crime scene or on articles collected at such a site.

The first microscopes, such as those used by Antonie van Leeuwenhoek (1632–1723), consisted of a tiny single convex lens. Because lenses with short focal lengths provide the greatest magnification, van Leeuwenhoek's lenses were nearly spherical and often no bigger than the head of a pin. With them, he could produce images nearly 200 times the size of the object being examined.

A convex lens has a focal point, the point where parallel light rays, such as those from a very distant object like the sun, are brought together. The distance from the center of the lens to its focal point is called the focal length of the lens. A lens held close (less than one focal length) to the object being examined produces a magnified but virtual image, as shown in Figure 3a. The image is called virtual because it can be seen only by looking through the lens; it cannot be viewed directly or "captured" on a screen, as a real image can. Light rays from points on the object are refracted (bent) by the lens and appear to originate from points farther apart than the same points on the object, as shown by the dotted lines.

The first figure shows how a real image is produced when an object, such as a photographic slide, is placed at a point more than one focal length from a convex lens and illuminated. Notice that the image is inverted relative to the object.

A detective's hand lens is larger and magnifies much less than did van Leeuwenhoek's. However, any material deemed worthy of greater magnification can be examined with a compound microscope, which can produce magnifications of 1,000 times or more. The compound microscope, invented by Robert Hooke (1635–1703), consists of two lenses mounted at opposite ends of a hollow cylinder, as shown in the second figure. The convex lens at the lower end of the cylinder—the objective lens—is placed close to the object being examined and is adjusted until it is slightly more than one focal length from the object being viewed. When it is illuminated, it produces a magnified, inverted real image of the object. This real image is then viewed through the upper or eyepiece lens. The length of the tube is carefully made so that the distance between the lenses is such that the real image formed by the objective lens lies slightly less than one focal length from the eyepiece lens. Since the real image lies inside the focal point of the eyepiece lens,

the second image, which the viewer sees through the eyepiece, will be a greatly enlarged virtual image of the real image. If the objective lens provides a magnification of 10, and the eyepiece lens does the same, the viewer will see an image 100 times larger than the object. If the eyepiece lens is changed to one that provides a magnification of 40, the viewer will see an image that is 400 times the size of the object.

Magnetoresistive Microscope

In 1996, scientists at the University of California at San Diego devised an instrument known as a magnetoresistive microscope that allows a viewer to see the tiny bumps and dips caused by changing magnetic fields that appear on a variety of magnetic media such as floppy discs and audiotapes. The instrument moves a piece of tape repeatedly back and forth under a computer's hard-drive read/write head where the direction and the strength of the magnetic field on the tape are determined at millions of points. A computer can then convert the data into sound. In 2000, scientists at the National Institute of Standards and Technology (NIST) modified the device for purposes of forensic science. They believe that it may be possible to use the magnetoresistive microscope to recover data erased from a tape, such as the information missing from the 18.5-minute gap in one of the Watergate tapes.

Stereoscopic Microscope

A stereoscopic microscope, as its name implies, provides the forensic scientist with a three-dimensional view of the object being examined. Since one eye sees the object magnified from a slightly different angle than the other, a three-dimensional view is made possible. The stereoscopic microscope is the most commonly used magnifying device in a crime laboratory.

A stereoscopic microscope is a combination of two compound microscopes, one for each eye. It has two eyepiece lenses connected to a single objective lens so that the viewer uses both eyes to examine the magnified image. This type of microscope also has prisms that invert the image, providing

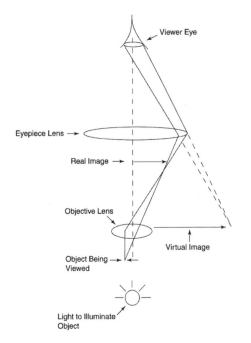

A compound microscope has two lenses. The magnification is the product of the magnification of each lens.

viewers with right-side-up images. The images in an ordinary microscope are inverted so that inversion must be taken into account when one is moving the slide or object on the stage to look at another part of the field. With the stereoscopic microscope, the user can manipulate, dissect, or probe the magnified object more easily.

The major disadvantage of a stereoscopic microscope is that in order to provide a large depth of focus and a wide field of view, the magnification is considerably less. Most of these instruments provide a magnification of only 10 to 125 times. Because the distance between the objective lens and the specimen under observation is large, it is the preferred instrument to use for examining material that may need to be moved and turned while it is being viewed. It is ideal for identifying trace evidence that may be mixed with other matter collected by a vacuum at a crime scene. The instrument is also useful in seeing trace evidence such as hairs, fibers, stains, and

marks that may be found on clothing, knives, tools, and human remains.

Cases

A microscope played a vital role in solving the shooting death of George Marsh, a wealthy retired soap manufacturer who was last seen alive on April 11, 1912. His body was found the next day on the salt flats near Lynn, Massachusetts. The only significant evidence was a gray overcoat button with an attached piece of cloth.

Newspaper requests for information about the crime from the public led to a report that a light blue automobile had been seen near the Marsh home on April 11. Upon learning of the light blue car, a woman who operated a boardinghouse told police that she had rented a room to a young man driving a car of that color. She also reported that the man, who had signed in as Willis A. Dow, spent a considerable amount of time peering through binoculars from a window that overlooked the back of George Marsh's home. Still another landlady said that she had rented a room to a Willis A. Dow who had left his buttonless overcoat in his room.

At the Lowell Textile School laboratory, Edward H. Baker and Louis A. Olney examined the button that had been found near Marsh's body under a microscope. They found that the cloth attached to the button matched the cloth in the overcoat that Dow had left in his room.

Later, a trawling fisherman told police of his discovery of a .32 pistol in the water near the place where Marsh's body had been discovered. The gun's serial numbers enabled police to trace the weapon to a William A. Dorr in Stockton, California. Dorr's aunt was a relative of George Marsh, and she, taken by the amorous attention she had received from her nephew, had made Dorr the sole beneficiary of her trust. The trust, police soon learned, was administered by George Marsh, who provided the aunt with only enough money to meet her living expenses. As long as Marsh was alive, Dorr could not help his aunt spend any of the accumulated capital.

With his aunt's approval, Dorr traveled to Massachusetts to confront Marsh and per-

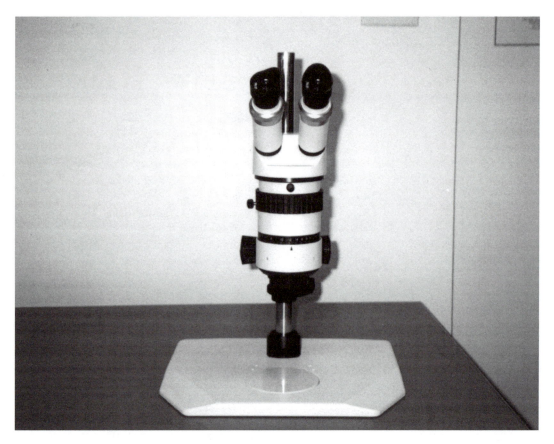

Stereoscopic microscope. Courtesy of David Exline, ChemIcon Inc., Pittsburgh, PA.

suade him to release his control over the trust. In a letter to his aunt, he told her of Marsh's demise. Aware of the relationship among Dorr, his aunt, and Marsh, police tapped her phone. When Dorr called from nearby, police traced the call and arrested him as he chatted with his aunt. He confessed to the murder but claimed that it had been done in self-defense. A jury disagreed, and he was executed on March 24, 1914.

In the case of the 1986 woodchipper murder, in which Richard Crafts was convicted of killing his wife, Helle, and then sending her frozen body through a woodchipper, microscopic examination of hair evidence played a role in solving the crime. Hairs found on Helle Crafts's hairbrush were very similar to those found along the bank of the river where Crafts was believed to have disposed of his wife's body. Harold Deadman, an expert in hair and fiber analysis, found that hairs taken from the river bank had the same characteristics as hair from Helle Crafts's hairbrush when they were examined microscopically. This alone would not prove that both hair samples were the same. However, the statistics turned against Richard Crafts when Deadman stated that hair taken from both places had very unusual characteristics, including a rare protrusion from the shaft of the hair known as "shouldering." *See also* Ballistics; Comparison Microscope; Electron Microscope; Spectrophotometry.

References

Aiken, C.G.G., and D.A. Stoney, eds. *The Use of Statistics in Forensic Science*. New York: Ellis Horwood, 1991.

Gaudette, B.D. "A Supplementary Discussion of Probabilities and Human Hair Comparisons."

Journal of Forensic Sciences, 27(2), 279–289, 1982.

Saferstein, Richard. *Criminalistics: An Introduction to Forensic Science.* Upper Saddle River, NJ: Prentice Hall, 1998.

Weiss, P. "Magnifier May Crack Crimes, Crashes." *Science News.* July 8, 2000, p. 20.

Magrath, George Burgess (1870–1938)

George Burgess Magrath was an American medical examiner in Boston who sought attention in an effort to educate the public about the value of forensic science. One way he attracted attention was to arrive at a crime scene in a vehicle with its siren screaming. To further attract attention, he wore colorful clothing and flaunted his flaming red hair, which he replaced with a wig as he grew older and more bald.

Boston was ahead of New York City in switching from the coroner to the medical-examiner system. New York City did not make the switch until 1918. Medical examiners, such as Magrath, because they were performing autopsies on all unnatural deaths, were uncovering crimes that would have gone undetected. Such autopsies had not been performed by coroners.

In 1916, while her husband, Frederick Small, was away, Mrs. Florence Small was found dead in her home, which had been destroyed by fire. Her body had fallen through cheap floorboards in her bedroom into several inches of water in the basement. The water prevented her body from burning completely and thus preserved evidence suggesting foul play. Magrath performed an autopsy and discovered that Florence Small had been murdered prior to the fire. The contents of a midday meal in the victim's stomach were undigested, indicating that she had died prior to the fire that erupted in the evening. Also, she had a cord around her neck and a .32-caliber bullet in her skull, and it was clear from bruises that she had been badly beaten. An unusual resin on the body explained the fire's intense heat. The resin was slag from a welding compound that burns at more than 5,000 degrees Fahrenheit.

Other investigators found an alarm clock with wires and spark plugs attached as well as a .32-caliber pistol that belonged to Frederick Small. They learned, too, that kerosene had been delivered to Frederick Small shortly before he left home on the day of the fire. Witnesses revealed that Small was an inventor with sufficient knowledge to build an explosive that could be ignited with a timing device. As a result of Magrath's careful autopsy and the investigation that followed, Small was found guilty of murdering his wife.

Magrath was also the medical examiner in the famous case in 1920 of Nicola Sacco and Bartolomeo Vanzetti, two men who were executed for gunning down a paymaster and a guard as they were delivering a large sum of money. It was Magrath who extracted the bullets from the victims. *See also* Arson; Ballistics; Comparison Microscope.

References

Evans, Colin. *The Casebook of Forensic Detection: How Science Solved 100 of the World's Most Baffling Crimes.* New York: Wiley, 1996.

Thorwald, Jürgen. *The Century of the Detective.* New York: Harcourt, Brace & World, 1965.

Male Rape. *See* Rape.

Mann Act. *See* Chaplin, Charlie: Paternity Case.

Manslaughter. *See* Homicide.

Mant, Keith (1919–2000)

Keith Mant was a British forensic pathologist whose investigations provided evidence for those prosecuting Nazi war criminals after World War II. In November 1945, Mant was appointed officer in charge of the pathology section of the war-crimes investigating team in northwestern Europe. In 1946, he became head of the Special Medical Section of the War Crimes Group. He exhumed the bodies of Allied airmen who had been buried in marked and unmarked graves by German military units. Most of the airmen had died when their planes crashed; however, he found 49 who were victims of *Genickschuss*, which literally means "neck shot." These men had died as the result of being shot by a gun placed at the base of the back of the skull. This method of execution was commonly used by the German military during the war.

After 1946, Mant investigated concentration camps. At Ravensbrück, where women and children were held captive, he discovered, through interviews and physical evidence obtained from exhumed bodies, that children had been shot in their legs and examined for the onset of gangrene so that different methods of treatment could be studied.

In 1948, Mant joined Keith Simpson at Guy's Hospital in London, where he became a leading forensic pathologist as chief of the department of forensic medicine at the hospital's medical school. When Simpson retired in 1972, Mant became head of the hospital's pathology department. Mant was involved in many cases where forensic science played a role. In 1973, he proved that a man at a building site had died not of a heart attack, as had been thought, but from a stray bullet fired at an army base nearly two miles away. In 1977, his analysis of evidence revealed that a woman who had swallowed an overdose of aspirin before crawling into a freezer had died of hypothermia before the excess aspirin had taken effect.

Mant often traveled to Richmond, Virginia, to visit friends in the chief medical examiner's office. One of his friends was Patricia Cornwell, who frequently sought his advice on plots for stories she was writing. He died on October 11, 2000. *See also* Camps, Francis; Cornwell, Patricia; Medical Examiner; Pathology.

Reference

Martin, Douglas. "Keith Mant Dies at 81; Pathologist Helped Convict Nazis." *New York Times*, Nov. 26, 2000.

Markov, Georgi (c. 1929–1978)

Georgi Markov, a Bulgarian who defected during the cold-war conflict between the United States, the Soviet Union, and its satellite nations, was the victim of one of the most bizarre poisonings in the annals of forensic science. Markov fled to England and was employed by the British Broadcasting Corporation (BBC) and Radio Free Europe. A noted commentator and writer in his homeland, Markov had stung the Sofia regime by a series of embarrassing revelations of political and personal corruption. Cold warriors were also abuzz about Markov's memoirs, which were a work in progress that promised even more scandalous insights.

On September 7, 1978, on his way to work at the BBC, Markov passed a crowded bus stop near London's Waterloo Bridge. The radio personality was jostled by a heavyset man and felt a sting in his right thigh. After a grunted apology, the man hopped into a cab and sped off. Markov continued on his way, but the puncture got larger, the leg stiffened, and he complained of pain. At home that evening, Markov spiked a high fever of about 104 degrees Fahrenheit and began to vomit. The next morning, he was admitted to St. James Hospital, where doctors diagnosed the disorder as flu. A dramatic rise in the patient's white-cell count from 10,600 at admission to 26,300 within a 24-hour period convinced physicians that this was not the flu, but they could not offer any other diagnosis. The sick man was not nearly as confused as the medical staff. To anyone who would listen, Markov loudly proclaimed that he had been poisoned by an umbrella-toting Communist agent. Three days later, Markov was dead, and for lack of a better explanation, poison was listed as the cause of death. Scotland Yard began an investigation.

During a routine postmortem examination, doctors noticed the reddened area around a puncture wound Markov had insisted was the entry point of the poison. The wound was excised and sent to Scotland Yard's lab for analyses. A perplexed scientist removed a microscopic pellet, about the size of a pinhead, from the wound. Examination under an electron microscope revealed that the sphere was made of a platinum-iridium alloy, a material that would not be rejected by the body. Further analysis showed that two small holes, about 0.4 mm in size, had been bored into the pellet. Even more curious was that the sphere was covered with sugar.

Scotland Yard scientists concluded that the pellet must have held a sophisticated biotoxin, one they were unfamiliar with, and that the sugar coating had held the poison in the sphere and had dissolved once it was in the body, releasing the deadly poison. The scientists estimated that only half a milligram had been necessary to kill Markov. Operating under this hypothesis, scientists were able to eliminate a number of poisons due to their lack of potency in such small amounts. After a close review of the symptoms, they settled on ricin, a derivative of the castor-oil plant, as the poison. The plant grew in abundance in Bulgaria. When scientists injected a pig with comparable amounts of the poison, the pig died within 24 hours.

Once the scientific evidence was made public, another Bulgarian, Vladimir Kostov, came forward with a similar account of an assault on him in Paris. Although he became very ill, Kostov survived. Scotland Yard surmised that the assassins had failed to use enough poison.

The Markov case was never definitively solved. In February 1992, the former head of Bulgarian intelligence, General Stoyan Savor, died under mysterious circumstances as he awaited an inquiry into the Markov murder. No Markov files were ever found in the Bulgarian state archives. However, following the downfall of Communism in Bulgaria, the new government revealed that Markov had been assassinated. *See also* Poisons.

References

Evans, Colin. *The Casebook of Forensic Detection: How Science Solved 100 of the World's Most Baffling Crimes.* New York: Wiley, 1996.
———. "The Poisonous Umbrella." *Time,* Oct. 16, 1978.
Willey, Fay. "Death by Umbrella." *Newsweek,* Sept. 25, 1978.

Marsh Test. *See* Arsenic Poisoning.

Mason, Perry

Perry Mason is a fictional investigating lawyer in a series of mystery novels by Erle Stanley Gardner. Mason is the unbeatable lawyer who breaks down seemingly indisputable cases and does not hesitate to use extralegal methods to secure evidence in preparing for court. In the first Perry Mason novel, *The Case of the Velvet Claws,* Mason could be accused of assault and battery; in *The Case of the Fiery Fingers,* he resorts to illegal wiretapping. Mason actually tampers with evidence in *The Case of the Curious Bride* by changing the sounds made by a doorbell. He then has a friend testify that the original sounds made by the doorbell are those to which Mason has changed them. In an effort to help people in trouble, Mason is not afraid to defy police and authority.

Perry Mason is helped by his loyal secretary, Della Street, and a private investigator named Paul Drake. Other characters in the series are Hamilton Burger, the district attorney who always loses to Mason in court, and police lieutenant Arthur Tragg, who is the arresting officer in many cases.

People who knew him said that Gardner, a successful courtroom lawyer for over 20 years, was very similar to the character Perry Mason that he created. Gardner was a dramatic and innovative defense lawyer who used legal tricks and forgotten statutes to defend his clients. Gardner wrote about his character, Perry Mason, "The character I am trying to create for him is that of a fighter who is possessed of infinite patience. He tries to jockey his enemies into a position where he can deliver one good knockout punch" (Hughes, 1978, p. 103). Before Gardner started writing the Perry Mason stories, he was a pulp-fiction writer, and this is evident in his early Perry Mason plots. The character matures as Gardner's style of writing changes. Early Perry Mason plots are full of fast-paced action, while later stories show a more subdued, calculating courtroom attorney. Book critic Francis Nevins wrote that Perry Mason "is a tiger in the social-Darwinian jungle, totally self-reliant, asking no favors, despising the weaklings who want society to care for them, willing to take any risk for a client no matter how unfairly the client plays the game with him" (Reilly, 1991, p. 358).

Six movies were made based on the Perry Mason stories. A radio show ran for 12 years, and a popular television program featured actor Raymond Burr as Perry Mason. In addition, several made-for-television movies were adapted from the Perry Mason books. *See also* Gardner, Erle Stanley.

References

Hughes, Dorothy B. *Erle Stanley Gardner: The Case of the Real Perry Mason.* New York: William Morrow, 1978, p. 103.

Nolan, William F. Erle Stanley Gardner (1889–1970). http://erlestanleygardner.com/nolan.htm. Accessed Aug. 21, 2001.

Reilly, John M., ed. *Twentieth Century Crime and Mystery Writers,* 2nd ed. New York: St. Martin's Press, 1991.

Senate, Richard. *A Brief Biography of Erle Stanley Gardner.* http://erlestanleygardner.com/senate.htm.

Steinbrunner, Chris, and Otto Penzler (eds). *Encyclopedia of Mystery and Detection.* New York: McGraw-Hill, 1976.

Mass Murderers. *See* Serial Killers.

Mass Spectrometry

A mass spectrometer is used in forensic laboratories to identify substances found in the evidence collected at a crime scene. In a mass spectrometer, a gas enters a chamber in which it is bombarded with high-energy electrons that knock electrons off the gaseous molecules or atoms, converting them to positively charged ions. The ions then drift through an opening that leads to a space between two charged plates. The electric field between the plates accelerates the ions across the plates to another opening. Charged particles passing through that opening enter a magnetic field perpendicular to their motion. There the particles experience a force that is perpendicular to both their motion and the direction of the magnetic field. The force causes them to follow a curved path. The radius of the curve depends on the particle's mass and speed. For particles with the same charge, the paths of lighter particles are more sharply bent than those of heavier ones. In fact, since the strength of the magnetic field and the voltage that is used to accelerate the particles across the electric field are fixed, the particles' mass-to-charge ratio is proportional to the square of the radius of the arc they follow in the magnetic field.

In a crime lab, chemical substances believed to be clues are first passed through a gas chromatograph to separate the pure substances in what is likely a mixture. After being separated according to their retention time, the substances then enter a mass spectrometer, one after another. Because the curvature of their paths in the spectrometer's magnetic field depends on their mass-to-charge ratio, the charged fragments follow different paths before striking a detector. There each fragment's mass-to-charge ratio is measured and counted. The resulting computer readout provides a pattern showing the abundance of each fragment as a function of its mass-to-charge ratio. Because the computer has stored in its memory the pattern for a great many known substances, it can match the patterns produced by the sample being investigated with thousands of patterns from known substances. For example, substances removed from a suspected drug smuggler can be tested by gas chromatography and mass spectroscopy to see if the substances are heroin, cocaine, or some other drug.

Time-of-Flight Mass Spectrometry (TOF/MS)

In time-of-flight mass spectrometry (TOF/MS), all ions with equal charge emerge from an ion trap with the same kinetic energy. As a result, the mass (m) of the ions can be determined by the time (t) it takes them to travel a fixed distance (d). Because the particles all have the same kinetic energy—½ mv^2—where v is the velocity (d/t), it follows that $m(d/t)^2$ is a constant (K). Since $m(d/t)^2 = K$, $t^2 = md^2/K$. Because d is a fixed distance, it too is a constant. Consequently, $m = t^2K'$, so the mass of the particle can be determined by the time of flight. The Lawrence Livermore National Laboratory (LLNL), in Livermore, California, has a Forensic Science Center that conducts forensic science research. The lab has developed a time-of-flight mass spectrometer that can be taken into the field to analyze air near a crime scene for trace amounts of chemical evidence.

Secondary-Ion Mass Spectrometry (SIMS)

In secondary-ion mass spectrometry (SIMS), a beam of oxygen or cesium ions is used to bombard a sample, which releases secondary ions from its surface. Electrostatic lenses focus the ions on the basis of the mass per charge. Identification of the secondary ions is obtained by comparing them to standard samples. SIMS has been used to analyze industrial accidents where workers have been overcome by toxic fumes of unknown origin. In currency suspected of being counterfeit, SIMS can not only detect elements not found in authentic U.S. paper money but can compare the relative concentrations of elements as well.

The Case of the Toxic Fumes

On February 19, 1994, toxic fumes from a dying woman, Gloria Ramirez, 31, in a Riverside, California, emergency room felled 23 of the 37 members of the staff who tried to save her life. The effects on the workers ranged from dizziness to, in the case of one medical resident, hepatitis, pancreatitis, and avascular necrosis. As the staff struggled unsuccessfully to save the woman's life, they noticed that an oil-like film covered her body and a fruity, garlicky odor emanated from her mouth. A sample of her blood had the odor of ammonia, and there were tiny manila-colored particles in her blood.

Neither a Riverside County hazardous-materials team nor the county coroner's office could find any unusual substances in Gloria Ramirez's body. As a result, the Riverside County coroner called on the Forensic Science Center at the Lawrence Livermore National Laboratory for help. Led by Brian Andersen, the center's director, the scientists decided to analyze the chemical compounds in the various tissue samples provided by the coroner as well as any gases that might have collected above the samples in their containers. To analyze the samples, they used a gas chromatograph and mass spectrometer. The analysis revealed the presence of an amine, a derivative of ammonia, that might account for the odor of ammonia detected in the emergency room. Analysts also discovered the presence of nicotinamide, a B vitamin that is also used by drug dealers to augment the volume of more expensive substances. Finally, they found that the victim's tissues contained unusually high concentrations of dimethyl sulfone [$(CH_3)_2SO_2$], a product of sulfur containing amino acids.

Chemists at the Lawrence Livermore National Laboratory's Forensic Science Center theorized that the victim, Gloria Ramirez, who suffered from cervical cancer, had rubbed DMSO—dimethyl sulfoxide [$(CH_3)_2SO$]—on her skin. (DMSO can be used to reduce pain.) The DMSO would account for the oily film on her body and the garlicky odor. When Ramirez collapsed, the emergency-room staff administered oxygen. The chemists believed that the oxygen combined with the DMSO to produce dimethyl sulfone, which, in turn, combined with more oxygen to form dimethyl sulfate [$(CH_3)_2SO_4$]. Dimethyl sulfate vapors can be deadly because they can cause convulsions, paralysis, and coma and damage vital body organs. At body temperature, dimethyl sulfate readily decomposes, but the chemists believed that when a nurse drew blood from the victim, the cooler blood allowed dimethyl sulfate to accumulate in the syringe. Some of the vapor escaped into the room, producing the toxic effects that affected the emergency-room staff. *See also*

Chromatography; Counterfeiting; Spectrophotometry.

References

Grant, Patrick, David Chambers, Louis Grace, Douglas Phinney, and Ian Hutcheon. "Advanced Techniques in Physical Forensic Science." *Physics Today*, Oct. 1998, pp. 32–38.

Saferstein, Richard. *Criminalistics: An Introduction to Forensic Science.* Upper Saddle River, NJ: Prentice Hall, 1998.

Stone, Richard. "Analysis of a Toxic Death. (Dying Patient in Emergency Room Releases Toxic Fumes)." *Discover Magazine*, April 1, 1995, pp. 66–75.

McGurn, Jack "Machine Gun." *See* St. Valentine's Day Massacre.

Medical Examiner

The medical examiner is a public official who has the responsibility of investigating all sudden, unexplained, unnatural, or suspicious deaths. He or she also performs autopsies in these cases and assists investigators in criminal homicides by providing information about the events leading to a death and its immediate cause.

Although medical examiners now usually do what coroners used to do, death-investigation practices still vary among jurisdictions. Some use a medical examiner, others use a coroner, and still others use a combination of both. The person in charge of death investigations may be uniform throughout a state or may vary from county to county within a state. There are some differences between a medical examiner and a coroner. A medical examiner is usually an appointed licensed physician, whereas a coroner is usually elected, may not be a physician, and must only fulfill a minimum age and residency requirement.

According to the Centers for Disease Control (CDC), medical examiners or coroners investigate about 20 percent of the 2 million deaths in the United States each year. The guidelines as to which deaths should be investigated vary from jurisdiction to jurisdiction, but most require that certain deaths must be investigated. These include deaths that are due to homicide, suicide, or accidental causes such as car crashes and drug ingestion. Also, most require that sudden or suspicious deaths, deaths from sudden infant death syndrome (SIDS), and unattended deaths should be investigated. Any deaths that are caused by an agent or disease that threatens the public health must be investigated. If a person dies at work or while he or she is in custody or confinement, his or her death is usually investigated, and deaths of people to be cremated are investigated by most jurisdictions.

Death investigations vary from case to case, and some are more thorough than others. A comprehensive death investigation has several components. The police, the attending physician, or a family member makes an initial report to the medical examiner or coroner. There is a determination of the circumstances surrounding the death. A postmortem exam or autopsy is completed. Lab tests are done to determine the presence of drugs, toxins, or infectious agents. Finally, there is certification of the cause and manner of death.

For years, medical examiners, the detectives of death, labored in near obscurity, and it is only within the last half century that their work has been publicly recognized. Thomas Noguchi, the chief medical examiner of Los Angeles County, was labeled the "coroner to the stars" for his work on several high-profile cases, including those of entertainer Marilyn Monroe, singer Janis Joplin, and actor William Holden. One of his more bizarre cases was that of an aspiring starlet who was found dead with a bullet-size hole in her head. The ensuing autopsy produced a forensic dilemma. There was no bullet in the wound, and there was no exit hole. Later, while he was Christmas shopping, Noguchi began to think about the shape of spiked heels of women's shoes. Somewhat apprehensively he told police of his theory, and they skeptically agreed to search the crime scene. Shortly thereafter, detectives called to report that they had found a woman's high-heeled shoe

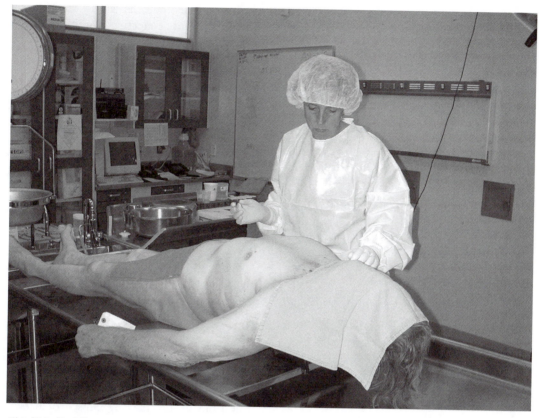

Chief Medical Examiner Dr. Mary Jumbelic of the Syracuse State Police Crime Laboratory, New York, beginning an autopsy. Courtesy of Dr. Marc Safran and Dr. Mary Jumbelic.

with dried blood on it. The blood type was the same as the victim's, and the heel's cross-section exactly matched the entrance wound.

Milton Helpern served for 20 years as chief medical examiner of New York City while acquiring a national reputation as a forensic pathologist. Called by one writer the "Sherlock Holmes of the microscope," Helpern believed there were no perfect crimes, only untrained or poorly trained investigators and medical examiners. Helpern's testimony and research were instrumental in the conviction of Dr. Carl Coppolino for the murder of his wife. A microscopic examination of the victim's body revealed a small puncture wound that proved to be the entry point for a poison. The poison immobilized Coppolino's wife and caused her to suffocate. As a collaborator on the definitive text *Legal Medicine, Pathology, and Toxicology*, Helpern rated a

dedication from mystery writer Erle Stanley Gardner.

Michael Baden, a student of Helpern and successor to his mentor as New York City medical examiner, was appointed to head the Senate Select Committee on Assassinations forensic pathology team. The committee was charged with investigating the deaths of Martin Luther King and President John F. Kennedy. Baden also helped establish the New York State Police Forensic Science Unit, which serves as a consultant to the state's many coroners.

Henry Lee, a former chief medical examiner for the state of Connecticut and coauthor of *Famous Crimes Revisited*, has been involved in investigating the Waco bombings, the O.J. Simpson case, and the Jon-Benet Ramsey murder. In addition, he and Michael Baden are involved in the reinvestigation of DNA evidence in the Albert DeSalvo (Boston Strangler) case. *See also* Coroner; Helpern, Milton; *Quincy*.

References

Baden, Michael M. *Unnatural Death: Confessions of a Medical Examiner.* New York: Ivy Books, 1989.

Black's Law Dictionary. 7th ed. Bryan A. Garner, editor-in-chief. St. Paul, MN: West Publishing, 1999.

CDC Medical Examiner and Coroner Information Sharing Program. http://www.cdc.gov/nceh /pubcatns/1994/cdc/brosures/me-cbro.htm.

MSBC TV News. *Examining the Examiner.* 1998. http://www.msbc.com/onair/nbc/dateline/ shockcoroner.asp.

Noguchi, Thomas T. *Coroner.* New York: Simon & Schuster, 1983.

Mengele, Josef. *See* Forensic Anthropology.

Methadone. *See* Opium.

Mickey Finn. *See* Chloral Hydrate.

Microscope. *See* Magnifiers.

Microspectrophotometry. *See* Spectrophotometry.

Missing Persons

Missing persons are individuals who have been reported to the police as missing by relatives, friends, or coworkers. When such a report is made, information about the missing person is filed by name, date, and physical appearance. The police check to make sure that the person has not been arrested, hospitalized, or found dead before they proceed any further. When the circumstances indicate that the missing person may be the victim of a violent crime, copies of his or her fingerprints are obtained, if possible, and placed on file with city, state, and federal authorities. Medical histories and a copy of a dental chart are also kept on file. In large cities, there are missing persons bureaus within the city police department. Detectives within this division will interview the relatives and friends of a missing person and follow up any leads.

Dental and medical records can help identify a body that has already decomposed to the point that fingerprints cannot be obtained. In one such case a decomposed body was identified by comparing X-ray plates of stomach ulcers taken before death with blood clots found on the remains. To help identify bodies or remains that may be those of a missing person, the FBI National Crime Information Center cross-references its Missing Person File and its Unidentified Person File.

When an unidentified body is found, a detailed physical description is made. Scars, tattoos, or any other unique attributes are noted to help identify the body. The body is photographed and fingerprinted. The fingerprints are checked using local, state, and federal files. If an identification cannot be made, the information is placed in the Unidentified Person File. In the case of a person who is found but cannot identify himself or herself because of amnesia or for some other reason, a description of the person, his or her photograph, and fingerprints are recorded and compared with data available in the Missing Person File.

When the missing person is a child, the necessary information, description, and possibly a photograph are given immediately to police in the area so a search can begin. Many times a child will be found within close proximity to his or her home. Unfortunately, the child is not always found. A missing child may have been abducted by a noncustodial parent, a kidnapper, a child molester, or worse. Some missing children, especially teenagers, are runaways.

Technology has helped in searching for missing persons. Computer imaging, broadcast faxing, e-mail, and the Internet have replaced sketches and posters tacked to telephone poles. Age-progression technology has played a significant role in the search for long-term missing children. A child reported missing at age 10 will look very different at age 20. Age-progression technology allows a photograph of a 10-year-old to be modified using photographs of parents and siblings together with data on facial growth and maturation to develop features that can be projected into a photograph of the same child at age 20. Both the original photograph and the age-enhanced picture can be distributed to police departments throughout the country. In September 1999, the National Center for Missing and Exploited Children, with the support of the Special Projects Unit of the FBI, reported the recovery of 209 missing children through the use of age-progression technology.

Some people will never be found. One probable example is that of Teamsters Union boss James Hoffa, who disappeared in July 1975 and has never been found. It is likely that he was kidnapped and murdered, but he is still listed as missing. *See also* Cause of Death; FBI Crime Laboratory; Fingerprints; Forensic Anthropology; Time of Death.

References

Söderman, Harry, and John J. O'Connell, revised by Charles E. O'Hara. *Modern Criminal Investigation.* 5th ed. New York: Funk & Wagnalls, 1962.

Theoharis, Athan G. (ed.) *The FBI: A Comprehensive Reference Guide.* Phoenix, AZ: Oryx Press, 1999.

M'Naughten Rule

The M'Naughten rule was originally a British legal statute formulated in 1843 by a panel of that country's prominent doctors and lawyers that detailed the standards of an insanity defense in determining criminal responsibility. The demands for some legal clarification stemmed from the 1843 case of Daniel M'Naughten. The original M'Naughten rule was modified a century later as forensic psychologists and psychiatrists came to a better understanding of the human mind.

M'Naughten, a Glasgow, Scotland, woodworker, believed that the Tory party, led by Sir Robert Peel, was trying to kill him. In a case of mistaken identity, the craftsman shot and killed Peel's secretary. At his trial, M'Naughten disavowed his crimes and claimed insanity. The jurors found him not guilty and he was acquitted, though he was hospitalized. As a result of the acquittal, the British legal establishment sought a clarification of what today has become the insanity defense. The particulars of the ruling may vary slightly in the United States, but the British parameters were also adopted by the American legal system.

There are two universal standards used to determine criminal responsibility, and meeting one of the two criteria can ensure a verdict of not guilty due to insanity. First, the simple right-and-wrong test remains a foundation of the insanity plea. In the eyes of the law, a defendant has to be able to distinguish right and wrong at the time of the felony. Failing this, attorneys can argue that the accused was sufficiently diminished mentally not to appreciate the nature and quality of the act. This standard could not anticipate the new findings concerning physiological, psychological, and environmental factors that influence behavior. About a century after M'Naughten, scholars from the American Law Institute formulated a broader definition in light of the new learning. They added a broad new clause stating that if the accused lacked the capacity to conform his behavior to the requirements of the law, he could use the insanity defense. *See also* Insanity Defense.

References

Burton, William C. *Burton's Legal Thesaurus.* 2nd ed. New York: Macmillan, 1992.

Renstrom, Peter G. *The American Law Dictionary.* Santa Barbara, CA: ABC-CLIO, 1991.

Molineux, Roland (1868–1917)

Roland Molineux, a wealthy New York City socialite and the son of the distinguished and politically powerful Civil War veteran General Edward Leslie Molineux, was a vengeful poisoner at the center of one of the city's most famous criminal cases. Handwriting analysis, acquisition of a unique poison, and the circumstances surrounding the deaths of two people closely associated with him led to an emotional and costly three-month case in which he was convicted and sentenced to death.

Educated as a chemist, Molineux reportedly managed the family business, a dye factory in Newark, New Jersey. In reality, he spent a good deal of his time at the Knickerbocker Club, a high-society men's refuge at 45th and Madison. His sense of entitlement and eccentricities did not endear Molineux to his fellow members. Nonetheless, since he was the national amateur horizontal-bar champion, his athletic prowess and social standing gave Molineux considerable influence at the establishment. Molineux was in love with a young debutante, Blanche Cheeseborough, who was being courted by another club member, Henry C. Barnet. Barnet persisted in pursuing Cheeseborough even after he knew of Molineux's interest. In late 1898, Barnet died under peculiar circumstances. In a state of delirium prior to his death, Barnet raved about a bottle that had arrived in the mail. The official cause of death was listed as diphtheria, but there were persistent rumors of poison. Molineux and the debutante were married shortly after the tragedy.

Barnet's death did not end Molineux's troubles at the Knickerbocker Club. Henry Cornish, the club's athletic director, aggravated the pretentious Molineux because he never seemed deferential enough to his betters. The long-simmering rivalry was fueled by the public embarrassment Molineux felt when he was bested in a weight-lifting contest he instigated between the two men. The self-absorbed loser went to the club directors demanding the immediate dismissal of Cornish. They refused, and in a fit of pique Molineux resigned his membership.

On December 23, 1898, Cornish received a bottle of Bromo Seltzer in the mail. The athletic director considered the gift a good humored prank from a club member warning about the pitfalls of holiday celebrations and absentmindedly took the bottle home. Three days later, his landlady, Mrs. Katherine Adams, awoke complaining of a headache. The compassionate boarder offered her a Bromo Seltzer. She drank the medicine but complained of the bitter taste. A sip convinced Cornish that something was wrong. Within minutes the woman was retching on the floor in convulsions as the deathly ill Cornish struggled to find some help. She died within the hour, though Cornish did survive the ordeal.

The death of Barnet and the attempted murder of Cornish appeared inextricably linked from the beginning of the investigation. A club doctor, recalling the unusual circumstances of Barnet's death, rushed to analyze the Bromo Seltzer taken by Cornish and Adams. He discovered that the remedy was laced with deadly cyanide of mercury. Authorities searched Cornish's office and discovered the gift's original wrapping. Handwriting experts singled out Molineux as the sender. This, combined with the well-known antagonism between the two men, led to the indictment of the socialite a mere six weeks after his marriage.

The sensational trial dominated turn-of-the-century New York news. Over 500 potential jurors were interviewed to seat 12. The cost of the trial, on both sides, was considered staggering. The prosecution provided not only motive for the attempted murder of Cornish, which had tragically killed his landlady, but also established means. Investigators uncovered the fact that Molineux had recently ordered a quantity of cyanide of mercury, ostensibly for use at the factory. In a critical mistake, prosecutors openly speculated about Barnet's death and its similarity to the Cornish calamity. After an eight-hour deliberation, the jury found Molineux guilty, and he was sentenced to death.

While he was waiting on New York's death row, Molineux began writing and became a minor New York celebrity based on his jail-house memoirs, *The Room with the Little Door*. Meanwhile, his family began an intensive appeals process. Their efforts succeeded when, in 1902, after three years in jail, Molineux was granted a retrial. The new trial was granted on the grounds that the previous judge had allowed the jury to hear evidence about the Barnet case that had no relevance to the Adams/Cornish case. The defense portrayed Molineux as a victim of popular prejudice and jealousy due to his social position. The success of his book led to the picture of a sensitive, talented writer. Molineux's lawyers also presented a key witness, Anna Stephenson, the wife of a policeman who was at the post office and saw the package being mailed. She could not identify the sender but knew for sure that it was not the defendant. Despite the fact that her testimony was discredited due to her poor vision, the jury took only four minutes to acquit Molineux.

After his release, Molineux was widely recognized as a featured legal analyst for a number of New York daily newspapers. In 1913, he suffered a nervous breakdown and was committed to an asylum, where he died in 1917.

References

Churchill, Allen. *A Pictorial History of Crime, 1849–1929.* New York: Holt, Rinehart & Winston, 1964.

Nash, Jay Robert. *Bloodletters and Badmen.* New York: M. Evans & Co., 1973.

Sifakis, Carl. *The Encyclopedia of American Crime.* New York: Smithmark, 1992.

Monomania

Monomania is considered partial insanity and is a psychosis or mental disorder characterized by thoughts confined to one idea or area of thought. It has been used as an insanity defense in murder cases.

One very early case in which monomania was used as a defense was the trial of Dr. Abner Baker, Jr., in 1845. This trial is of interest because it occurred very close to the beginning of the legal acknowledgment of insanity as a defense. It took place just two years after the M'Naughten ruling in England, which instituted the insanity plea based on the idea that the defendant did not know

right from wrong, and just one year after the establishment of the Association of Medical Superintendents of American Institutions for the Insane, which was the first professional organization of psychiatrists in the United States. Baker was accused of killing his brother-in-law, Daniel Bates. His lawyer argued that Baker had suffered from monomania and should not be held accountable for the murder because of his mental illness. However, this crime occurred in a small town in Kentucky, where this defense was not well received, and Baker was found guilty and executed. *See also* Insanity Defense; M'Naughten Rule.

References

Dorland's Illustrated Medical Dictionary. 27th edition. Philadelphia: W.B. Saunders Co., 1988.

White, R. "The Trial of Abner Baker, Jr., MD." http://www.cc.emory.edu/AAPL/bull/J183–223.htm.

Montero, Britt. *See* Buchanan, Edna.

Morgue

A morgue is a public building that has refrigerated chambers in which unidentified bodies are kept pending their identification and the investigation of their cause of death. *See also* Eyewitness Account.

Moritz, Alan R. (1899–1986)

American Alan R. Moritz, known as the Sherlock Holmes of Medicine, developed a technique to make the follicle of hair more visible for forensic investigation. In 1937, he brushed a thin layer of nail polish mixed with amyl acetate on a slide and placed a stretched hair on it. When the layer of polish had dried, he removed the hair. This left an impression of the hair on the slide that was more defined than a direct microscopic inspection of the hair follicle. During his career, Moritz was a professor of legal medicine at Harvard University and director of the Institute of Forensic Medicine at Case Western Reserve University. He also worked with Samuel Gerber, who gained fame as the coroner involved in the controversial Sam Sheppard murder case and was a member of the Warren Commission that investigated the assassination of John F. Kennedy. *See also* Hair Evidence; Trace Evidence.

References

Thorwald, Jürgen. *The Century of the Detective.* New York: Harcourt, Brace & World, 1965.
———. *Crime and Science.* New York: Harcourt, Brace & World, 1967.

Mormon Forgery Murders

The Mormon forgery murders, which precipitated the largest criminal investigation in Utah history, were cracked by forensic document experts and bomb investigators. The murders took place on October 15, 1985, when the normal calm of the Salt Lake Utah Valley was shattered by two bomb explosions that killed two seemingly unrelated citizens. A second bomb on the following day created an image of Salt Lake City as the "Beirut of the West" and left the population wondering what was happening. At the center of the turmoil was the victim of the third bomb, Utah celebrity and historical-document dealer Mark Hoffman, later referred to as the most skilled forger the country had ever seen. His case involved greed and the desire to protect a reputation gained through fraud and deceit. All of this was tied to the legendary and mythical origins of the Church of Jesus Christ of Latter-Day Saints, the Mormons.

As one of the fastest-growing and most prosperous religious denominations in the world, Mormonism has been haunted by its early history. Almost from the date the church was established by the Prophet Joseph Smith, there were questions about the faith. Particularly troublesome were the allegations that the Prophet was a charlatan, plagiarizer, treasure hunter, and believer in folk magic. Most devastating was the assertion that the Book of Mormon, considered the third part of the Holy Scripture and the foundation of the Latter-Day Saints, was merely the product of Smith's superstitious

imagination, not divine inspiration. The church's early advocacy of polygamy only added fuel to the assaults. By the mid-1960s and into the 1970s, there was an aggressive Mormon scholarship that began to investigate the origins of the church, its doctrine, and its history. Sensitive to the challenge, the church believed that it was its duty to protect the faithful from the persistent slanders of the modern era. The church's sensitivity provided the basis for Mark Hoffman's intrigue in what amounted to blackmailing his own church.

Mark Hoffman was a sixth-generation Mormon who turned a youthful hobby in historical documents into a very lucrative business. Combining faith and business, his specialty was documents that related to Mormon church history. Described by one author as a "scholarly country bumpkin," Hoffman had a likable personality, though some observed that he could be secretive and eccentric. In the early 1980s, Hoffman uncovered and supplied a number of documents to the church, especially those that applied to the early years of Mormonism. The Mormon elders were keen on suppressing anything that even remotely hinted at fraud in the church's founding. As a result of his acquisitions and contacts, Hoffman was increasingly given unparalleled access to the highest levels of the church hierarchy. By the mid-1980s, his reputation extended beyond Utah, "genius" was used with his name, and Hoffman was considered a major phenomenon in the historical document world.

Hoffman's greatest contribution to the Mormon archives was the discovery and subsequent sale to the church of the Salamander Letter. The letter, from an early Smith disciple, embarrassingly showed that the Prophet was involved in folk magic and superstition. Further, these had a major impact on the founding of the church and the writing of the Book of Mormon. The church, hoping to conceal Smith's involvement in folk magic, went to great lengths to be certain that the manuscript was authentic, and even the FBI Crime Laboratory agreed that it was genuine.

Hoffman told the church that he was on the trail of a large group of documents from Texas, the McLellin Collection. Hoffman believed that these anti-Mormon letters, written by a onetime friend of Joseph Smith, would be available shortly, and he took a large cash advance to facilitate the purchase. Then came the bombings.

The first bomb victim who was killed was Steve Christensen, a bibliophile who had a lifelong interest in Mormon history and served as a middleman for the church in purchasing Hoffman's manuscripts. The second person killed in this series of bombing deaths was a friend and associate of Hoffman, Kathy Sheets. The third victim, who was only wounded, was Hoffman himself.

Christensen's death, combined with what was believed by Mormon officials to be an attempt to assassinate Hoffman, led to rumors that some extremist right-wing Mormon faction, angry at Hoffman's discoveries, was responsible for the murders and assault. Hoffman did little to dispel the notion, but law-enforcement officials were reluctant to subscribe to the popular sentiment. In fact, bomb experts from the Bureau of Alcohol, Tobacco, and Firearms (ATF) were convinced that Hoffman himself was the terrorist, although they could not supply a motive.

Interviewed in the hospital with regard to the bomb that had wounded him, Hoffman said that there was a mysterious package in the car, and as it fell from the seat of the car, it exploded. Based upon the scene and the condition of the car, ATF experts believed that the package had been on the driver's seat or right next to it. They believed that this bomb's detonation had been a mistake, that Hoffman had been delivering the bomb, and that his clumsy handling had nearly cost him his life. Other inquiries uncovered a witness who had seen a stocky man in the elevator, prior to the first explosion, with a package addressed to Steve Christensen. The witness could not offer a more complete description, but did recall that the man was wearing a high-school letterman's jacket. A jacket similar to the description given was found at Hoffman's home.

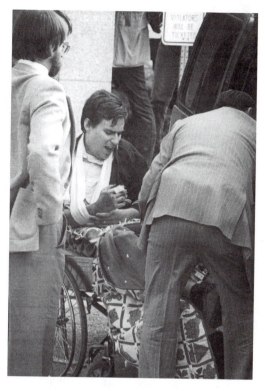

Mark Hoffman at the hospital after the car bombing that injured him. © *Salt Lake Tribune.*

As the investigation progressed, it became obvious that Hoffman had amassed a staggering personal debt of more than a million dollars. Finally, there was also a strong suspicion that some of the material from the homemade bombs had been purchased by Hoffman using the alias Mike Hanson. All of this was circumstantial, and detectives still could not come up with a motive for the killing of Kathy Sheets. The prosecutors' case suffered a blow when Hoffman took a lie-detector test and passed. The suspect's friends and colleagues rallied around him, believing that he was being persecuted by a beleaguered police department.

The Salamander Letter was Hoffman's downfall. The constant validation of the letter bothered George Throckmorton. As the only practicing forensic document examiner in the state and a criminologist at the state crime lab, Throckmorton raised the question of the authenticity of the letter. Throckmorton rightly insisted that it was impossible to accurately date paper and ink, as most of the authorities had done. The best the experts could say was that both the paper and ink were consistent with those in use at the time the letter had allegedly been written. Yet the authorities repeatedly cited paper and ink as evidence of the validity of the manuscript. Anyone could mix a similar ink and buy nineteenth-century paper. Throckmorton proposed a more extensive forensic analysis of the letter and other Hoffman items. Salt Lake City prosecutors had never considered forgery a motive, but given the public firestorm about Hoffman's innocence, they agreed. The Mormon church, anxious to validate its document collection, also agreed to cooperate.

To aid Throckmorton, Utah authorities enlisted William Flynn of the Arizona state crime lab and president of the Southwestern Association of Document Examiners. In December 1987, the two men, sequestered at the Mormon church headquarters, began an extensive examination of 81 documents connected to Mark Hoffman. For four straight days, the men put documents through 17 different tests. In the end, they could not prove forgery. There was one niggling characteristic, however, that made the Hoffman collection different from other nineteenth-century pieces the two scientists examined. Under a microscope, the ink in the Hoffman documents showed small cracks, what the examiners called alligator skin, which appeared only in the Hoffman documents. In a textbook, Flynn discovered a recipe for ink used by nineteenth-century forgers. He found that when certain additives were applied to the recipe and the ink was heated to give it an aged appearance, the characteristic alligator skin appeared. With proof that the Salamander Letter was a forgery, the detectives began an exhaustive search of the provenance or background of each of Hoffman's documents. They discovered more lies about the documents' pedigrees and how Hoffman had acquired them. In fact, detectives estimated that about half of the 81 Mormon church documents were forgeries.

Faced with the overwhelming circumstantial evidence and the fact that the prosecutors had a plausible motive, namely, disclosure of the Hoffman forgery factory, Hoffman's lawyers sought a plea bargain. Hoffman pleaded guilty to two counts of second-degree murder and two counts of felony theft by deception. His sentence would be 1 to 15 years for felony theft and life in prison for murdering Steve Christensen. Prosecutors made it clear that parole was unlikely. In return for not going to trial for murder, Hoffman agreed to a complete disclosure about the crime.

Just as police theorized, Hoffman had killed Christensen because he was pressuring the dealer to either come up with the McLellin Collection, which did not exist, or return the loan he had secured with the church's help. It was clear to Hoffman that Christensen was suspicious and very skeptical. Hoffman needed time to get money from his other fraudulent transactions or to create the documents to pacify the church. The murder of Kathy Sheets, the second bombing, was a diversion for the police and sought to direct attention away from Hoffman. The police were also correct about the third bomb: Hoffman had bungled it. Despite the condition of the plea bargain, Hoffman never fully cataloged his forgeries. Ironically, Hoffman forgeries now bring a premium price on the document markets. *See also* Disputed Documents.

References

Lindsey, Robert. *A Gathering of Saints*. New York: Simon & Schuster, 1988.

Naifeh, Steven and Gregory White Smith. *The Mormon Murders*. New York: New American Library, 1989.

Sillitoe, Linda, and Allen Roberts. *Salamander*. Salt Lake City: Signature Books, 1988.

Morphine. *See* Opium.

Mudgett, Herman Webster (1860–1896)

U.S.-born Herman Mudgett, one of the most notorious felons in the annals of American crime, is considered the first mass murderer in American history. A forensic search of the killer's home that confirmed his murderous spree led police sources to estimate that he had killed 20 to 100 people, mostly young women, in his "Murder Castle" in Chicago. In a death cell confession, Mudgett admitted to 27 killings and judged himself a "moral idiot" devoid of remorse and sentiment. His criminal activities extended beyond murder. He was an accomplished swindler as well.

Beset by debts, the handsome, suave, and well-dressed swindler moved to Chicago in 1890. Abandoning his wife, Mudgett assumed his alias, H.H. Holmes. As Dr. Holmes, Mudgett established himself as an indispensable assistant at Mrs. E.S. Holton's pharmacy on the city's South Side. As the business improved, the widow Holton and her daughter left town and supposedly sold the pharmacy to Mudgett. Police later speculated that the two women were among Mudgett's earliest victims.

The successful businessman astounded his neighbors by purchasing land across the street from the pharmacy and building a three-story structure, ultimately labeled the "Murder Castle" by Chicago police authorities. Mudgett explained his investment by saying that he intended to rent out rooms during the forthcoming Columbian Exposition. The pharmacy owner used a number of different construction crews to ensure that no one fully understood the nature and function of his hundred-room house of horrors. The top floor of the castle doubled as Mudgett's living quarters and office. The ground floor was reserved for shops and businesses. The middle floor and basement staggered the disbelieving Chicago police when they finally raided the building.

Most rooms had gas connections and doors wired to activate a buzzer in the owner's chambers. There were windowless and soundproof rooms, a maze of secret passages, peepholes everywhere, false partitions, and a system of trapdoors and chutes to a basement disposal area. One vaultlike room had sheet-iron walls lined with asbestos and showed signs of scorching by fire. Human bones were found in an oversized stove. It

was the basement that provided the most grisly evidence of Mudgett's sinister activities. There was a vault of quicklime in the cellar floor and a zinc-lined cedar vat of acid. There was also a furnace big enough and hot enough to serve as a crematorium and a surgical table with a number of skeletons under it. Human bones were buried in the floor or just lying around the basement. Police suspected that Mudgett either destroyed the remains of his victims or sold cadavers or cleaned skeletons to medical schools.

The discriminating murderer lured young women to the castle through newspaper ads promising employment or advantageous marriages. As a sign of good faith, the doctor requested that his clients bequeath their wealth and possessions to him in case of an accident. They relocated to his residence, and when they died, he inherited their estates. No doubt the excitement and preoccupation generated by the exposition aided Mudgett in covering up his crimes.

Ironically, it was not the elaborate house, the schemes to finance its construction, or the suspicion generated by the missing women that exposed Mudgett in 1894. Rather, it was the usually reliable insurance swindle gone bad. Held in a St. Louis jail under suspicion of fraud, Mudgett confided an elaborate fraud plan to his cellmate, a bank robber named Marion Hedgepeth. Mudgett asked the criminal if he could recommend a lawyer to aid in his release and one who would participate in the scheme. The grateful Mudgett promised Hedgepeth $500 in return for the name of Jephtha Howe, a St. Louis attorney. Upon his release, Mudgett went to Philadelphia to meet his coconspirator, Benjamin F. Pitezel, who had taken up residence in the city as B.F. Perry.

The plan was a simple one that Mudgett had used successfully earlier in his career. Pitezel/Perry, posing as a prosperous chemist, took out a $10,000 insurance policy. Mudgett was responsible for planting a dead body in his accomplice's lab and setting it ablaze. A grieving Mudgett would then identify the badly burned body as that of his friend, the chemist, and collect on the policy. Somehow the two partners had a falling out, and Mudgett killed Pitezel and put in his claim. Fidelity Mutual Life Association paid the claim, and the matter was presumed closed until Hedgepeth, angry at not receiving his money, told the warden about Mudgett's plan.

Pinkerton detectives were called in to investigate and ultimately arrested Mudgett for murder. Ever the innocent, Mudgett admitted his duplicity and acknowledged Pitezel's involvement but insisted that the dead body found in the lab had been obtained from an unidentified doctor. Mudgett insisted that Pitezel was still at large. The erstwhile doctor was convicted of fraud and remained in custody as authorities mounted an extensive manhunt for the second man in the plot. Detective Frank Geyer led the investigation that determined that Mudgett had not only killed Pitezel but also two of his children sent east to identify the body of their father and claim the insurance money. Mudgett's legal difficulties in Philadelphia led Chicago police to the "Murder Castle." Despite the summer-long analysis of the castle, Mudgett was never charged with the atrocities in Chicago. After a five-day trial in Philadelphia, he was convicted of multiple murders in the Pitezel case and was sentenced to death by hanging.

Mudgett sold his story to W.R. Hearst but received no money. He was executed by hanging in Philadelphia on May 7, 1896. Even in death Mudgett was different. To protect his body from grave robbers and medical science, he requested to be buried 10 feet down in a casket encased in cement. *See also* Serial Killers.

References

Blundell, Nigel. *Encyclopedia of Serial Killers.* North Dighton, MA: JG Press, 1996.

Duke, Thomas. *Celebrated Criminal Cases of America.* Reprint. Montclair, NJ: Patterson Smith, 1991.

McNamee, Tom. "Monster of 63rd Street Was Terrifying Serial Killer of the 1890s." *Chicago Sun-Times,* May 7, 1996.

Schechter, Harold. *Depraved: The Shocking True Story of America's First Serial Killer.* New York: Pocket Books, 1994.

N

Neutron-Activation Analysis

Neutron-activation analysis, a technique that involves bombarding a sample of matter with neutrons and then measuring the energy of the gamma rays released, has been used in forensic science to detect trace elements in paint, drugs, soil, hair, gunpowder residues, and metals. Samples of different copper wires, for example, may appear to be identical. By using neutron-activation analysis, which can measure fractions of a part per million, one can determine whether two samples came from different sources or from the same source.

When atoms of many heavy elements are bombarded with neutrons, some of them capture a neutron and become a different isotope of the same element. For example, the nucleus of most gold atoms has 79 protons and 118 neutrons. If such an atom captures a neutron, it becomes an isotope of gold with 79 protons and 119 neutrons. This atom is unstable and releases a beta ray (an electron) and a gamma ray that has a discrete amount of energy.

Unlike many other analytical methods, neutron activation is a nondestructive procedure that can detect and differentiate among nanogram quantities of different elements. However, the neutrons used to bombard the sample come from the heart of a nuclear reactor, and very few crime laboratories have access to a reactor or the sophisticated devices required to detect and identify the gamma radiation.

There had been a persistent rumor that despite the switch to iron-alloy ("white") pennies in 1943 as a means of conserving copper for the war, a few copper pennies had also been minted. The U.S. Treasury Department consistently denied that any such pennies had been made in that year. In 1963, a private coin collector confronted the Treasury with a few 1943 pennies that he believed were authentic 1943 copper pennies. He demanded that they be analyzed to determine whether or not they had the same components as 1942 pennies, which were known to be made of copper.

To avoid damaging the coins, they were placed in a nuclear reactor in order to make them radioactive. They were then examined using neutron-activation analysis. The gamma-ray emission confirmed the collector's suspicion. The coins exhibited the same concentrations of copper and trace elements as did 1942 pennies. The rumors had been true. Someone in the mint had secretly made a few pennies with the same metals that were used in 1942.

In 1965, neutron-activation analysis was used in a case involving a Canadian named René Castellani who was accused of killing his wife by giving her small doses of arsenic. In an effort to find out whether the concentration of arsenic in the victim's body corresponded to the frequent but intermittent bouts of illness she had been experiencing, forensic chemist Norm Erickson sliced hair samples from the victim's exhumed body into lengths of 1.0 cm, 0.5 cm, and, finally, 0.25 cm—the lengths that hair normally grows in a month, two weeks, and one week, respectively. Erickson's results from neutron-activation analysis showed beyond reasonable doubt that Esther Castellani's periods of illness occurred at the same times that the concentrations of arsenic in her hair were peaking. *See also* Cause of Death.

References

American Jurisprudence Proof of Facts Annotated. Vol. 15. San Francisco: Bancroft-Whitney Co., 1964, 115–148.

Lane, Brian. *The Encyclopedia of Forensic Science.* London: Headline, 1992.

Saferstein, Richard. *Criminalistics: An Introduction to Forensic Science.* Upper Saddle River, NJ: Prentice Hall, 1998.

Norris, Charles (1867–1935)

Charles Norris, New York City's first medical examiner after the city abandoned the coroner system in 1918, introduced the principles of forensic science to New York and, with grit and determination, created a model medical-examiner system. Shortly before his death, he won his long fight to have a chemical and toxicological laboratory added to the New York police facilities.

After Norris received his medical degree from Columbia University, he went to Europe, where he studied forensic medicine for two semesters in Kiel and one semester in Göttingen. He went on to Berlin, where he worked under the famous pathologist Rudolf Virchow. From 1894 to 1896, he collaborated with Eduard von Hofmann and Alexander Kolisko in Vienna. At the time, Europe was far ahead of the United States in forensic medicine, and Norris received excellent training from these men.

On his return to the United States, he was professor of pathology and head of the chemical and bacteriological laboratory at New York's Bellevue Hospital from 1904 to 1918. When he became the medical-examiner for New York City in 1918, he hoped to incorporate the knowledge and insight he brought back from Europe. The coroners, who had been replaced by the new medical-examiner system, fought for their old jobs through politics. There were repeated attempts by politicians to cut the budget of the new office. Norris, who was a wealthy man, at times used his own money to buy laboratory equipment.

Norris was able to bring some of his ideas to fruition. He established a central medical examiner's office in Manhattan that had a telephone switchboard open 24 hours a day. He opened a branch office in each of the New York City boroughs and added an assistant pathologist in each morgue. Other assistant pathologists did fieldwork, and there was always someone ready to go to the scene of a crime.

Norris was a physically large man, weighing over 200 pounds, who had a dynamic personality. He used his presence to help get what he wanted, from police cooperation to political backing for his offices. There were 15,000 to 20,000 reports of suspicious deaths a year in New York during Norris's term. His offices performed 4,000 to 6,000 autopsies a year, so there was always a need for new materials and equipment.

His expert testimony put many men behind bars and saved a few innocent men. In one case, Joseph B. Elwell, a well-known bridge authority, had been found dead in his apartment, and the death had been ruled a suicide. However, Norris discovered that in the position that the revolver must have been held, Elwell could not have pulled the trigger. The murder was never solved. In another strange case, Norris gave testimony that saved a husband from being charged with his wife's death. In 1919, Bessie Troy was discovered dead on a sidewalk. Relatives accused her husband, Michael Troy, of throwing her from a window to cover up her murder. However, Norris testified that Bessie Troy had been alive when she hit the sidewalk.

At this time there were no medical schools in America that offered courses in forensic medicine. This infuriated Norris, for there was nowhere in the country that a doctor could go for training in this specialty. It was not until a year before his death from a heart attack that he succeeded in creating the Department of Forensic Medicine at the New York University College of Medicine. *See also* Coroner; Medical Examiner.

References

Saferstein, Richard. *Criminalistics: An Introduction to Forensic Science.* Englewood Cliffs, NJ: Prentice Hall, 1990.

Sifakis, Carl. *The Encyclopedia of American Crime.* New York: Smithmark, 1992.

Thorwald, Jürgen. *Crime and Science.* New York: Harcourt, Brace & World, 1967.

O

Odontology (Forensic Dentistry)

Odontology is the study of the anatomy, growth, and diseases of teeth. In forensic science, an odontologist is a dentist who uses his knowledge of teeth to help police solve crimes or identify victims who cannot be identified by visual means or by fingerprints because they have been badly burned, decomposed, disfigured due to trauma, or reduced to skeletal remains.

As any physical anthropologist will state, teeth survive the perils of time better than any other body tissue. The same is true of fillings, caps, bridges, and dentures, which are resistant to both chemical and physical decomposition. Since dentists have X-ray data of their patients' teeth as well as records showing the location of fillings, caps, root canals, and bridges, identification procedures may include, through postmortem X rays, portions of the teeth embedded in the jaw as well as those separated from the body. That two people have the same dentition is possible but highly unlikely. Consequently, dental remains often provide a very useful means of identifying the remains of an unknown body. This is particularly true following mass disasters such as airplane crashes when identification is needed not only for the peace of mind of relatives but for positive confirmation of death needed for funeral purposes, death certificates, and the execution of wills and insurance policies.

Teeth, Age, and Records

Because the emergence of teeth in youngsters follows a rather universal chronological pattern, it is quite easy to identify the age of death among young victims to within a month or two. A child's first set of 20 teeth begins to emerge at 6–9 months, when the lower central incisors break through the gums, and continues until the second molars are cut by the age of 2. The emergence of the second set of teeth, which eventually constitutes the set of 32 normally found in adults, begins with the "six-year-old molars" and continues until the third molars (wisdom teeth) are cut at somewhere between 17 and 25 years of age. After age 25, age determination by dentition is more difficult. However, with time, teeth wear down, gums recede, the pulp area of the teeth is reduced in volume, roots are resorbed, and root tips become translucent. As a result, age at death based on dentition of adults is possible, but is no better than ± 4 years.

One major difficulty in using dental remains for identification is that more than 200 different methods are used throughout the world to chart teeth. In the United States, the teeth are numbered starting with the upper right third molar, which is labeled "1,"

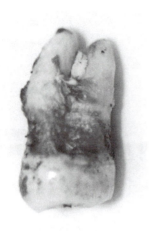

A tooth used in a forensic criminal investigation. Courtesy of the Division of Forensic Science, Commonwealth of Virginia, Department of Criminal Justice Services.

across the upper jaw to the upper left third molar, which is labeled number 16, down to the lower left third molar (17), and across to the lower right third molar, which is number 32. The records also include the five visible surfaces of each tooth. Diagrams of these surfaces for each tooth serve to record the location of any dental work.

In 1986, the American Dental Association initiated a program encouraging dentists to bond a tiny coded microdisk onto each child's upper molar. The 12-digit code provides the information needed to identify the individual for as long as the molar remains. With the early fluoride treatments common today, people seldom lose teeth.

Cases

One of the first cases to use dental means to identify an unknown skeleton took place in Britain during World War II. While workers were cleaning debris from a damaged church, they found a female corpse consisting of a skeleton and some dried skin. Because the head and legs had been severed and the body had been doused with lime, authorities believed that she was a victim of murder, not German bombs.

On the basis of an estimate of the victim's height and a list of missing persons from the London area in which she was found, police checked her dentition with a dentist who had treated Rachel Dobkin. Police knew that Dobkin had been in a dispute with her husband regarding child support. Her dentist, Dr. Barnett Kopkin, provided records that showed that the dentition of the victim matched Dobkin's dental records. He also reported that in extracting two teeth from Dobkin's lower left jaw, portions of the root had broken off and had remained in the bone. An X ray of the victim's skeletal jaw revealed the roots of the extracted teeth. The dental evidence, together with a match between a photograph of Rachel Dobkin and the recovered skull, led to the conviction and hanging of her former husband, Harry Dobkin, in 1943.

In the 1986 woodchipper murder case, in which Richard Crafts sent his wife's frozen body through a woodchipper, police were able to retrieve two teeth caps among nearly 3,000 bits of human tissue. Of the 50,000 forensic tests conducted on these remains, one was particularly significant; forensic dentist Dr. C.P. Karazulas was able to use dental records to show that one of the caps came from the mouth of Helle Crafts. His work, confirmed by another forensic dentist, Dr. Lowell Levine, was instrumental in convicting Richard Crafts of murdering his wife.

On August 30, 2000, police removed a body from a 1990 Jaguar that had burned early that morning on Squaw Island near Hyannis, Massachusetts. When firemen responding to a 911 call from a nearby resident had arrived at about 3 A.M., they had found the car engulfed in flames so hot that the car's tires had melted. The charred remains of the driver made identification difficult, but a forensic pathologist was able to identify the body using dental records.

On January 3, 1998, Annemarie Cusano disappeared. She was the mother of twin daughters, a resident of Shelton, Connecticut, and an employee of Executone Information Systems in nearby Milford. Through cellular phone records, police were able to trace her to a room in Hartford rented by Gregory McArthur, the last person with whom she had been seen prior to her disappearance. Police believe that it was her part-time work with an escort service that led her to McArthur's room, where investigators found a bag of bloody clothing. McArthur, who had a criminal record involving robbery, larceny, and sexual assault, was extradited from a Massachusetts prison to Connecticut on September 1, 2000, to face murder charges for Cusano's death. McArthur led police to a wooded area outside Hartford where the remains of a body were uncovered. Based on a comparison of the skeleton's dentition with Cusano's dental records, the state's chief medical examiner confirmed that the remains were those of Annemarie Cusano. *See also* Bite Marks; Dobkin, Harry; Forensic Anthropology; Webster, John: Murder; Woodchipper Murder.

References

"Connecticut Police Find Body of Missing Mother." Associated Press, *Cape Cod Times*, Sept. 3, 2000.

Evans, Colin. *The Casebook of Forensic Detection: How Science Solved 100 of the World's Most Baffling Crimes.* New York: Wiley, 1996.

Fisher, Barry A.J. *Techniques of Crime Scene Investigation.* 6th ed. Boca Raton, FL: CRC Press, 2000.

Jeffrey, Karen. "Man Dies in Hyannisport Car Fire." *Cape Cod Times*, Aug. 31, 2000, p. 1.

Oklahoma City Bombing (1995)

The Oklahoma City bombing on April 19, 1995, was the worst and deadliest terrorist attack ever committed on American soil. Forensic scientists were soon on the scene attempting to discover who had carried out the attack.

A rental truck parked outside the Alfred P. Murrah Federal Building exploded at 9:02 A.M. The blast was heard miles away. The explosion left an 8-foot-deep, 30-foot-wide crater, which became the initial focal point of the investigation. The explosion tore off the front of the nine-story structure and damaged 200 other buildings. The damage was estimated at about $500 million. In human terms, the price was even more startling and tragic. Killed were 168 people, including 19 children from a day-care center. Over 400 others were injured, primarily from flying glass.

Despite the difficulties associated with sealing off a bomb site where there was a round-the-clock search for survivors and bodies, forensic scientists were able to locate traces of ammonium nitrate in the rubble, which provided a clue as to the composition of the bomb. Bomb experts from the FBI estimated that the bomb contained a number of 55-gallon plastic drums filled with a mixture of ammonium nitrate fertilizer and fuel oil. The extent of the damage suggested that the bomb's power was increased by the addition of hydrogen or acetylene.

The pivotal piece of evidence was supplied by a twisted truck axle that provided the truck's identification number. Knowing the truck's identification number, investigators found the local Ryder rental outlet in Junction City, Kansas, from which the truck had originated. They obtained a description of the two men who had rented the truck and began the task of going from motel to motel on the chance that the men might have spent the night prior to the bombing in the area. Luckily, the owner of the Dreamland Motel identified one of the composite sketches as Timothy McVeigh, who had stayed at the motel on April 14. McVeigh gave his address as Decker, Michigan. Clerks at the motel remembered the Ryder truck and recalled that McVeigh had checked out on Tuesday, April 18. A database search revealed that a man named McVeigh had been arrested on April 19 about 90 miles from Oklahoma City approximately 90 minutes after the explosion. The officer who stopped McVeigh for a minor driving violation noticed that McVeigh had a concealed gun, a shoulder-holstered pistol. Upon inspecting the gun, the trooper found that it was loaded with hollow-point bullets, known on the street as "cop killers." McVeigh was taken into custody and held for two days until a bail hearing. Five minutes prior to the hearing, the FBI asked that he be held for further questioning. On April 21, the 27-year-old McVeigh was charged with malicious damage to federal property while the FBI set out to build a case against its prime suspect.

McVeigh's similarity to the composite sketch based on the memory of the truck renter proved to be but a small part of the circumstantial evidence the government amassed against the alleged bomber. Chemical analysis of McVeigh's clothes and car revealed traces of the explosives used in the bomb. On April 26, Terry Lynn Nichols, an army buddy of McVeigh's, who was wanted for questioning, turned himself in to police.

A search of Nichols's Kansas farm turned up a blasting cord used in the bomb as well as blasting caps of the type used in the bomb. The search for evidence also uncovered a receipt for ammonium nitrate fertilizer from a local farm store. McVeigh's thumbprint was found on the paper. Investigators surmised that a trail of storage-locker rentals allowed the two men to hide the ton of fertilizer used in the bomb. At his bail hearing, Nichols testified that three days prior to the bombing, McVeigh told him that something big was going to happen. Bolstering all this was the highly anticipated testimony of Michael Fortier, another army friend of McVeigh. Fortier reported that he had cased the Murrah site with McVeigh several months prior to the bombing.

Government agents also examined McVeigh's past. They discovered that he had had a troubling childhood and adolescence involving abandonment, abuse at school, and an obsession with guns. Despite having served in Operation Desert Storm in 1991, McVeigh had failed to qualify for a Special Forces assignment and had left the army frustrated and bitter. The suspect was also a fringe follower of the heavily armed, antigovernment militia subculture that was convinced that a conspiracy of bankers, politicians, media moguls, and other elites were planning to take over the country. Their concerns were fueled by the government's 1993 assault on the Branch Davidian compound at Waco, Texas. The persistent belief

that the government deliberately slaughtered the Branch Davidians provided considerable grist for the underground movement. Investigators were quick to recognize that the Oklahoma City catastrophe took place on the second anniversary of the Waco tragedy and that the Murrah Building contained an office of the Bureau of Alcohol, Tobacco, and Firearms, the agency responsible for the Waco incident.

Ultimately, prosecutors singled out rage against government and the paranoia of the extreme right as the motive for McVeigh's assault. On May 10, the government also charged Terry Nichols as an accomplice in the bombing. Prosecutors conceded that Nichols might not have planted the bomb or been present in Oklahoma City on April 19, but they were convinced that he knew about McVeigh's plans and helped construct the bomb.

The period prior to the trials saw a number of machinations, some designed by defense lawyers to create reasonable doubt about the government's conspiracy theory and some regarding the impossibility of a fair trial in Oklahoma. The issue of pretrial publicity led to a change in venue in February 1996, and the defendants were also granted separate trials. Potentially most damaging to the prosecution's case was criticism of the FBI Crime Laboratory. One of the lab's own, FBI forensic chemist Frederic Whitehurst, accused the lab of mishandling evidence, basing conclusions on crime-scene experience rather than precise data, and lack of scientific objectivity. Whitehurst claimed that since the lab worked only for prosecutors, results were predisposed to proving a suspect's guilt. Attorney General Janet Reno commissioned a panel of international forensic scientists to investigate the charges. They found sufficient substance to Whitehurst's claims to get three FBI lab supervisors reassigned, but while the final report did note some sloppy lab work and other lapses, it concluded that none of the facility's conclusions and findings had been compromised.

At the trials of McVeigh and Nichols, jurors found the prosecutors' cases compelling and both men were found guilty. McVeigh was sentenced to death and Nichols to life in prison without parole. McVeigh was executed by lethal injection at 7:14 A.M. on June 11, 2001. *See also* FBI Crime Laboratory.

References

CNN.com. The Execution of Timothy McVeigh. http://www.cnn.com/SPECIALS/2001/okc.

Jones, Stephen and Peter Israel. *Others Unknown: The Oklahoma City Bombing Case and Conspiracy.* New York: Public Affairs, 1998.

Serrano, Richard A. *One of Ours: Timothy McVeigh and the Oklahoma City Bombing.* New York: W.W. Norton, 1998.

Opium

Opium, a strong narcotic, is the juice harvested from the seed capsules of opium poppies (*Papaver somniferum*), which is widely grown throughout Southeast Asia. It dries to form a brownish, gummy substance that is made into crude opium by rolling it into balls that are then covered with leaves. One of the challenges faced by forensic scientists is the identification of opium and narcotic substances derived from it. Ever since narcotics became illegal in the United States, a significant number of prison inmates have been convicted of either using or selling narcotics or other illegal drugs.

Laudanum, or tincture of opium, was widely used in nineteenth-century England as a relaxant or to add an extra "kick" to alcoholic beverages. Opium reached the United States in the middle of the nineteenth century when railroad companies imported Chinese workers to lay the tracks that would join the American coasts. A quarter of a century later, there were more than 200 opium dens in San Francisco alone. Smoking opium was popular among both the rich and the poor, but as its addictive properties became known, morphine and, later, heroin became the drugs of choice. Though heroin is no less addictive than opium, the source of both morphine and heroin, the American Medical Association approved heroin for general use

in 1906, and it became a common ingredient in a variety of patent medicines. They were not made illegal until about the time of World War I. Opium is the source of most narcotic drugs, which pharmacologists define as substances that bring relief from pain and produce sleep. Such drugs are legally available only through a doctor's prescription.

Morphine

Morphine, a white crystalline solid, is an alkaloid of opium. Alkaloids are basic, nitrogen-containing compounds found in a number of dicotyledonous plants. They are often bitter tasting and may have pharmacological action. Friedrich Serturner, a German chemist, succeeded in extracting morphine from opium in 1806. It was the first alkaloid, but certainly not the last, to be extracted from the poppy extract. Others include atropine, caffeine, cocaine, codeine, nicotine, quinine, and strychnine. All end with the letters *ine*.

Morphine is a narcotic used in medicine to relieve pain. Its analgesic property made it widely used in the medical community before its addictive nature—the result of its euphoric effects—led physicians to restrict its use. Patients suffering from terminal cancer do benefit from its emollient quality, as do others in extreme pain.

The magnitude of a fatal dose varies; however, 325 milligrams will generally lead to respiratory failure following euphoria, drowsiness, and nausea. A common symptom of a morphine overdose is the contraction of the pupils until they are no more than pinpoints and do not respond to light or dark.

Heroin

Morphine reacts with acetic anhydride or acetyl chloride to form heroin. Because heroin is much more soluble in water than morphine, it is easily prepared on the street for intravenous injection. Such injection produces an almost instant euphoria accompanied by drowsiness and a sense of well-being. The effect usually wears off after three or four hours. Heroin is commonly diluted with quinine, which, like heroin, has a bitter taste.

The addict who buys heroin on the street is never certain of the heroin's strength. Such uncertainty can lead to overdosing and death.

Codeine

Codeine is also found in opium, but is more commonly prepared from morphine. It is a milder narcotic than morphine and is widely used as a cough suppressant. Because it is a much weaker narcotic than morphine or heroin, it is less attractive to drug addicts.

Methadone

Methadone is one of the narcotic drugs not derived from opium or morphine. It is widely used to treat drug addicts because it seems to inhibit the craving for heroin without producing significant side effects.

The Dr. Robert Buchanan Case

In 1892, Anna Buchanan, who ran a very profitable brothel, died of a cerebral hemorrhage according to the New York coroner who signed her death certificate. As a result of her death, her husband, Dr. Robert Buchanan, inherited $50,000, which was a significant sum at that time. One of her pimps, also a former lover, identified only as Mr. Smith, demanded that the coroner's office open an investigation. He was overheard by Ike White, a reporter for the *New York World*, who took up the cause. White discovered that Buchanan had returned to Nova Scotia, where he remarried his former wife only three weeks after Anna Buchanan's death. His discovery served to whet his suspicion and led him to interview others who had worked at the brothel. One of them recalled a comment that Robert Buchanan had made during the trial of Carlyle Harris, who had murdered his wife by giving her an overdose of morphine. Buchanan called Harris an amateur because his wife's pinpoint pupils led investigators to suspect morphine poisoning. Buchanan had boasted that he knew a way to avoid such telltale evidence.

When White interviewed the nurse who had attended Anna Buchanan just prior to her death, the woman remembered that Buchanan had placed drops in his wife's eyes several times during her short illness. Could the drops have contained atropine, which, because it causes pupils to dilate, could have hidden the effect of morphine? White knew that he was onto something, and he launched a newspaper campaign demanding that Anna Buchanan's body be exhumed for examination. The police finally responded by exhuming the body. Toxicologist Rudolph Witthaus found clear evidence of morphine in the victim's tissues and agreed with White that the usual pinpointing of the pupils could have been prevented by atropine drops. At the trial, the prosecution administered a lethal dose of morphine to a cat followed by drops of belladonna, which contains atropine. It was clear that the pinpointing effect was prevented by the belladonna. Buchanan was found guilty of murder and was executed on July 2, 1895, just 26 months after Carlyle Harris was executed for the same crime.

A More Recent Drug Case

In June 1988, Hong Kong authorities received word from Australian police that one of their undercover operations had discovered a drug syndicate that was planning to use a yacht to smuggle a large shipment of heroin from Hong Kong to Australia. Hong Kong police responded with their own undercover agents who, posing as sailors aboard the yacht, arrested 3 pushers and confiscated 43 kilograms (95 pounds) of heroin. Undercover agents then faked the delivery of yacht-borne heroin in Sydney, Australia, where they arrested 16 more people. *See also* Atropine; Drugs.

References

Cyriax, Oliver. *Crime: An Encyclopedia*. North Pomfret, VT: Trafalgar Square, 1996.

Evans, Colin. *The Casebook of Forensic Detection: How Science Solved 100 of the World's Most Baffling Crimes*. New York: Wiley, 1996.

Lane, Brian. *The Encyclopedia of Forensic Science*. London: Headline, 1992.

"Now for Another Opium War." *The Economist*, Nov. 19, 1988, p. 40.

Orfila, Mathieu Joseph Bonaventure (1787–1853)

Spaniard Mathieu Joseph Bonaventure Orfila is considered by many to be the father of forensic toxicology, which deals with poisons and drugs and their effects. Orfila wrote the first scientific treatise on the detection and effects of poison, establishing forensic toxicology as a legitimate scientific discipline.

Although his family wanted him to be a seaman, a difficult voyage at age 15 convinced the child prodigy to pursue his academic interests in chemistry and medicine. He studied in Spain and then went to Paris, where he became a medical doctor. Orfila then scratched out a living as a chemistry tutor and lecturer in the French capital. During this period, he also began chemical experiments with poisons, most notably arsenic. In 1813, at 26, he published the first of three books on poison that were a summation of all that was known about poisons at the time. The encyclopedia, *Treatise on Poisons or General Toxicology*, immediately became the seminal work on the topic. Orfila cataloged the symptoms of all kinds of poisons, the appearance of the body when it was under the influence of poison, the psychological side effects, and, most significantly, how to detect poison in the body. Orfila showed that even if there was no poison in the stomach, other organs, like the liver, spleen, or intestine, could provide residual evidence of poisoning.

Orfila played a significant role in the Lafarge case. While Charles Lafarge was on a business trip to Paris, his wife, Marie, sent him a cake laced with arsenic. When he returned alive, she added the same substance to his milk. The family's maid told police that she had seen white flakes floating on the milk's surface. Later, she saw Marie Lafarge adding pinches of the white powder to her husband's food. Following the death of Charles Lafarge on January 16, 1840, the maid delivered some of the powder to the police, and Marie Lafarge was soon under arrest.

At the trial, a local chemist testified that he had found traces of arsenic on the glass that held the milk Charles Lafarge had drunk and on the box that held the white powder the maid had seen Marie Lafarge adding to food, as well as in Charles Lafarge's stomach. The defense called on Orfila to testify. Orfila disputed the chemist's tests, claiming that they were inconclusive with regard to arsenic. Through Orfila, the prosecution learned about the Marsh test and requested him to apply the test to the evidence available. Applying the Marsh test to the remains of Charles Lafarge, Orfila found arsenic in the stomach, liver, heart, brain, and intestines. This new evidence led to Marie Lafarge's conviction and a sentence of hard labor for life. The finding revolutionized forensic detection and catapulted Orfila to professional security. Not only was he recognized as an expert witness, but he gained a promotion at the University of Paris. In 1831, he was made dean of the medical school faculty. *See also* Arsenic Poisoning; Poisons.

References

Block, Eugene B. *Science vs. Crime: The Evolution of the Police Lab.* San Francisco: Cragmont, 1979.

Cyriax, Oliver. *Crime: An Encyclopedia.* North Pomfret, VT: Trafalgar Square, 1996.

Osborn, Albert S. (1858–1946)

It was through the efforts of Albert S. Osborn and other experts of the late nineteenth and early twentieth centuries that evidence involving disputed documents was accepted by courts in the United States. Disputed documents are materials that contain written markings, made either by hand or by machine, whose source or authenticity may be in doubt. Investigations pertaining to documents include the examination of handwriting to see if it matches that of a suspect, analyzing the paper, ink, or print found in a document, examining charred paper under infrared light to make print visible, and other forensic techniques used to analyze documents. Osborn was the author of *Questioned Documents*, which, although it was written in 1910, is still used as a reference.

One case in which Osborn was involved was the "Yule Bomb Killer." The case was significant because investigators were able to determine the killer from just a few scraps of packaging paper in which the bomb had been wrapped.

A package had been delivered to John Chapman's home in Marshfield, Wisconsin, a few days after Christmas 1922. As Chapman stood nearby, his wife, Clementine, began to open the package, which exploded, killing her and injuring Chapman. From scraps of the packaging paper not destroyed in the explosion, investigators found part of a hand-written address and a postmark that indicated that the package had been sent from a mailbox belonging to Thorval Moen, who lived on Route 5 not far from Marshfield. After Moen had been cleared, the investigators turned their attention to the handwritten address, which had some interesting features. Marshfield had been misspelled as Marsflld. The address on the bomb package had been written with an ink that had a unique mixture of ingredients. The writing was also clumsy, indicating that the writer was probably not well educated. John Tyrell, an examiner of questioned documents, felt that the misspelling suggested that the person was from another country. When the misspelled word was pronounced phonetically, it sounded Swedish.

Chapman had been involved in a dispute over a drainage ditch that would cut through a piece of land belonging to John Magnuson, a Swedish farmer. Investigators found pieces of wood similar to those used in the bomb in Magnuson's barn. When Magnuson was asked to provide a handwriting sample, he misspelled Marshfield as Marsflld. Investigators also found the unique combination of inks in a pen at Magnuson's home.

Osborn was brought in as a handwriting expert. He testified that Magnuson had written the address. He found at least 14 similarities between the handwriting of Magnuson and the handwritten address. His testimony was combined with the testimony of a professor of the Swedish language, J.H. Stromberg, who stated that a Swedish immigrant would likely pronounce and write Marshfield as Marsflld or Marsfld because the Swedish language does not have the *sh* sound or the combinations *ie* or *ei*. On the basis of these findings, along with other evidence such as wood and metal similar to that in the bomb that was found in his workshop, Magnuson was convicted of the murder and sentenced to life in prison.

The most famous case in which Osborn was involved was the Lindbergh kidnapping case. Osborn was one of the handwriting experts who testified that Bruno Richard Hauptmann was the person who wrote the ransom notes to the famous aviator Charles Lindbergh seeking money from him for the return of his young son, Charles, Jr. *See also* Disputed Documents; Lindbergh Kidnapping and Trial.

References

Cyriax, Oliver. *Crime: An Encyclopedia*. North Pomfret, VT: Trafalgar Square, 1996.

Evans, Colin. *The Casebook of Forensic Detection: How Science Solved 100 of The World's Most Baffling Crimes*. New York: Wiley, 1996.

Sifakis, Carl. *The Encyclopedia of American Crime*. New York: Smithmark, 1992.

Oursler, C. Fulton (1893–1952) *See* Abbot, Anthony.

Overlying

Overlying most commonly occurs when a parent who is sleeping close to her or his baby smothers the child. A part of the parent's body, usually inadvertently, comes to lie over the child's mouth and nose. Overlying is of concern to forensic scientists because smothering is sometimes intentional and, of course, in such cases constitutes murder, but proof of intent to kill is extremely difficult to obtain. Deliberate smothering of one adult by another, often when the victim is asleep, is a common means of homicide.

Although the danger of overlying always exists when an adult sleeps in the same bed as a baby, it is more likely to happen if the adult is medicated or drugged and, therefore, less likely to awaken. Overlying is less common than sudden infant death syndrome (SIDS), which was formerly called "crib death." Again, some deaths reported as SIDS may actually be cases of homicide, but proof is difficult.

In many cultures, parents commonly sleep with their young children. Such a practice was widespread in Western society before cribs were introduced—a cultural change that was never studied or shown to be safer for the child. SIDS has declined since parents have been warned to place babies on their backs. Babies who sleep on their stomachs are, it is believed, more likely to be smothered when their heads become buried in blankets or clothing.

A 1999 report based on a review of death certificates from 1990 to 1997 by the Consumer Product Safety Commission indicated that 15 deaths in the United States each year can be attributed to cosleeping of adults and infants. The commission recommended that babies sleep on their backs in a crib that has a firm, tight-fitting mattress and that all soft bedding, pillows, and stuffed animals be removed from the crib.

Deliberate overlying, which was used on adults by the Edinburgh body snatchers William Burke and William Hare, came to be known as "Burking." To ensure that the bodies would be suitable for medical school dissection, Burke or Hare would place a hand over the victim's mouth and nose while simultaneously placing his body on the victim's chest to prevent the diaphragm from contracting. *See also* Sudden Infant Death Syndrome (SIDS).

References

Lane, Brian. *The Encyclopedia of Forensic Science.* London: Headline, 1992.

Sears, William. *SIDS: A Parent's Guide to Understanding and Preventing Sudden Infant Death Syndrome.* Boston: Little, Brown, 1995.

"Untold Dangers in the Family Bed?" *The Community Perspective (Newsletter)*, Growing Families International, Spring 1997. http://redrhino.mas.vou.edu/Ezzo/cosleeping.

P

Paint Evidence

Paint evidence associated with possible crimes is usually in the form of tiny chips of paint, a common type of trace evidence associated with hit-and-run accidents and burglaries. Forensic investigators routinely look for such evidence on victims, near the scene of a crime, on suspect cars, and on suspect burglar tools or the burglars themselves, who often remove small patches of paint when they are breaking into a building or a safe. In the case of paint chips left by a car, the color, make, and model of the vehicle can often be determined at a forensic laboratory. Under a microscope, the color, texture, and sequence of layers can be ascertained. It may even be possible to find a fit, similar to pieces of a jigsaw puzzle, between pieces collected at the point of a hit-and-run accident and a car or between a burglar's tool and the paint remaining on a window frame. Paint samples subjected to infrared spectrophotometry, pyrolysis gas chromatography, and other tests may provide patterns that will allow investigators to compare samples from a crime scene with those associated with a suspect or his or her vehicle. Since 1974, the Law Enforcement Standards Laboratory at the National Bureau of Standards (now the National Institute of Standard and Technology) has collected and sent paint samples from all makes and models of cars manufactured in the United States to crime labs throughout the country.

The case of Malcolm Fairley illustrates the use of paint chips in solving crimes. In 1984, an area north of London was terrorized by a masked, left-handed, bisexual burglar and rapist known as the Fox who carried a shotgun taken from one of his victims. After one of his attacks, police searching the fields around the victim's home found tire tracks and part of a sheet the criminal had taken from the home, as well as his shotgun hidden under some leaves. They also found a flake of yellow paint on a tree at a height nearly four feet above the ground. Police set up a surveillance expecting the burglar-rapist to return for his gun, but he never came.

It was the flake of paint that eventually led to his capture. A forensic laboratory identified the color of the paint flake as Harvest Gold, a color used only by Austin. Further analysis showed that Austin had used this particular paint only on Allegros made between 1973 and 1975. As police searched for previously arrested burglars who might own an Austin Allegro, they came upon Malcolm Fairley washing his yellow Allegro. On the rear body of the car, at a height of 45 inches, they found that a chip of paint was missing. On the seat of the car was a wrist watch. When police handed the watch to Fairley, he

Table 7
Comparison of the Concentration of Elements Found in 12 Samples of Paint

Sample	Titanium (%)	Aluminum (%)	Sodium (ppm)	Copper (ppm)	Manganese (ppm)
1	8	0.7	190	1,100	11
2	21	0.4	190	12	170
3	19	0.3	110	100	1.6
4	20	0.3	110	86	260
5	11	0.1	70	130	2.4
6	15	0.2	290	72	570
7	8	0.1	60	270	6.0
8	9	0.9	220	1,200	12
9	12	0.1	50	95	0.9
10	15	0.2	80	160	4.6
11	11	0.2	70	160	0.7
12	8	0.1	40	16	4.6

Note: Sample 1 was a paint fleck found at a crime scene. Sample 8 was taken from the suspect's car. The other ten samples of blue paint used on cars collected at random.

Source: Adapted from Figure 5 of Robert and Joan Watkins, "Identification of Substances by Neutron Activation Analysis," *American Jurisprudence Proof of Facts: Annotated Vol. 15*, Jurisprudence Publishers, Inc., 1964, p. 130.

put it on his right wrist, showing that he was left-handed. A search of Fairley's home revealed other articles associated with his burglaries. Faced with the evidence against him, the Fox confessed, was convicted, and was sentenced to life in prison.

In a case involving a hit-and-run driver, a fleck of blue paint found at the crime scene was subjected to neutron activation analysis. To a high degree of certainty, the composition of the blue paint on the suspect's car matched that of the crime-scene sample. The results of the analysis, showing the crime-scene sample (1), the paint on the suspect's car (8), and ten other samples, are provided in Table 7. *See also* Automobile Accidents; Chromatography; Magnifiers; Trace Evidence.

References

American Jurisprudence Proof of Facts Annotated. Vol. 15. San Francisco: Bancroft-Whitney Co., 1964, 115–148.

Saferstein, Richard. *Criminalistics: An Introduction to Forensic Science*. Upper Saddle River, NJ: Prentice Hall, 1998.

Wilson, Colin. *Clues! A History of Forensic Detection*. New York: Warner Books, 1991.

Zonderman, Jon. *Beyond the Crime Lab: The New Science of Investigation*. New York: Wiley, 1990.

Palmer, Dr. William (1824–1856)

Dr. William Palmer, known as "Palmer the poisoner," was one of the most successful serial killers of the nineteenth century. Only the exhumation of an alleged victim and analysis of the remains by one of the early forensic experts on poisons brought the killer to justice.

Motivated by greed, the Englishman went on a nine-year murder spree that resulted in at least nine homicides. In 1855, he was arrested and convicted of poisoning a friend in a case with overwhelming circumstantial evidence. By age 17, the doctor-to-be was already an inveterate liar and had been dismissed from two pharmacy positions. The first dismissal was for embezzlement; the second may have been for running an illegal abortion clinic. During his medical training, Palmer allegedly fathered 14 illegitimate children. Foreshadowing his future activities was the suspicious death of one of the children after a visit from the father.

Palmer qualified as a physician in 1846 at London's St. Bartholomew's Hospital. The following year he married Ann Brooks, the daughter of a moderately wealthy widow, though the doctor would ultimately be disappointed in his wife's financial status. Palmer essentially confined himself to searching for the good life on vacations he could ill afford. He spent most of his time at the racetrack, leaving his floundering practice to an assistant. Eventually, he became involved with loan sharks and moneylenders who pressured him for payments.

As the 1850s approached and Palmer's debts increased, there was a distinct rise in the mortality rate among his family, friends, and acquaintances. The deaths were always accompanied by some unusual financial considerations. In 1849, Palmer invited his mother-in-law to spend some time with his family. The elderly matron had reservations, telling anyone who would listen that she would never survive the visit. After a week, the woman suddenly died. Her daughter, and by extension Palmer, inherited a small bequest.

Sudden deaths related to Palmer continued to occur with startling frequency. A fellow gambler unexpectedly succumbed during a visit to the Palmer household. The bookie's account ledger, listing Palmer as a creditor, was lost, and his winnings disappeared without a trace. In close succession came the death of another creditor, Palmer's uncle, and even a number of his own children. In 1853, Palmer dutifully insured his wife. She died after Palmer had paid only one premium. Buoyed by the success of his scam, Palmer insured his alcoholic brother, who obligingly drank himself to death, leaving his bereaved brother with a tidy sum of money. The insurance company, convinced that there was some irregularity in the death, refused to pay the benefit. The determined Palmer then took out a policy that insured a friend, George Bates, who promptly died. The ensuing investigation uncovered a witness, a boot boy, who had seen Palmer pour something into Bates's drink before he became ill.

On November 21, 1855, a desperate Palmer accompanied his friend and horse owner John Parsons Cook to the Shrewsbury races. The two friends watched as Cook's horse won. They then retired to a local tavern to celebrate the horseman's good fortune. Cook downed some brandy and immediately jumped to his feet shouting about a burning in his throat. Palmer drained the few remaining drops of the afflicted man's drink and pronounced the liquor untainted. Cook was taken to a nearby inn and attended by his doctor friend, but his condition deteriorated rapidly. He vomited regularly and was afflicted by severe spasms. So violent were the convulsions that just prior to his death Cook's heels actually touched the back of his head. Upon the horrible death of his friend, Palmer quickly produced a document in which the deceased promised his doctor a considerable amount of money. Cook's suspicious stepfather demanded an autopsy.

Palmer, as a doctor and friend of the dead man, was allowed to observe the autopsy. From the start, the observer's actions aroused suspicion. He inadvertently bumped an attending physician, which caused the contents of Cook's stomach to fill the body cavity. At one point in the procedure, the doctor was observed attempting to leave the room with the specimen jar containing the dead man's stomach. Failing in this, Palmer tried to bribe the lab assistant who was removing the jars.

Palmer's bizarre actions only focused the harsh light of suspicion on him. People recalled the number of deaths associated with the doctor, and his defense was not helped by the revelation that he had forged a check while Cook lay dying. The document promising Palmer a large sum of money was also exposed as a fraud. The authorities ordered the exhumation of both Palmer's brother and wife. Alfred Swain Taylor, professor of medical jurisprudence and an authority on poisons, was called to examine the bodies. Cook's stomach was also sent to the forensic expert for study. The expected strychnine residue was not discovered. The only commonality Taylor found was a high level of antimony, which was hardly enough, by itself, to convict Palmer of anything. The circumstantial evidence and the strong belief that Palmer was guilty led to a spectacular trial. After a little over an hour and a half of deliberation, the jury returned a guilty verdict. He was hanged on June 14, 1856, but to the end he never confessed. *See also* Serial Killers.

References

Blundell, Nigel. *Encyclopedia of Serial Killers.* North Dighton, MA: JG Press, 1996.

Cyriax, Oliver. *Crime: An Encyclopedia.* North Pomfret, VT: Trafalgar Square, 1996.

Wilson, Colin, with Damon Wilson and Rowan Wilson. *World Famous Murders.* London: Robinson, 1993.

Palmprints

Although palmprints are not routinely recorded, they can, like fingerprints, be used as a means of identification. Consequently, forensic investigators will look for palmprints as well as fingerprints at a crime scene.

In 1942, an English pawnbroker was beaten to death by burglars. Although evidence at the crime scene was scarce, one of the burglars had left a palmprint on a safe in the store. Detective work led to the questioning of two men who had been seen examining a revolver in a pub on the day the crime was committed. A check of their palmprints showed that one of the men had opened the safe. The man with the matching palm quickly confessed but claimed that his companion had committed the murder. The judge rejected the plea that only one of the two was guilty of murder because both had pursued a common purpose in the robbery and murder. Both men were convicted and hanged.

In 1948, a palmprint led to the solution of another murder in England. A woman had been killed while she was walking her dog on a golf course. The killer had left a portion of his palmprint on a tee iron used to mark the 17th tee. He had used the marker to strike his victim. Police proceeded to take palmprints from all the some 7,000 people who lived or worked in the area. Their persistence

paid off. They found that the print on the marker matched that of a 17-year-old youth who lived with his parents near the golf course. He confessed and was quickly convicted and sentenced to life in prison. His life was spared only because he was under 18 years of age.

In 1973, Gloria Carpenter, a resident of Modesto, California, was found submerged in her bathtub. The cause of death appeared to be drowning following a heart attack, but the coroner found no water in her lungs. He also found a thin line around her neck. She had been strangled and raped. An investigation led police to suspect Jimmy Wayne Glenn. Glenn admitted that he had been drinking with Carpenter at a bar on the night of the murder, but claimed that he had never entered her apartment. A faint palmprint on the rim of the bathtub matched Glenn's and proved that he had been not only in Carpenter's apartment, but at the scene of the crime. He was found guilty and sentenced to life in prison.

Beginning in April 1986, a serial killer terrorized southwest London, England. His victims were all elderly people, both men and women, whom he strangled, robbed, and sexually assaulted. In some cases, he turned nearby photographs away from the crime or covered them as if he believed that the people in them could serve as witnesses to his crimes. By the end of July, "the Stockwell Strangler," as he came to be known, had killed eight people. In one case of attempted murder, the attacker was repulsed when Fred Prentice, a 73-year-old living in a home for the aged, managed to press an alarm button near his bed, causing the attacking strangler to flee.

At one murder scene, a fingerprint was recovered from a plant pot. At another, a palmprint was discovered. The two prints matched those of Kenneth Erskine, a small-time burglar who had served a prison sentence. Police learned that while Erskine had no permanent address, he was receiving social security payments. They staked out the place where the payments were being sent and soon had the suspect in custody. Erskine admitted burglarizing the apartments but denied murdering anyone. When police pointed out that the burglaries were at the same locations as the murders, Erskine said that he might have committed them, but he had no memory of doing do. In court, police presented the fingerprint and palmprint evidence, as well as the fact that Erskine had made bank deposits on dates that immediately followed each murder and robbery. They also had Fred Prentice's eyewitness identification. Prentice had successfully distinguished Erskine from others at a lineup. The jury found Erskine guilty of eight murders and one attempted murder. He was sentenced to eight life sentences for the murders and one 12-year sentence for attempted murder.

More recently, Boulder, Colorado, police used palmprints to show that there was no connection between a sexual assault on a 14-year-old girl in 1997 and the Christmas 1996 slaying of Jon-Benet Ramsey. Despite similarities between the two cases, including the fact that both girls attended the same dance academy and lived within two miles of one another, palmprints found at the two crime scenes were not from the same hand. *See also* Drowning; Fingerprints.

References

"Cops Rule Out Ramsey in Sex Case." *AP Online,* Sept. 13, 2000. bigchalk.com, inc. http://www.crimelynx.com.ramsey.html.

Evans, Colin. *The Casebook of Forensic Detection: How Science Solved 100 of the World's Most Baffling Crimes.* New York: Wiley, 1996.

Haime, Kevin. "The Stockwell Strangler: A Crazed Serial Killer Was Murdering Senior Citizens in Southwest London." *Ottawa Sun,* August 13, 2000, p. 16.

Wilson, Colin. *Clues! A History of Forensic Detection.* New York: Warner Books, 1991.

Paraffin Testing

Paraffin testing involves applying melted paraffin to a suspect's hands with a paintbrush. After the paraffin hardens, it is removed and tested by adding a solution of diphenylamine. The presence of blue flecks in the paraffin is indicative of gunpowder residues. When a gun is fired, gunpowder residues that contain nitrates and nitrites are blown back onto the hand of the person doing the shooting. For many years, the paraffin test, also known as the "dermal nitrate test," was used to test for nitrate residues on the hands of suspects believed to have recently fired a gun. The test is no longer used because many other substances in the environment, such as fertilizers, explosives, tobacco, urine, and cosmetics, contain nitrates and nitrites; consequently, the paraffin test lacks specificity. Furthermore, since the test relies on the oxidizing nature of the nitrate ion, other oxidizing agents can produce a similar effect.

The more common approach today is to swab the suspect's hands with dilute nitric acid and then test the swabs for barium and antimony using absorption spectrophotometry. Another approach is to use tape to remove any tiny particles of powder from the suspect's hands. The tape can then be examined with a scanning electron microscope, where the characteristic size and shape of the primer-powder particles can be identified. *See also* Electron Microscope; Gunshot Wounds; Spectrophotometry.

References

DiMaio, Vincent J.M. *Gunshot Wounds: Practical Aspects of Firearms, Ballistics, and Forensic Techniques.* New York: Elsevier Science, 1985.

Saferstein, Richard. *Criminalistics: An Introduction to Forensic Science.* Upper Saddle River, NJ: Prentice Hall, 1998.

Parkman, Dr. George: Murder. *See* Webster, John: Murder.

Paternity Testing

Paternity testing involves the use of DNA or blood samples so that forensic scientists can settle questions of paternity. A man accused of fathering a child may be able to prove his innocence through blood tests or DNA testing. Occasionally, babies are accidentally switched in a hospital, and parents take someone else's baby home.

Long before DNA testing became available, blood was used to try to determine paternity. (See discussion on "Inheritance of Blood Type and Paternity" in Blood entry.)

Table 6 (in Blood entry) also shows the genes that the individual may transmit through his or her gametes (sperm or egg cells). Chromosomes separate during the formation of gametes so that only one member of each pair of gene-bearing chromosomes enters a sperm or egg cell; consequently, only one gene for A-B-O blood type will be in any one gamete. When the gametes join to form a zygote, the cells that result from later cell division will carry two genes for blood type, one on each of two chromosomes. A person who is homozygous (genes on both chromosomes are the same) for type-A blood can transmit only the A gene to offspring. A person who is heterozygous (genes are not the same on both chromosomes) for type-A blood can transmit either the A or the o gene to offspring. All people with type-O blood are homozygous because the gene is recessive. If either the **A** or the **B** gene for blood type were present, the person would have type-A or type-B blood, not type-O. A person with type-AB blood will carry both the **A** gene and the **B** gene.

Consider the case of a person heterozygous for type-A blood who is married to a person heterozygous for type B-blood. The person with type-A blood has a 50–50 chance of transmitting either the **A** or the **o** gene to his or her offspring. The person with type-B blood has a 50–50 chance of transmitting either the **B** gene or the **o** gene. Since a child receives one gene for each gene pair from each parent, the genotypes of the offspring from this marriage could be **AB**, **Ao**, **Bo**, or **oo**. Their children have an equal probability of belonging to any of the four blood types. The possible combinations of the genes (genotypes) for offspring of this marriage are shown in Table 8.

Suppose parents who both have type-O blood find that the baby they brought home from the hospital has type-A blood. They know that the baby is not theirs. Since both carry only recessive genes, they could not

Table 8
Possible Blood-Type Combinations in Offspring of Heterozygous Parents

		Possible Gametes from Genotype Bo	
		B	o
Possible Gametes from Genotype Ao	A	AB	Ao
	o	Bo	oo

produce a child that carries the dominant **A** gene. On the other hand, if one of the parents had type-A blood, the baby could be theirs, but A-B-O blood typing alone cannot prove that the baby is definitely theirs.

A woman with type-O blood claims that a man with type-A blood is the father of her illegitimate child who has type-B blood. The man's lawyer points to the fact that neither the man nor the woman carries the gene for type-B blood; consequently, the man cannot be the child's father. However, if the man accused of paternity has type AB blood he could be the father, but A-B-O blood typing alone cannot establish his paternity with certainty.

Charlie Chaplin Case

A famous case involving paternity occurred in 1943 when a young female actress named Joan Barry accused actor-comedian Charlie Chaplin of fathering her child. The blood tests, which were monitored by a three-doctor panel, showed that Chaplin had type-A blood. Barry's blood was type A, and the baby's was type B. The evidence was conclusive: Chaplin could not have fathered the child. Nonetheless, Barry's lawyers were successful in having Chaplin convicted because at that time blood tests were not considered to be conclusive under California laws.

With advances in DNA testing, blood types have become less widely used in establishing paternity. However, blood typing is much less expensive than DNA testing, so blood tests are often done first to see if blood typing can negate any claims of paternity, as was the case with Charlie Chaplin.

Bill Cosby and Paternity

In 1997, Autumn Jackson was convicted of extortion. She had tried to obtain $40 million from comedian Bill Cosby by threatening to go public with her claim that Cosby was her father. The FBI had secretly recorded Jackson negotiating with Cosby's lawyers about "hush" money. Cosby admitted to having had an affair with Jackson's mother, Shawn Upshaw, but he denied that he was her father. He told the jury that shortly after Autumn Jackson was born, her mother told him that he was the father. Cosby denied paternity but refused to submit to a blood test because he was afraid that it would be leaked to the press and have a negative effect on his career. Nevertheless, Cosby gave Upshaw about $100,000 and paid for her children's education. The court ruled that Cosby's affair was irrelevant to the case, which had to do with the extortion charge against Jackson. Following Jackson's conviction, Cosby's attorney, Jack Schmitt, stated on CNBC's *Rivera Live* television show that Cosby was undergoing a DNA test to prove once and for all that he was not Jackson's father. Jackson's attorney, Robert M. Baum, said that his client would not submit tissue for DNA testing prior to being sentenced. He also noted that tests will not change the fact that Jackson was raised believing that Cosby was her father.

Paternity Tests Reveal Children Switched in Hospital

In 1998, Paula Johnson, in an effort to obtain child support for her three-year-old daughter, Callie Marie, obtained a court order requiring her former boyfriend to submit to a paternity test. The results of the test showed that the former boyfriend, Carlton Conley, was not the child's father. More shocking was the fact that a second test revealed that Johnson was not Callie's mother. Somehow, three years earlier at the University of Virginia Medical Center in Charlottesville, her baby had been switched with another child born at approximately the same time. An investigation failed to discover how the switch had occurred.

Authorities at the hospital used its records to track down Johnson's biological daughter. She was Rebecca Grace Chittum in Buena Vista, Virginia, about 75 miles west of Johnson's home in Ruckersville. She was living with her grandparents because her parents, Kevin Chittum and Whitney Rogers, had been killed in a traffic accident. Initially, both the grandparents and Johnson agreed that the children should remain with the people who had raised them during the first three years of their lives. However, the situation soon changed. Johnson and the Chittum and Rogers families sued one another for custody and visitation rights. In March 2000, a circuit court approved a settlement in which the girls remain with the families who raised them. Both sides, however, will have visitation rights. Johnson has filed lawsuits against the hospital and the company that manufactures bracelets for newborns. An earlier lawsuit by Johnson against the state of Virginia, the hospital, and 17 doctors and nurses was dismissed. The Chittum and Rogers families accepted a $1 million settlement from the state on behalf of Rebecca Chittum in 1998. Johnson rejected a similar settlement, believing she could obtain more through the courts.

What Constitutes Fatherhood?

Must a man pay child support for a child he did not father? State courts in Florida have consistently ruled that once a man is married, he is responsible for any child born to his wife. However, in recent years, since DNA testing has become available to determine paternity, the courts have issued conflicting decisions because some courts have freed men of financial responsibility for children proven not to be theirs.

In *Anderson v. Anderson*, the Florida Supreme Court will face the question, What constitutes fatherhood? Michael Anderson's attorney believes that men should not have to pay child support for children they did not father. DNA testing revealed that Anderson was not the father of the daughter born to Cathy Anderson during their brief marriage. Cathy Anderson's attorney argues that a man who takes the marriage vows is responsible for any child born to his wife regardless of that child's biological origin. If men are forgiven support payments because of DNA testing, more and more children will be left without financial support. Furthermore, Michael Anderson could have had himself and his wife's daughter tested for genetic matching prior to the divorce, but he did not. As a result of his lack of action, he is legally responsible for the child born to his wife during their marriage. In December 1999, the Florida District Court of Appeals ruled in favor of Cathy Anderson. *See also* Blood; Chaplin, Charlie: Paternity Case; DNA Evidence.

References

Follick, Joe. "In Child Support Case, Ex-Husband Plays DNA Card." *Tampa Tribune*, Aug. 19, 2000.

Guthrie, Bruce, Linda Kramer, Ellen O'Conner, and Jane Sims. "Up Front: The Heartbreak Kids: Two Fair, Blue-eyed Girls Find Themselves at the Center of a Tragic Baby-switch Case." *People*, Aug. 17, 1998.

Sanminiatelli, Maria. "Insurer's Payoff on Switched Girls Up in Air." *Washington Times*, Aug. 8, 2000, p. C3.

Woodruff, Bob, Mark Mullen, and Blake Asha. "Bill Cosby Testifies in Extortion Trial." *ABC World News This Morning*, Jul. 16, 1997.

Zoglin, Richard, reported by Charlotte Faltermayer. "Justice: From Here to Paternity: Bill Cosby Offers a Blood Test, and Autumn Jackson Gets Another Father Figure or Two. Is This a Sitcom?" *Time*, Aug. 11, 1997, p. 32.

Pathologist. *See* Pathology.

Pathology

Pathology is the study of the causes, processes, development, consequences, and manifestation of diseases. A pathologist is a doctor who specializes in interpreting various aspects of disease by examining body tissues and fluids. A clinical pathologist works in a laboratory where he or she carries out procedures to identify and quantify foreign organisms and organic substances in body fluids and tissues.

A forensic pathologist carries out autopsies in an effort to detect the cause of sudden, violent, or unusual deaths, particularly those associated with crimes such as murder. By doing so, the pathologist provides medical evidence that may be used in court. *See also* Cause of Death; Time of Death; Toxicologists; Toxicology.

Reference

Saferstein, Richard. *Criminalistics: An Introduction to Forensic Science.* Upper Saddle River, NJ: Prentice Hall, 1998.

Poisons

A poison is a chemical that produces harmful effects in living tissues. Before the development of chemistry provided analytical methods for detecting poisons, poisoning was a common means of killing someone. Poisons were the bullets of earlier times.

Almost any substance is toxic if it is taken in large quantities or if it enters the body in the wrong way. For example, water, which is essential to life, can cause death by drowning if it enters the lungs rather than the stomach. In addition to man-made chemicals such as cleaning compounds and medicines, which can be dangerous if they are ingested in improper amounts, there are natural poisons produced by plants and animals to ward off predators or kill prey. Eating belladonna seeds or daffodil bulbs can be fatal. Rattlesnakes, black-widow spiders, and other animals produce poisonous venoms that can cause nausea, vomiting, paralysis, respiratory failure, convulsions, and even death.

Poisons reach body tissues in different ways. Carbon monoxide, like any gas, is inhaled into the lungs. The gas is found in the exhaust of automobiles and is very commonly used to commit suicide and less frequently to commit murder. In the lungs, it combines with the hemoglobin in red blood cells. Normally, hemoglobin unites with oxygen and carries this vital gas to the cells of the body where the gas is readily released. The compound formed by hemoglobin and carbon monoxide is much more stable and decomposes slowly. As a result, carbon monoxide reduces the capacity of the red blood cells to distribute oxygen, and cells suffer from oxygen starvation. Concentrations as low as two parts per thousand will produce loss of consciousness in 30 minutes and death in about three hours. Higher concentrations will cause death in less time.

Some poisons can be absorbed through the skin, but most are swallowed and enter the digestive tract. More than 90 percent of all poisonings occur in the home, and the most common cause is the ingestion of poisonous liquids or solids by young children. However, the incidence of accidental poisonings among children age five or less has declined significantly since the introduction of childproof caps.

Often described in morbid fashion in mystery novels, poisoning accounts for less than 5 percent of all homicides, and toxicologists are usually able to detect the poison during an autopsy if there is reason to suspect poisoning. Because poisons often produce symptoms similar to those of ordinary illnesses, doctors may conclude that a death was due to natural causes when poison was actually involved. Without an autopsy, murders due to poisoning may be attributed to natural causes.

Poisons may be classified as (1) those, such as carbon monoxide and cyanide, that interfere with normal biochemical processes in the body; (2) corrosive materials, such as mustard gas and chloroform, that pierce the stomach wall; (3) substances that produce widespread damage, such as arsenic and other heavy metals, and alkaloids such as morphine and strychnine; and (4) substances that leave no trace of their presence, such as ricin, a derivative of the castor bean, which is a deadly poison that causes clumping of the red blood cells and leads to death through an imbalance of ions.

The deadly ricin was believed to be the poison used in the 1978 murder of Georgi Markov, a Bulgarian who had defected from that country's Communist regime eight years earlier. While Markov was walking in London, he felt a sharp pain in his right thigh. He turned to see a stranger with an umbrella mumble an apology and hurry off. Upon arriving home, he found a red mark on his thigh where he had felt pain. The next day he was in the hospital with a fever and vomiting. X rays failed to reveal anything in his thigh, but the region surrounding the puncture wound was inflamed. Despite large doses of antibiotics, Markov's condition became critical, and he died three days later.

An autopsy revealed a tiny pellet less than 2 mm in diameter beneath the puncture in his thigh. Authorities believed that the assailant's umbrella held a gun that had been used to drive the pellet into Markov's flesh. Since the pellet could have held only a minute quantity of poison, toxicologists tried to determine what substance, in such a small amount, could have caused the symptoms exhibited by the victim. They soon ruled out all toxins except for one—ricin. To test their hypothesis, they injected a full-grown pig with the same amount of ricin believed to have been held by the metal pellet. Within 24 hours the pig was dead. An autopsy revealed tissue damage that was virtually the same as that found in Markov's body. The evidence was convincing but not conclusive because there was no trace of poison. If ricin was used, it was probably for that very reason. It breaks down in the body, leaving only natural body products in its wake. The Bulgarian Embassy denied any assassination plot. However, more than a decade later, following the end of communism in Bulgaria, the new government admitted that Markov's assassination had been planned by Communist leaders in Sofia.

Three scientific tests had been conducted by October 2000 on a lock of Ludwig van Beethoven's hair. The researchers who had analyzed 20 strands of hair from the head of Beethoven concluded that he suffered from, and may have died of, lead poisoning. The concentration of lead in his hair was many times higher than normal. The excess lead may have caused many of the health problems that plagued him, such as digestive disorders, stomach pain, irritability, and depression. There was no indication of excess mercury, which led the research team directed by Dr. William Walsh of the Health Research Institute in Naperville, Illinois, to question some scholars' belief that Beethoven was afflicted with syphilis. Syphilis, at the time of Beethoven's death in 1827, was treated with compounds of mercury. *See also* Arsenic Poisoning; Opium; Strychnine.

References

Cyriax, Oliver. *Crime: An Encyclopedia.* North Pomfret, VT: Trafalgar Square, 1996.

Evans, Colin. *The Casebook of Forensic Detection: How Science Solved 100 of the World's Most Baffling Crimes.* New York: Wiley, 1996.

Grolier Multimedia Encyclopedia. Compact disc. Danbury, CT: Grolier Electronic Publishing, 1995.

Wilson, Colin. *Clues! A History of Forensic Detection.* New York: Warner Books, 1991.

Polygraph Test

A polygraph is sometimes used by forensic investigators to try to determine whether someone is telling the truth. People often give away telltale signals when they are lying. They may blush, avoid eye contact, perspire, or exhibit other telltale body language. Cesare Lombroso noticed such reactions and designed the world's first lie detector, which he called a hydrosphygmograph. The device consisted of a water-filled glass bulb sealed with a rubber membrane that enclosed the subject's fist. The bulb was connected to an air column where changes in the volume of the subject's fist due to heartbeats could be clearly seen and recorded.

When a thief, Bersonne Pierre, was arrested and searched for a railroad robbery, another person's passport was found in his pocket. Lombroso used Pierre to test his hydrosphygmograph. When Pierre was questioned about the railroad heist, his heart rate remained steady as he denied any knowledge of it, but questions related to the passport led to an increased heart rate with larger pulses.

Lombroso told police that Pierre was not involved in the railroad robbery but had stolen the passport. Further investigation showed that Lombroso's analysis based on his machine was correct.

Modern polygraphs (lie detectors) are improved versions of one invented by John Larson in 1921. Larson's device was a sphygmomanometer to which he attached a needle so that blood pressure could be recorded on a roll of soot-darkened paper that turned beneath the needle. Larson tested his polygraph on August Vollmer, the chief of police of Berkeley, California, who had suggested to Larson that lying might produce measurable changes in blood pressure. Vollmer, who watched the meter as he responded "yes" or "no" to Larson's questions, was mightily impressed. Every time he lied, the needle jumped, registering a spike on the record.

Later, Larson's lie detector was improved. It became known as a polygraph because it now measured a number of changes associated with the body's autonomic nervous system, which controls involuntary, unconscious actions of smooth muscles and glands. In addition to blood pressure, the modern polygraph can record changes in pulse, breathing rate, blood flow, brain waves, and galvanic skin response, which is a measurement of the skin's electrical resistance. Perspiration, which contains salt, will reduce the skin's electrical resistance, causing an increase in the electrical current between electrodes attached to a subject's fingers.

Most polygraphs measure just three indicators while a subject is asked a series of questions. Rate and depth of breathing are measured by straps that wrap around the chest. Blood pressure is detected by a sphygmomanometer, and galvanic skin response is measured by electrodes attached to the fingertips. Insulated wires connected to these devices lead to inked styluses that record the data on rolls of moving paper.

The polygraph is based on the theory that questions about incriminating behavior will stimulate the autonomic nervous system and produce physiological responses that can be detected. However, in 1930, Albert Riedel realized that the polygraph was far from perfect in detecting lies. A tired subject whose adrenalin was depleted might lie without providing a detectable response. Other extraneous stimuli such as heat, noise, or an uncomfortable position could produce responses unrelated to the questions used to test an individual's veracity. A question thought to be neutral by the examiner might evoke a dramatic response from a subject, or a key question designed to produce telltale physiological responses by a guilty party might elicit no response because of the subject's cultural experience.

To make the polygraph a reliable instrument, Riedel established protocols for testing. Subjects should be asked sequential questions for no more than three minutes; frequent breaks should occur to avoid fatigue. Questioning should be done in a comfortable setting without any distractions. Questions should be unambiguous and consist of a mixture of relevant and irrelevant queries, and the social and cultural background of the subject should be considered when the examiner is framing questions. When the precautions established by Riedel are used, a pattern usually emerges after an hour of questioning. An innocent person will be more likely to respond to extraneous questions designed to elicit a response than to questions relevant to the crime. On the other hand, a guilty subject exhibits more pronounced reactions to questions pertaining to the crime and less to irrelevant questions.

Nevertheless, there are people who can fool the polygraph. A very nervous but innocent person may react as if he or she is guilty, and guilty parties who have superb control of their emotions may appear innocent. Aldrich Ames, a CIA agent convicted of spying for Russia, was able to beat polygraph tests in 1986 and again in 1991. Chester Wegner, who had murdered three women, passed polygraph tests in 1960 by taking aspirins washed down with cola. Richard Crafts, the infamous woodchipper murderer, passed a polygraph test before police discovered that his wife's body had been passed through a woodchipper. In the hands of a trained expert, the polygraph is probably about 95 percent accurate, but since lie-detector data are not always reliable, polygraph tests are not accepted as evidence in many courts unless both the defense and the prosecution agree to accept the results or the judge requires it. *See also* Larson, John; Lombroso, Cesare; Vollmer, August.

References

Cyriax, Oliver. *Crime: An Encyclopedia*. North Pomfret, VT: Trafalgar Square, 1996.

Gardner, Robert, and Dennis Shortelle. *From Talking Drums to the Internet: An Encyclopedia of Communications Technology*. Santa Barbara, CA: ABC-CLIO, 1997.

Science and Technology Illustrated. Chicago: Encyclopaedia Britannica, 1984, vol. 15, pp. 1804–1805.

Wilson, Colin. *Clues! A History of Forensic Detection*. New York: Warner Books, 1991.

Zonderman, Jon. *Beyond the Crime Lab: The New Science of Investigation*. New York: Wiley, 1990.

Psychic Detectives

Psychic "detectives" are people who claim that they are able to use paranormal methods to solve crimes. Approximately 35 percent of city police departments have used a psychic at least once in an attempt to solve a crime or find a missing body or article. Rural police have used psychics about half as frequently. But neither urban nor rural psychics can claim an outstanding success rate.

A Dutch psychic who called himself Peter Hurkos (his real name was Pieter Van der Hurk) was typical of people who claim to have psychic powers. For a period of 30 years beginning at the end of World War II, Hurkos did well on stage when he was in control, but often failed to exhibit paranormal powers when he was called upon to help police solve crimes. Hurkos claimed to have a psychometric capacity—the ability to obtain information about the owner of an object by holding the object.

Because of a glowing reputation as a psychic sleuth that he helped foster with the press, Hurkos was often hired by police departments to help them solve crimes. In 1964, he identified the Boston Strangler as a man weighing 130–140 pounds with a pointed nose, a scar on his left arm, and one who loved shoes. Responding to Hurkos's description, police arrested a shoe salesman. For a while, the murders associated with the strangler stopped; then another murder occurred, and Albert DeSalvo, who later confessed to the crimes, was arrested. DeSalvo's detailed knowledge of the crimes attributed to the Boston Strangler, which allegedly only the murderer could have known, convinced authorities that DeSalvo, not the shoe salesman, was the guilty party. Like many psychic detectives, Hurkos often made several ambiguous predictions about the person(s) who had perpetrated a crime. If the crime was solved, the incorrect assessments were neglected and the correct and ambiguous ones were interpreted to fit the facts and offered as evidence of paranormal powers.

In a 1994 test designed by British psychologists Richard Wiseman, Donald West, and Roy Stemman, the performance of three psychic detectives was compared to that of a control group consisting of three college students who laid no claims to having psychic powers. All the subjects were shown three items: a bullet, a scarf, and a shoe. Each item had been associated with a crime. The subjects in the experiment were video taped individually as they held the objects and were asked to talk about them. After commenting, each subject was given three response sheets, one pertaining to each object. On each sheet there were 18 statements, 6 of which were true of the crime associated with the object. The subjects were asked to mark the 6 statements that were relevant to the crime. While the psychics made nearly twice as many statements as the students, there was no difference in the accuracy ratings of the two groups on either part of the test. Following the tests, the subjects were told about the crimes associated with the objects. Interestingly, the psychics, after hearing the truth about each crime, believed that their responses agreed very closely with the actual

offense. They emphasized the responses that had been reasonably accurate and seemed to forget the many incorrect statements and assessments they had made.

The results of this test agreed closely with other experiments that have been conducted to evaluate the paranormal ability of those who claim to be psychic detectives. For example, Dr. Martin Reiser, working with the Los Angeles Police Department, mixed guns that had been used as murder weapons with guns that had never been fired. Psychics who claimed to have psychometric powers were unable to distinguish one weapon from another.

In a less well controlled experiment, skeptics and a Philadelphia television station invented a story about a missing girl and then released "details" of her life. More than 100 people claiming to be psychics called the girl's father to offer information about her location and condition. None of them realized that she was not really missing.

While the press tends to bring psychic successes to the public's attention, the many failures are often not reported. A few are. A Missouri sheriff had 6.4 million gallons of water drained from an abandoned quarry because a psychic had dreamed that a missing local teenager was at the quarry's bottom. No missing body was found, but police did discover a 1950 Hudson automobile and the bones of an unknown adult. *See also* DeSalvo, Albert.

References

Christopher, Milbourne. *Mediums, Mystics, and the Occult.* New York: Crowell, 1975.

Cyriax, Oliver. *Crime: An Encyclopedia.* North Pomfret, VT: Trafalgar Square, 1996.

"I See . . . Yes! A Backhoe." *New York Times Magazine,* Aug. 25, 1996, p. 8.

"Pursuing the Paranormal." *CQ Researcher.* Washington, DC: Congressional Quarterly, Mar. 29, 1996 vol. 6, no. 12, p. 280.

Randi, James. *An Encyclopedia of Claims, Frauds, and Hoaxes of the Occult and Supernatural.* New York: St. Martin's Press, 1995.

Wiseman, Richard, Donald West, and Roy Stemman. "Psychic Crime Detectives: A New Test for Measuring Their Successes and Failures." *Skeptical Inquirer*, Jan./Feb. 1996.

Psychological Profiling

Psychological profiling is the process of developing a personality sketch of a criminal. It is based on the belief that criminals leave psychological and behavioral clues at the crime scene and is used only in crimes of extreme violence such as rape, child molestation, murder, and arson.

A psychological profile is created by the careful analysis of evidence, both physical and intuitive. A profiler analyzes the investigator's report, including the crime-scene photographs, the autopsy report on the cause and manner of death, weapons used, whether there was postmortem activity, and whether there was torture before death. He or she examines detailed witness interviews and a time line of the victim's travels immediately before the crime. The profiler also seeks background information on the victim's age, gender, race, physical description, marital status, sexual adjustment, intelligence and achievement, occupation, personality, lifestyle characteristics, mental and medical history, use of alcohol or drugs, information on personal relationships, and data on the victim's racial, ethnic, and social traits, as well as his or her neighborhood. The profiler also looks at the crime scene for evidence of motivation and factors such as rage, hatred, love, or fear. Once all the evidence is analyzed and interpreted, it may reveal both the victim's and the offender's behavior during the crime. Based on the behavior of the offender, the profiler tries to determine characteristics of the offender such as race, age, gender, size, marital status, sexual preferences, occupation, and socioeconomic status. Psychological profiling is not an exact science. There is a substantial amount of intuitive guesswork, but profiling has been quite accurate in consolidating the list of suspects for a crime.

In the United States, profiling was pioneered by psychiatrist James Brussel and is carried on today mostly by the Behavioral Science Unit of the FBI. Brussel believed that if one can study a person and make accurate predictions about that person's behavior, one should be able to reverse the process and examine a person's actions and infer what type of person he or she is. He used his technique to develop a profile of the "Mad Bomber," who planted bombs in public places over a period of 16 years. By studying the crime scenes and letters from the bomber, Brussel drew a psychological sketch of a single male in his 40s to 50s, neat in appearance, who was a skilled mechanic. He was an unsocial introvert, could become violent if criticized, and suffered from progressive paranoia. Brussel believed that the bombings were in retaliation for being fired or severely reprimanded. Since paranoia takes time to develop, Brussel thought that the incident that brought on the bombings had occurred 10 to 20 years earlier. His profile proved eerily correct when George Metesky, the "Mad Bomber," was arrested in 1956.

The Unabomber was a more recent case in which psychological profiling played a role in helping investigators find their culprit. For 17 years, from 1978 to 1995, the Unabomber was involved in 16 bombings, 3 deaths, and 23 major injuries. The profile developed by the FBI in 1993 was found to be accurate when the Unabomber, Theodore Kaczynski, was finally arrested in 1995. Their profile was that of a well-educated, obsessive-compulsive loner, hostile to technology, who liked to work with wood and was probably an environmentalist.

In developing a profile for the Unabomber, the FBI Behavioral Science Unit used a computer program called Profiler. The program was developed by compiling information from comprehensive interviews with convicted serial killers, their families, doctors, psychiatrists, neurologists, and social workers during the years 1979–1983. All the information about a case is loaded into a computer program, and, on the basis of this data, Profiler gives a description of the criminal's age, sex, race, marital status, intelligence, occupation, family upbringing, and other personality traits likely to be found in the offender. Profiler helps the investigators develop a list of suspects. The program can even suggest ways to question potential suspects based on the suspect's probable emotional vulnerabilities. From the research, investigators have determined that there are basically two varieties of killers, the disorganized killer and the organized killer. A disorganized killer may not have planned to kill the victim. A sudden rage or fear of being identified could have precipitated the murder. An organized killer is usually someone who has carefully planned the murder.

Psychological profiling has been featured in recent movies and television programs. The two movies about the fictitious murderer Hannibal Lector, *Silence of the Lambs* and *Hannibal*, featured Jodie Foster (in *Silence of the Lambs*) and, later, Julianne Moore (*Hannibal*) as a profiler from the FBI's Behavioral Science Unit (BSU) interviewing a very dangerous killer played by Anthony Hopkins. The television programs *The Profiler* and *Millennium* dealt with psychological profiling. *Millennium* was set in the supernatural realm of profiling, while *The Profiler* was grounded in the real world. *See also* Forensic Psychiatry; Forensic Psychology.

References

Decaire, Michael. *Criminal Profiling: An Informal Introduction and Discussion.* http://www.suite101.com/article.cfm/forensic_psychology/21333.
———. *Criminal Profiling in Fiction.* http://www.suite101.com/article.cfm/forensic_psychology/24469.
Lane, Brian. *The Encyclopedia of Forensic Science.* London: Headline, 1992.
Turvey, Brent E. *Criminal Profiling Overview.* http://corpus-delicti.com/profile.html.
Zonderman, Jon. *Beyond the Crime Lab.* New York: John Wiley & Son, 1990.

Pudd'nhead Wilson

Pudd'nhead Wilson, a novel about the antebellum South, is the story of David Wilson, a fictional forensic scientist created by the pen of Mark Twain (Samuel Clemens). The story begins as Wilson arrives in Dawson's Landing south of St. Louis on the Mississippi River, where he hopes to establish a law practice. Unfortunately, soon after his arrival, he is nicknamed Pudd'nhead by the dull-witted residents, who cannot grasp his sardonic humor.

Wilson tries unsuccessfully for years to establish himself as a lawyer. To make a living, he works as an accountant and surveyor, but continues to seek legal clients. As a hobby, which is essential to Twain's plot, Wilson collects fingerprints. During the course of two decades at Dawson's Landing, he collects the fingerprints of nearly every resident, white and slave, on glass plates that he carefully preserves and studies.

A major character in the novel is a very light-skinned (one sixteenth black), sharp-minded mulatto slave named Roxana who bears a son, fathered by one of the town's leading aristocrats, on the same day that her master's wife also gives birth to a boy. A short time later, the master's wife dies, and Roxana is given the responsibility of caring for both infants. Since her master is preoccupied with business affairs, he cannot distinguish one child from the other except by the clothes they wear. Recognizing the opportunity to save her son from the terror of slavery, Roxana exchanges him (Chambers) for her master's son (Thomas Driscoll), a switch that is never perceived by her master.

Two decades later, Wilson's opportunity as a lawyer arises when he defends Italian twins accused of killing Judge York Driscoll, uncle and now guardian of the false heir, Thomas Driscoll, a deceitful, conniving, and vicious young man who has learned through his mother of his true identity. All evidence points to the twins. They were discovered at the crime scene, and one of them had threatened to kill the judge. However, there are fingerprints on the knife used as the murder instrument, and witnesses saw a young woman leaving the judge's house shortly after the murder.

Wilson searches his files of fingerprints but can find no woman's fingerprints that match those on the knife. However, as luck would have it, Tom Driscoll drops by in a mood to irk Wilson, who he feels has lost the case.

Inadvertently, Tom, seeing Wilson engaged in examining fingerprints, notes that one plate contains the prints of Roxana as well as his and those of Chambers. Suddenly, Wilson recognizes, but does not reveal to Tom, the prints that match those on the knife. He sees, too, because he took the fingerprints of both boys before and after they were switched by Roxana, that Tom is really Chambers, the child of a slave. The true Tom, the judge's heir, has been raised as a slave.

The next day, in a dramatic courtroom scene, Wilson shows that the prints on the knife do not match those of either of the Italian twins, but they do match those of the man thought to be Thomas Driscoll. He then explains how the boys were switched by Roxana. On the basis of the evidence, the twins are freed, and the true heir receives his inheritance and freedom, but cannot escape the slave culture in which he was raised. The villainous false heir, Tom, who confesses his guilt, is sold down the river, where he becomes a slave, leaving his mother distraught.

When Twain wrote this novel in the 1890s, fingerprints had not been established as a vehicle for crime detection by police departments. Consequently, he must have been well aware of the research that had been done on fingerprinting and of the potential that fingerprints offered in establishing criminal guilt. *See also* Fingerprints.

References

Ever the Twain Shall Meet. joseph@telerama.lm.com.

Nettels, Elsa. "The Tragedy of Pudd'nhead Wilson." *Benet's Reader's Encyclopedia of American Literature.* New York: HarperCollins, 1991, p 886.

Twain, M. (Samuel Clemens). "The Tragedy of Puddn'head Wilson." *Century Magazine*, 1894.

————. *The Unabridged Mark Twain, Vol. 2.* Ed. Lawrence Teacher. Philadelphia: Running Press, 1979.

Q

Quincy

Quincy was the title of a one-hour, weekly mystery series on NBC Television from 1976 to 1983. Jack Klugman starred as Quincy, a Los Angeles medical examiner who used forensic science to solve crimes. Quincy, who had no first name, was aided in his investigations by his lab assistant, Sam Fujiyama, played by Robert Ito. When he was not in his lab or around town, Quincy could usually be located at his friend Danny's restaurant. Quincy's beleaguered boss Dr. Astin (John S. Ragin) was constantly arguing with his subordinates over involvement in cases. Further, Quincy spent a good deal of his time tracking down leads outside the lab. Other regulars included Gary Walberg as police lieutenant Monahan and Val Bisoglio as Danny Toya,

owner of the restaurant that was Quincy's home away from home.

To understand the nature of the work and the general atmosphere of the lab, Klugman spent time at the Los Angeles coroner's office, attended autopsies, and inspected wounds. To further ensure medical and scientific authenticity, the program used technical advisors from the forensic field. A number of the program's subjects were generated from the medical examiner's office. The show dealt with topics that included unqualified coroners, mistreatment of the elderly, disreputable plastic surgeons, and treatment of autistic children as well as the usual crime themes. Quincy's real-life counterparts recognized the character's professionalism but felt that he became too involved in cases and, in so doing, lost his scientific objectivity.

R

Rape

Rape is sexual intercourse without the consent of one of the individuals involved in the act. According to *The World Almanac and Book of Facts, 2000,* in 1998 there were 93,100 reported cases of forcible rape in the United States, which was an incidence of 34.4 per 100,000 people. This represents a significant decline from the 109,060 cases reported in 1992, when the incidence of rape was 42.8 per 100,000 inhabitants. Most, but not all, court cases involve a woman who was forced to engage in sexual intercourse by a man.

Rape is not limited to forcible penetration of the vagina. Penetration of any body orifice may constitute rape. For example, if one male forcibly sodomizes another, it is considered rape. Rape does not require evidence of forceful resistance; fear of personal injury is sufficient. Furthermore, in some states, a man's wife may charge him with rape. In view of the fact that in the United States a woman is battered every 15 seconds, it is not surprising to learn that marital rape is common and is often the cause of murder in self-defense. Rape of a fraudulent nature is also possible. For example, a doctor may claim to be providing medical treatment, or a man may impersonate a woman's husband.

For two decades prior to 1992, rape was increasing at a faster rate than any other crime. Part of that growth may be attributed to more frequent reporting of the crime by its victims. Increased frankness about sexual matters that took place during those years coupled with the feminist movement undoubtedly made victims more willing to report the crime. Prior to that time, women who were raped were less likely to go to the police. Embarrassment, reluctance to undergo cross-examination by a defense lawyer, and insensitive treatment by police and medical examiners caused many victims to suffer silently—and suffer they did. The psychological aspects of rape—shame, humiliation, fear, rage, confusion—can lead victims to develop a sense of vulnerability, an inability to control their lives, a fear of going near the area where the incident occurred, insomnia, eating disorders, and other effects of a long-term nature.

Feminists strongly oppose the decriminalization of prostitution and unfettered circulation of pornographic materials because they claim that both contribute to an increase in the incidence of rape. Whether they do or not is open to debate, but the feminist movement has certainly changed attitudes about rape. Police have become more empathetic toward rape victims, and many departments now have special female personnel who handle rape cases with sensitivity and concern.

Medical examiners now recognize the emotional and psychological effects of rape and are more sympathetic toward those who have been victimized. Victims should be examined not only for evidence but for injuries, pregnancy, and any sexually transmitted diseases. They should also be given access to psychological counseling.

Evidence

If a rape victim reports the crime soon after it occurs, enough evidence may be found to convict the assailant. Victims should not shower or clean away blood or semen before reporting the crime. Those reluctant to report rape should remember that the rapist will probably attack others unless he is apprehended. A medical examination of the victim may reveal semen or blood left by the rapist, but lack of semen does not prove that rape did not occur. Other evidence might include bruises and bleeding, as well as hairs, fibers, and skin left by the rapist. The examination should include vaginal, oral, or anal swabs, depending on the nature of the rape. The pubic region of both victim and suspect, if there is one, should be combed for foreign hairs and hairs that can serve as controls. Blood and saliva samples should be obtained from the victim and from any suspects. Any clothing worn by the victim or a suspect should be placed in large paper bags and taken to the laboratory. The garments may contain seminal stains, blood, hair, fibers, and other bits of evidence that will prove useful to prosecutors.

Semen leaves a stain with a stiff crusty appearance. Such stains, even if they are not visible, are readily detected and identified using the acid phosphatase color test. Acid phosphatase is an enzyme secreted by the prostate gland that contributes to the seminal fluid. A piece of wet filter paper rubbed over the suspected area will absorb acid phosphatase. If acid phosphatase is present, the addition of a drop of sodium alpha-naphthylphosphate and Fast Blue B solution will produce a purple color within 30 seconds.

Of course, the presence of live spermatozoa, which can be detected under a microscope at a magnification of 400 times, is a clear indication of seminal fluid; however, spermatozoa disintegrate readily outside the body. Furthermore, the criminal could have been aspermatic as a result of a vasectomy or a very low sperm count. However, the presence of p30, a protein unique to semen, is a definite indicator. The protein can be detected with anti-p30 serum. If an extract of the stain and anti-p30 are placed on an electrophoretic plate, a distinct line will form between the two when a voltage is applied to the plate. Once semen is detected, DNA testing can be used to compare the semen with blood from a suspect.

Semen found in the vagina may offer evidence related to the time of the attack. If live sperm cells removed from the vagina are detected under the microscope, the attack took place within the last 4 to 6 hours. Evidence of p30 indicates that the attack took place within the last 8 hours, and nonmotile sperm cells may be evident for 3 to 6 days after being deposited. For this evidence to be useful, the victim must report any recent voluntary sexual activity.

Date Rape

A 1989 survey of 6,000 college students revealed that 15 percent of them had experienced intercourse without their consent. In most cases, the rapist was someone they knew; they had experienced what is known as date rape. Authorities estimate that about 80 percent of all rapes involve attacks by someone the victim knows.

During the 1990s, a number of cases of date rape involved what has come to be known as "the date-rape drug." Also known as "roofies," "rope," and "roach," the drug is related to Valium, but is much more powerful. It is manufactured by Hoffman–La Roche and is sold in a number of countries under the name Rohypnol as a remedy for extreme cases of insomnia. Because the drug is colorless, odorless, and tasteless, a rapist can slip it into a woman's drink and it will not be noticed. The rape is carried out after the victim passes out. The woman will recall nothing when she regains consciousness. To prevent such an attack, a woman should prepare her own drinks and watch or hold her glass at all times.

Another "date-rape drug," GHB (gamma-hydroxybutyrate), also known as "Easy Lay" because it can produce loss of consciousness and amnesia, originally developed as an anesthetic. It has also been used as an aphrodisiac, to control weight, or as a diet supplement by bodybuilders. It produces a sense of euphoria, but it can also produce seizures and death. In fact, as reported by Jim Suhr in an Associated Press story that appeared in the *Cape Cod Times* on January 31, 2000, it has caused at least 58 deaths and more than 5,700 overdoses in the United States since 1990.

To avoid date rape, Allison Bell recommends that a woman not be alone with a man she does not completely trust; that she make her policy on sex clear before the date; that she dress conservatively because men often think that a woman wearing tight or revealing clothing is seeking sex; that she not drink because alcohol inhibits logical thinking; that, if possible, she walk away from a situation that is getting out of control; and that if a firm and loud "No!" does not deter sexual aggression, the victim should yell for help.

Male Rape

Although male rape is commonly thought to be an act involving homosexuals, it is not uncommon among heterosexual males, particularly in prisons. The act is not limited to prisons, but an accurate determination of its incidence is difficult to ascertain because the crime is seldom reported. The humiliation suffered by its victims often leads to severe depression and fear. Their situation is further compounded by the fact that they have no one in whom they can confide. Most rape crisis centers are staffed by people concerned with female rape victims; they have no understanding of male rape.

Statutory Rape

Statutory rape is defined as sexual intercourse with anyone under a specific age or with a person who is mentally deficient or disordered or under the influence of drugs. Laws related to statutory rape assume that persons younger than a certain age lack the maturity to consent to a sexual act. The exact age depends on the legislation established within a particular state. It is usually between 14 and 18.

A Case Involving GHB

In 1996, shortly after returning from a dance club near her home in La Porte, Texas, Hillory Farias was suffering from nausea and a severe headache. Within 24 hours, the high-school volleyball player was dead. An autopsy showed that her body revealed no evidence of alcohol or other common drugs. Responding to gossip about a new club drug, gamma-hydroxybutyrate (GHB), toxicologists decided to make further tests and discovered that Farias had died of a GHB overdose. Investigators assumed that someone slipped GHB, which is tasteless, odorless, and water soluble, into the young woman's soda when her attention was elsewhere. As Farias's death revealed, the drug can be deadly.

False Rape

During the summer of 1999, two young women claimed that two men had raped them in a New York City nightclub bathroom. The men were arrested. A few days later, the women confessed that they had made up the crime, and the men were released. By that time, the men's names had appeared in newspapers, giving them a reputation as rapists. The men can sue the women for defamation of character, but should those who falsely accuse another of rape be tried in a criminal court and subjected to the same penalties as a person who commits perjury? When readers of *Cosmopolitan* were asked this question, 77 percent of the women and 89 percent of the men who responded thought that such false accusers should be tried as criminals.

An Associated Press release on November 4, 2000, revealed that two women at the University of Massachusetts at Dartmouth recanted claims that they had been raped on campus shortly after classes opened for the fall semester. One of the women admitted having consensual sexual relations. Fearing that she might become pregnant, which would lead her parents to discover that she was sexually active, she invented the rape story. The other woman offered no explanation. Neither woman was prosecuted by police authorities, but both were disciplined by the school's judicial review system.

John Doe, Rapist

At the request of Manhattan district attorney Robert Morgenthau, a New York grand jury charged an unknown rapist with two rapes committed in 1995 and a third in 1997. Police know the rapist as John Doe, a sexual predator who is believed to be guilty of at least 16 rapes, all but one on the Upper East Side of Manhattan. His true name is unknown, but his genetic code has been well established by DNA testing of the semen left at the scenes of four of his crimes. The decision to indict a John Doe on the basis of his DNA was based on the fact that the five-year statute of limitations on the 1995 rape had nearly expired when the grand jury acted. *See also* Blood; Crime Scene; DNA Evidence; Paternity Testing.

References

Bell, Allison. "Date Rape." *Teen Magazine*, July, 1997, pp. 68 ff.

Cyriax, Oliver. *Crime: An Encyclopedia*. North Pomfret, VT: Trafalgar Square, 1996.

Gorman, Christine. "Liquid X." *Time*, Sept. 30, 1996, p. 64.

Gose, Ben. "Brown University's Handling of Date-Rape Case Leaves Many Questioning Campus Policies; an Accused Student Faced Suspension Although No One Refuted His Claim That The Woman Initiated Sex." *Chronicle of Higher Education*, Oct. 11, 1996, p. A53ff.

Hawaleshka, Danylo. "A Zip in Time." *Maclean's*, Mar. 27, 2000, p. 58.

"Is It a Crime to Lie about Rape?" *Cosmopolitan*, Nov. 1999, p. 56.

Seligman, Jean, and Patricia King. " 'Roofies': The Date-Rape Drug: This Illegal Sedative Is Plentiful—and Powerful." *Newsweek*, Feb. 26, 1996, p. 54.

Suhr, Jim. "Manslaughter Trial Begins over 'Date-Rape' Drug." The Associated Press. *Cape Cod Times*, Jan. 31, 2000.

The World Almanac and Book of Facts, 2000. Mahwah, NJ: World Almanac Books, 1999.

Reeve, Arthur Benjamin (1880–1936)

American writer Arthur Benjamin Reeve created the fictional character Craig Kennedy, known as the American Sherlock Holmes because he used forensic-science techniques in solving crimes. Reeve studied law at Princeton University but ended up becoming a journalist. Most of his work featured the scientific sleuth Craig Kennedy. Detective Kennedy was very popular during the era of the silent-film serials. Reeve wrote the scripts to these serials, which featured a young, pretty leading lady named Elaine who was always getting into trouble. Kennedy would save her from the evil villains using his scientific miracles. A major achievement of Reeve's mysteries about Craig Kennedy was his use of psychoanalytic techniques to solve cases despite the fact that they were written more than 20 years before Freudian psychology gained national acceptance.

The Craig Kennedy mystery stories are not well known today because most of the scientific methods used in the stories are either outdated or are now known to be technically impossible. However, Reeve's popularity in the early twentieth century was apparent when, because of his futuristic ideas, he was asked to help establish a spy- and crime-detection laboratory in Washington, D.C., during World War I. *See also* Kennedy, Craig.

References

Penzler, Otto, Chris Steinbrunner, and Marvin Lachman, eds. *Detectionary*. Woodstock, NY: Overlook Press, 1977.

Steinbrunner, Chris, and Otto Penzler, eds. *Encyclopedia of Mystery and Detection*. New York: McGraw-Hill, 1976.

Reichs, Kathy (c. 1948–)

As one of only a few anthropologists certified by the American Board of Forensic Anthropologists, Kathy Reichs brings to her novels a unique authenticity that stems from having done the forensic procedures she describes and from playing an active role in criminal investigations. The professor of anthropology at the University of North Carolina and official forensic anthropologist for North Carolina and the Canadian province of Quebec wastes little time in establishing the setting of her books. *Déjà Dead*, Reichs's critically praised first novel, opens with Dr. Temperance Brennan gluing skull fragments together to establish a victim's identity.

Integral to the plot of her second novel, *Death du Jour*, is an arson that claims the lives of several Montreal cultists. Brennan's expertise is required to reconstruct the badly burned victims and ascertain a cause of death. Reichs worked on a similar arson-suicide-murder case and, on the basis of that experience, can readily describe the process of determining an identity and cause of death in such a situation.

Dr. Temperance (Tempe) Brennan, like her creator a forensic anthropologist who splits her time between Montreal and Charlotte, North Carolina, is Reichs's heroine. Brennan is a fortyish divorcée with a college-age daughter, Katy. When she is in Canada, she shares a downtown condo with her cat, Birdie. The daughter of an alcoholic, Brennan herself is fighting a battle against the bottle, though she has been sober for six years. The doctor is a bone specialist since skeletons often figure into causes of death, and she is employed by the police.

Other recurring characters in the Reichs ensemble include Pierre LaManche, Brennan's boss and director of the police laboratory. Luc Caudel, the protagonist's skeptical police colleague, is reluctant to consider any scientific information, much less that offered by an American woman. Homicide detective Andrew Ryan provides the romantic tension in the novels, though there are complications in their relationship.

For most forensic novelists, autopsy descriptions are a staple, and Reichs is no exception. In *Deadly Decisions*, a biker war provides a forensic nightmare for Brennan. When normal autopsy procedures cannot answer the questions posed by police authorities, Brennan is called to the case. The bombing deaths of two bikers who were identical twins temporarily stymies her because all that remains of the two is 11 pounds of pulverized flesh. There are no fingerprints, and DNA will not help in an identification since the men were identical twins.

Deadly Decisions also analyzes other aspects of forensic investigations. The discovery of a partial skeleton at biker headquarters further involves Brennan in an escalating turf war. Reichs traces Brennan's analysis of the skull as she determines the approximate age of the victim based on the incomplete fusion of the skull bones. The anthropologist turns student as a colleague hypothesizes how a biker was killed based on the blood splatters seen in photographs of the crime scene. The pattern of blood on the apartment ceiling was most likely caused when an attacker swinging a weapon abruptly terminated his stroke and reversed direction to begin the next blow. On the basis of forward blood splatter from an exit wound and back splatter from an entrance wound, the technician deduces that the victim was beaten to the edge of death, then propped up in a chair and shot in the head. One last clue to the murder is found in a void pattern created when an object blocks the path of the blood and is then removed. With this information, Brennan continues her search for the killer knowing that he removed something from the room.

Brennan's skeletal-trauma specialty provides some unique forensic observations in *Déjà Dead* when she encounters a possible serial murder. Frustrated in her attempts to convince the stubborn Caudel of that possibility, the doctor begins her own inquiry by studying and comparing the cuts on the bones of two seemingly unrelated murders. She amazes detective Ryan with a veritable dissertation on saws and the damage they can produce. The doctor demonstrates that a microscopic analysis of the troughs in the bone can determine the blade dimensions, size, spacing, and settings in the saw's teeth. The use of a power saw in a dismemberment can be determined by the polish of the cuts. The conclusion of the novel hinges on Brennan's analysis of bite marks. (Not personally conversant with odontology, Reichs consulted forensic dentists to assure accuracy in her novel.) In a photograph of a suspect's apartment, Brennan notes a partially bitten piece of cheese. Using computer magnification and manipulation, Brennan conclusively identifies the killer and earns a grudging nod from Caudel.

References

Gartner, Hana. "Gory and Grisly Life of Crime Writer Kathy Reichs." *CBC Infoculture Interview.* http://www.infoculture.cbc.ca/archives/bookswr/bookswr_07231999_reichs.html.

Mudge, Alden. "Meet the Season's Best Discovery: Kathy Reichs." *First Person Book Page.* http://www.bookpage.com/9709bo/firstperson1 html.

Rice Murder (1900)

The Rice murder initiated a memorable criminal investigation and prosecution when the autopsy of American merchant and philanthropist William Marsh Rice revealed poisoning symptoms. Born in Massachusetts in 1816, Rice migrated to Texas in 1838 and ultimately became one of the Lone Star State's most famous millionaires. The quiet, reclusive Rice's greatest ambition was to repay his adopted state by establishing an institution of higher learning that would also immortalize him. Rice achieved his goal by establishing the Rice Institute in Houston, but a bizarre murder plot intended to steal his bequest and the ensuing trial riveted the nation's attention on turn-of-the-century New York City.

The import/export firm of Rice and Nichols, Texas and Louisiana land acquisitions, and a variety of other concerns accumulated a fortune for Rice. When his second wife died

in 1896, the eccentric's plans for his school were placed in immediate jeopardy. Prior to her death, Rice's wife wrote a will that disposed of a goodly portion of the family wealth, which was considered community property under Texas law. Rice's own will, drawn up shortly after his wife's death, designated the institute as his major beneficiary. A bitter Rice contested the validity of his wife's will, claiming that he was not a Texas resident, just a frequent visitor, and therefore the community-property statute did not apply. At 81, Rice was ensconced in his apartment on Madison Avenue in New York City attended only by a cleaning lady and his 23-year-old secretary/valet Charles Jones. Rice was so confident in his secretary that he allowed the $55-a-week Jones to handle his daily financial affairs.

The task of investigating Rice's residency claims fell to a New York lawyer and transplanted Texan, Albert T. Patrick. In November 1899, Patrick called at Rice's home for an appointment, but the cantankerous owner refused to see him. Undeterred, Patrick called Jones offering money for a forged letter to Rice's attorney admitting his Texas residency and requesting an expeditious settlement because he knew that he could not win the suit. Realizing an opportunity to advance his own fortunes and understanding Jones's position in the Rice household, Patrick went beyond his bribery attempt and advanced a new scheme and partnership to defraud the ailing Rice.

In early 1900, Jones typed a new four-page will, and Patrick forged Rice's signature at the bottom of each page. The document left the bulk of the Rice estate to Patrick. The partners eliminated the original bequest to the school and, to ensure that the will would be uncontested, left Rice's relatives and the school's trustees more money than the initial document. Patrick convinced Jones that it would arouse suspicion if he received a large sum of money. The two partners agreed that Patrick would pay Jones $10,000 a year for life after the will passed probate.

The conspirators realized that leaving such

a large sum of money to a relatively unknown person would arouse suspicion. They immediately set out to establish an elaborate paper trail that would demonstrate a long-standing relationship between the two men. Jones typed several letters to Patrick on which the lawyer forged Rice's signature. The letters went beyond mere friendship to indicate that Rice relied on Patrick for business advice. In them, Patrick received power of attorney over the estate and was granted deposit and withdrawal privileges at the bank. One letter even directed Patrick to supervise Rice's cremation in the event of his death.

Two things forced the partners to reconsider their initial plan of gradually poisoning Rice. First, the financier did not cooperate. In fact, Rice appeared to be getting stronger. Second, and perhaps more significant, was that Rice informed Jones that he was planning to rebuild a Houston refinery, destroyed by fire, which would require a substantial investment. On September 22, 1900, Jones, using chloroform, murdered his unsuspecting employer while he slept. A death certificate procured that evening from the doctor who treated Rice by Jones and Patrick indicated that Rice died of natural causes, old age, mental fatigue, and weakness of the heart.

The scheme began to unravel almost immediately. Patrick produced a forged letter supposedly expressing Rice's desire for immediate cremation. Both men knew that any autopsy would surely uncover foul play. The mortician was willing but informed the desperate Patrick that it would take 24 hours to fire up the crematorium. At the same time, Jones was wiring Rice's relatives the sad news and informing them of the impending cremation. Rice's Houston attorney and nephew were immediately suspicious of the hasty arrangements and telegraphed New York not to do anything until they arrived in the city.

Sensing that things were not going well, Patrick instructed Jones to draw up checks valued at $250,000 payable to Patrick and dated the day prior to the death. The lawyer forged Rice's signature as he had done numerous times. Unfortunately for the murderers, Jones misspelled Patrick's name (Abert instead of Albert), which aroused the bank's suspicions. The bank called Rice's residence, and Jones assured it that the checks were valid. When the bank pressed for a personal confirmation from Rice, the secretary acknowledged that the millionaire was dead. A handwriting expert confirmed the bank's forgery concerns. The police were called in, and a panicked Patrick rushed to his office and tried to flush the incriminating evidence down the toilet. A clogged pipe caused a backup, and an angry janitor provided all the evidence the authorities needed for an arrest.

An autopsy revealed symptoms consistent with chloroform poisoning. A medical prescription for chloroform was necessary in New York State, but investigators discovered that Jones's brother had sent him some from Texas. In the face of the growing circumstantial evidence, Patrick concocted a you-first suicide pact with Jones. Jones attempted to cut his own throat with a penknife Patrick had smuggled into the jail. The attempt failed, and Jones spent two weeks in Bellevue Hospital recovering. During his recuperation, Jones provided a full confession and promised to testify against his partner in return for full immunity.

The January 1902 trial lasted ten weeks and hinged on the very convincing testimony of Jones. Patrick's defense unrelentingly tried to portray Jones as a confessed liar, murderer, and perjurer, but to no avail. Patrick was convicted and sentenced to death. Supposedly fueled by a wealthy brother-in-law's money, Patrick began a lengthy appeals process. In 1912, Governor John A. Dix, without consulting the New York district attorney, inexplicably gave Patrick a full pardon. Patrick moved west, resumed his law practice, and died in Tulsa, Oklahoma, in 1954. Jones committed suicide at age 80 in Bayview, Texas, in the same year. The ashes of William Marsh Rice were interred beneath his statue on the campus of Rice Institute, now called

Rice University, which did, in the end, receive his fortune.

References

Friedland, Martin L. *The Death of Old Man Rice: A True Story of Criminal Justice in America.* New York: New York University Press, 1994.

Muir, Andrew Forest. *William Marsh Rice and His Institute: A Biographical Study.* Houston, TX: Rice University, 1972.

Wolf, Marvin, and Katherine Mader. *Rotten Apples: Chronicles of New York Crime and Mystery.* (1689 to the Present.) New York: Ballantine Books, 1991.

Ricin. *See* Markov, Georgi; Poisons.

Rigor Mortis. *See* Time of Death.

Rogues' Gallery

A rogues' gallery, more colloquially known as "mug shots," is a collection of photographs of known criminals that is kept by a police department. It is used to help eyewitnesses or victims identify the one among several or many that police believe may have committed a crime. It was first developed by Alphonse Bertillon as a way to improve the French Sûreté's method of describing criminals. The photographs, which have become standard among police throughout the world, include both a full-face and a profile picture of the felon.

In an effort to reduce crimes related to gangs, police in many California cities have taken to photographing not only known gang members but those suspected of being members as well. In 1997, Claudio Ceja was stopped and photographed by Anaheim police for the fifth time. Each time, Ceja, who was an 11th grader in a city high school and held two jobs, told police that he was not a member of a gang. They simply ignored him and entered his picture in their own rogues' gallery—a gang-tracking computer database. Jessica Castro, president of United Neighborhoods, an Anaheim organization that helps people fight police abuse, told Ceja

that he had a right not to be photographed. Despite stating his rights the next time he was stopped, police took his photograph anyway.

When a group of young Vietnamese American women waiting for a ride in Garden Grove, California, were lectured and photographed by police, they contacted the American Civil Liberties Union. The ACLU filed a class-action lawsuit on the women's behalf. The case, which was settled in 1995, awarded the women $85,000, and the Garden Grove police agreed that they would no longer photograph people without having a reasonable basis for doing so. They also agreed to obtain written consent before photographing anyone suspected of belonging to a gang. One of the young Asian women, who later attended the University of California at Irvine, reported that police continue to harass young Asians despite the agreement. *See also* Bertillon, Alphonse; Forensic Photography.

Reference

Siegal, Nina. "Ganging Up on Civil Liberties." *Progressive*, Oct. 1997, pp. 28–31.

Rojas, Francesca: Murder Case (1892)

Francesca Rojas has the questionable distinction of being the first murderer to be indicted on the basis of fingerprint evidence. Criminological history was made in the unlikely place of Necochea, Argentina, in 1892.

On June 19, 1892, Francesca Rojas's two children, a boy aged six and a girl aged four, were beaten to death in their beds. The distraught mother alleged that a local ranch hand named Velasquez was responsible. Rojas told authorities that Velasquez wanted to marry her, but she had turned down his proposal. In response, he had threatened her and, in an act of revenge, had killed the children. The mother even said that she had seen the killer run from her home as she returned from work.

Velasquez was arrested and readily admitted that he had threatened Rojas, but denied any involvement in the murders. At first, Velasquez was subjected to intense questioning but maintained his composure and innocence. Police methods at that time had little regard for the rights of the accused, and with no confession, the police chief tried a new tactic. He had Velasquez staked out and tied down next to the bodies of the children for a night. There was no change in Velasquez's story. For a week, the worker was subjected to periodic beatings and torture, but he never wavered from his innocent plea.

Desperate to find the guilty party, the authorities heard rumors that Rojas had another younger boyfriend, who would not marry her because of the children. The inventive local police chief then tried to frighten Rojas into confessing by spending one night acting as the ghost of the children, haunting the mother. When the misguided attempt failed, the local police called on regional headquarters in La Plata for aid. Police investigator Alvarez arrived and began his investigation at the crime scene.

Alvarez found a bloody thumbprint on the frame of the door to the children's bedroom. Aware of the advances in fingerprinting developed by his chief, Juan Vucetich, Alvarez had the print removed and called Rojas to the police station. He had the mother fingerprinted, and under a magnifying glass, even his novice eye could tell that the prints were identical. Confronted with the evidence, Rojas broke down and confessed that she had beaten her children to death with a stone from the well that she had then thrown down the shaft. She wanted to marry her young lover, who was not interested in an immediate family. Whether the fingerprint evidence would have stood up in court without the confession is debatable, but her admission based on the evidence discovered by Alvarez led to a sentence of life in prison for Rojas. *See also* Fingerprints.

References

Evans, Colin. *The Casebook of Forensic Detection: How Science Solved 100 of the World's Most Baffling Crimes.* New York: Wiley, 1996.

Nash, Jay Robert. *Encyclopedia of World Crime.* Wilmette, IL: Crime Books, 1990, vol. 3.

Ruxton Murder (1936)

The Ruxton murder was a case involving Dr. Buck Ruxton (1899–1936), an English doctor who brutally murdered his wife and her maid. The dismemberment of the two women created a forensic puzzle few medical examiners had ever experienced. The 37-year-old physician was an Indian, Bukhtyar Hakim, who had anglicized his name to Buck Ruxton. After studying at the Universities of Bombay and London, Ruxton opened a practice in Lancaster, England. With him was his wife Isabella. Although the two had never legally married, they had been together since 1928, and she was generally regarded as his wife. By 1936, the couple had three small children.

The Ruxtons' common-law marriage was a difficult one. The two quarreled bitterly about her perceived infidelity. Ruxton was convinced that his wife was unfaithful, though there was no truth to his conviction. On more than one occasion, Ruxton threatened his wife before witnesses. Former employees spoke of revolvers in the bedroom, knifepoints at his wife's throat, and a generally threatening manner. Police were called to the Ruxton home once and found the doctor in a near fit and incoherent. There were times when Isabella Ruxton actually moved out of the house, only to be persuaded to return by her contrite husband.

The old insecurities bubbled over on September 7, 1935, after Ruxton had followed his wife on her outing with family friends and accused her of having an affair with their son, a local town-hall clerk. On September 14, 1935, Isabella Ruxton left Lancaster alone for her annual reunion with her sisters at Blackpool. She left them about 11:30 P.M. to return home but was never seen again.

On Sunday, September 15, Buck Ruxton called his housekeeper and told her not to come to work because Isabella Ruxton and her maid, Mary Rogerson, had gone on holiday to Edinburgh. Later that Sunday, the doctor went to a neighbor, who was also a patient, and asked for help cleaning the house. Ruxton had badly cut his hand opening a can of peaches for his children, and the house needed attention. He was particularly concerned about the bathtub, which was stained by a dirty yellow color up to about six inches from the rim. Bloodstained carpets, torn up for the remodeling, were piled in the back yard along with some soaked towels and a shirt, all the refuse of his accident. Ruxton said that he had tried to burn the waste, but it was all too wet. He offered some of the salvageable carpet pieces to his neighbor as a thank-you for the help.

Things resumed a normal pace in the next two weeks, though the servants all remarked on the foul smell in the house. Ruxton was hard pressed to explain his wife's absence. His story changed depending on the audience. First, it was the holiday to Edinburgh or Blackpool; then he spread the story that Mary Rogerson, the maid, was pregnant and Isabella Ruxton had taken her friend away to avoid the shame and embarrassment. The Rogerson family never accepted this and was very aggressive in seeking a more plausible answer for their daughter's disappearance. Finally, Ruxton concluded that, as he had always feared, his wife had run off with another man.

On September 29, 1935, a woman visitor in Moffat, Scotland, strolled across a bridge and happened to look down. There, protruding from a parcel, was a human arm. A police search turned up four bundles of human body parts wrapped mainly in newspaper, dating from August 2 to September 18, 1935. More grisly discoveries were made over time, as late as November 4, and at distances up to a half mile from the bridge. Despite the heavily decomposed state of the remains, a preliminary investigation by forensic experts Professor John Glaister and Dr. Gilbert Millar did give the police some initial information. The killer had gone to great lengths to conceal the identity of the victims. Teeth had been removed after death, fingertips were missing or mutilated, and the heads had been brutalized. Nonetheless, it was clear to the experts that these were two female bodies. They also believed that the murderer probably had medical knowledge

due to the precision of the dismemberments. The remains were sent to the University of Edinburgh's forensic lab, where a five-man team, including Glaister, Millar, and forensic pathologist Professor Sydney Smith, began a closer analysis and ultimately produced a remarkable volume of evidence.

For their part, the police began a highly competent and noteworthy investigation. The authorities could not estimate a time of death, but they could reliably estimate when the remains had been dumped in the river. They knew that rains had swollen the river on September 18 and 19, and since the parcels were stranded well above the normal water level, the probability was that the remains had been deposited during the flood stage. The most recent paper wrapping the body fragments was dated September 18, so police concluded that the dumping had taken place between September 15 and 19. Inquiries in the local area about missing persons produced no leads. Questions about unusual movements around the river or unfamiliar cars in the vicinity were also unproductive.

One of the wrapping newspapers was a special insert used in the Lancaster area to promote a carnival, and when an officer read of the missing Mary Rogerson and Isabella Ruxton, the investigation shifted to that area. Ten days after the remains were found, Buck Ruxton went to police protesting his innocence and offering an inspection of his home. The police questions and implications were hurting his practice, and he wanted the probe ended.

While the police pursued their leads, the forensic medical team produced astounding results. Among the 70 fragments of remains were three breasts, which confirmed the initial assumption that these were two female victims. Most of the soft tissue was missing, but the experts pieced together the bony material of the victims. The ghastly jigsaw puzzle produced one nearly complete body and one that was missing the upper torso. From the breasts, experts learned that one of the women had not had children. Glaister, by examining the skull sutures, determined the approximate age of the woman. Sutures in the skull seal themselves over the years and are usually closed by age 40. In one of the two skulls they were not closed, but they were almost closed in the other. Consequently, one of the women was close to 40 and the other under 30. X rays of the smaller, younger skull revealed that the wisdom teeth were not fully erupted, further narrowing the age to the early 20s. This younger woman was about 5'6" tall, had had teeth extracted sometime before death, and had suffered current bouts of tonsillitis. Police informed the scientific investigators that Mary Rogerson was 5'6" tall, had never married, and had had six teeth extracted for various reasons. In regard to identifiable marks, Rogerson had scars on her right thumb and abdomen from an appendix operation. Significantly, the torso of the younger body was never found, and it was missing the right thumb from the hand.

The more complete body of the older victim offered just as much forensic information. The woman had light brown hair, was about 5'5" tall and had had children. All of these points corresponded with the missing wife of Buck Ruxton. Isabella Ruxton had dentures, and the skull had missing teeth that indicated such dental work. She also had a bunion on her left big toe, and this was missing from the dismembered foot. Still not finished, the team made casts of the two victims' feet. One foot easily fit Mary Rogerson's shoes, and the foot from the other body slipped into Isabella Ruxton's shoe. Similarly, the scientists got some serviceable fingerprints from the younger body and palmprints from the older body that authorities had come to believe was Isabella Ruxton.

A final piece of evidence that clinched the forensic evidence and led to the arrest and trial of Buck Ruxton was a picture, probably one of the most famous photos in the history of the science of identification. A member of the forensic team took a photo of Isabella Ruxton and superimposed a negative of the skull of the older victim over it. It was a perfect fit, marking the first time that evidence such as this was used in a case. The medical evidence, though circumstantial, was supported by further data collected from the Ruxton home after the doctor's arrest on October 13, 1935. An inspection of the home uncovered fingerprints that matched the "unidentified" younger body, blood spatters on the stairwell and banister, and traces of blood and tissue in the bathroom drain. Of course, the carpets that Ruxton had pulled up were also introduced into evidence.

The trial began on March 2, 1936. Ruxton's lawyers knew how difficult their task would be in the face of so much evidence. They offered two explanations of the prosecution's evidence. First, the two bodies could not be proven to be Isabella Ruxton and Mary Rogerson. The medical experts, especially Glaister and Smith, were unshakeable in their testimony. They were positive that the bodies were those of the women in question. Second, the defense lawyers insisted that the blood remnants the authorities had found were the result of the doctor's injured hand or a miscarriage a year previous by Isabella Ruxton. Again, the scientists convincingly declared that there was too much blood and it was in places too unusual, like the upper wall of the bathroom, to have been caused by either of these incidents. After a short deliberation, the jury returned a guilty verdict. Dr. Buck Ruxton was executed on May 12, 1936.

References

Blundell, R. H., and Wilson Haswell. "Blood in the Drains: The Ruxton Killings." *Crimes and Punishment: The Illustrated Crime Encyclopedia*. Westport, CT: H. S. Stuttman, 1994.

———. "Buck Ruxton." *Famous Trials*, ed. James H. Hodge. Harmondsworth, Middlesex, England: Penguin Books, 1950.

Goodman, Jonathan. "The Jigsaw Murder Case." *The Mammoth Book of Murder Science*, ed. Roger Wilkes. New York: Carroll and Graf, 2000.

Wilson, Colin, with Damon Wilson and Rowan Wilson. *World Famous Murders*. London: Robinson, 1993.

S

Sacco, Nicola, and Bartolomeo Vanzetti.
See Ballistics; Comparison Microscope.

St. Valentine's Day Massacre

A legendary gangland slaying in Chicago known as the St. Valentine's Day massacre took place on February 14, 1929, when seven members of George "Bugs" Moran's mob were killed. The relatively new science of ballistics never conclusively settled the issue of who was responsible for the slaughter.

Though the case was technically never solved, there is no doubt that the killings were the work of rival crime boss Al Capone's henchmen. The murders were largely responsible for elevating Capone's name to national recognition and providing a perverse glamour to the rising gangsterism of the 1930s. The sheer brutality of Chicago's biggest news story of the decade sparked a public demand for action that led to a relentless investigation of the Chicago mobster until he was convicted and imprisoned for tax evasion some years later.

The year 1929 was not only the height of Prohibition, but also the climax of the long-simmering Capone-Moran feud. At stake in the battle was the lucrative Chicago North Side. Capone, as the reigning czar of the Windy City's grimy underworld, had systematically tried to eliminate any competition.

Only Moran, who really held his position by default, remained. Liquor hijacking, saloon bombings, and 38 gangland murders in 1928 alone marked the escalating violence between the two factions. Informed that there could be a bloodbath before control was resolved, Capone cryptically said, "I'll send flowers."

To distance himself from the growing feud, Capone retreated to his Florida estate at Palm Island. Early in 1929, bodyguard "Machine Gun" Jack McGurn approached his chief about a preemptive strike at Moran. McGurn had more than a passing interest in the North Side gang, having narrowly survived an ambush at the hands of the Gusenberg brothers, Moran's lead gunners. McGurn wanted retribution. Capone agreed to pay McGurn $10,000 but emphasized that his name should not be involved in any way. The killer agreed and methodically began to implement his ingenious plan.

In what became a trademark prelude to a Capone murder, McGurn rented rooms across the street from Moran's headquarters at 2122 North Clarke Street. Posing as night-shift cabbies, McGurn's thugs monitored the SMC Cartage Company for three weeks, establishing the habits of the opposition. The building served as the central receiving and distribution point for Moran's bootlegging operation, as well as a clubhouse for gang members. At the same time, Mc-

Gurn was assembling his hit team, chief of whom were Fred "Killer" Burke and the "Murder Twins" John Scalise and Albert Anselmi, Capone's most experienced and accomplished killers. With Capone ensconced in Florida, McGurn knew that he would be the next logical suspect and went to great lengths to establish an alibi. As execution day approached, on February 13, he made a show of checking into a Chicago hotel with his girlfriend.

The North Siders were set up by a fraudulent phone call on that same day. A liquor hijacker that Moran was familiar with called looking for a quick turnaround on a shipment he had just taken "off the river." He offered his prize to Moran for a very reasonable price, and the gangster took McGurn's bait. The delivery was set for 10:30 the next morning, Valentine's Day, at the Clarke Street location.

The following day, the Moran gang, including Moran's brother-in-law and "business manager," Frank Gusenberg, waited for the illegal booze. As Moran, who was late, rounded the corner to the warehouse, he noticed a policelike black Cadillac pull up to the door and four men get out, two uniformed police officers and what could only be two plainclothes detectives. Little did Moran know that the props for McGurn's charade had been stolen days before the raid. The four went inside on what appeared to Moran and his two companions to be one of the frequent police raids any self-respecting gangster was subject to in Chicago. Moran went to a nearby coffee shop to wait out the inconvenient arrival of the police.

Once inside, the uniformed "officers" herded the unsuspecting hoodlums to a wall. Believing that this was one of the normal nuisance raids of the Chicago Police Department, they readily turned over their weapons and offered no resistance. After the crew was disarmed, the executioners opened fire with two machine guns, a sawed-off shotgun, and a .45. Within 10 seconds the job was done. The coroner later estimated that each man was shot at least 15 times by the methodical side-to-side firing of the guns, which started at head level on the first pass, came back about waist level, and traced the original route along the floor. When the authorities arrived, Frank Gusenberg, slumped over a chair, was barely alive. He was rushed to the hospital, only to die three hours later, refusing to give any information to the investigators.

The investigation showed that this was plainly a murder assignment and not a robbery because many of the victims had large sums of cash in their pockets. An angry public, embarrassed by the city's growing criminal notoriety, offered $100,000 for information on the killers. Ironically, it was not Capone, since he was out of town, who was the initial suspect. Witnesses testified that they saw the police enter the building, but Chicago authorities had no record of any raid at the North Side address. The rumor persisted that the massacre had been perpetrated by rogue cops getting their revenge. Moran, when pressed by newsmen, blurted, "Only Capone kills like that." The continued rumors of a renegade police operation led several Chicago businessmen to finance the development of a crime lab at Northwestern University. Major Calvin Goddard, the nation's foremost expert in the new science of ballistics, was hired to clean up the situation and head the lab. Goddard tested every machine gun in the police arsenal and declared that none had been involved in the murder.

With the police eliminated, the search for a culprit shifted to the Capone gang. How could Capone be involved if he was in Miami? In fact, the boss of bosses went out of his way to be noticed in the Florida city. Police arrested and questioned Scalise, Anselmi, and McGurn but could turn up nothing. There was a momentary hope when one year after the massacre police discovered two machine guns in the possession of Fred "Killer" Burke. Goddard, the ballistics expert, was able to prove that the weapons had been used in the killings. Burke, however, was being tried in the murder of a Michigan policeman. He was convicted and sentenced to life. He never revealed anything about the crime. No one was ever indicted for the murders or even positively identified as part of the assassination squad.

Neither Capone nor McGurn could have anticipated the wave of reform that resulted from the crime, nor would they have believed that the murders would be so widely remembered today. The primary actors in the drama did not fare well. Scalise and Anselmi were beaten to death with a baseball bat by Capone himself when he discovered that they were plotting against him. Ironically, seven years after the Chicago slaughter, on Valentine's Day, McGurn was gunned down in a Chicago bowling alley. A makeshift Valentine's Day card was pinned to his chest. Capone was eventually convicted of tax evasion but only served part of his term. He was released when authorities learned that he was in poor heath. He died in 1947 of venereal disease. The warehouse on North Clarke Street became a tourist attraction, then an antique-furniture business, and was finally razed in 1967. *See also* Goddard, Calvin Hooker.

References

Bergreen, Laurence. *Capone: The Man and the Era.* New York: Simon & Schuster, 1994.

Churchill, Allen. *A Pictorial History of American Crime, 1849–1929.* New York: Holt, Rinehart & Winston, 1964.

Nash, Jay. *Blooodletters and Badmen.* New York: M. Evans and Co., 1973.

Sayers, Dorothy L. (1893–1957)

English writer Dorothy L. Sayers, who incorporated chemistry and other sciences into her crime-solving detective and mystery stories written during the 1920s and 1930s, probably is best known for her detective stories featuring the elegant and sophisticated amateur sleuth Lord Peter Wimsey, whose solutions to crime reflected insights reminiscent of forensic science. Sayers graduated from Somerville College in 1915, took top honors in medieval literature, and was one of the first Englishwomen to receive a college degree.

In 1930, she introduced Harriet Vane, a woman accused of killing her lover in the book *Strong Poison.* Wimsey falls in love with her as he tries to prove her innocence. In this story, Sayers uses her knowledge of arsenic and medicinal chemistry to catch the killer. She shows that arsenic is lethal in liquid form, but local natives can build an immunity to it when it is used in solid form. The locals took it to increase endurance, clear their complexions, and make their hair sleek. The key was not to drink for at least two hours after taking it so that it did not get absorbed quickly and produce poisonous effects. The real murderer was found to have arsenic in his fingernails and hair. It was also observed that he did not drink at dinner on the night of the murder. Wimsey noted, too, that the killer had sleek hair and and a clear complexion, indicating that he knew how to avoid arsenic's poisonous effects.

In another story, *The Incredible Elopement of Lord Peter Wimsey,* Sayers has Wimsey use shrewd reasoning and medical awareness to solve the mystery. An American woman is thought to be bewitched by the local people. Wimsey is able to determine that the woman actually suffers from myxedema, a thyroid-deficiency disease. By withholding her medication, her husband has reduced her to feeblemindedness. Wimsey smuggles the medication to her while her husband is absent and restores her to good health.

Sayers collaborated with Robert Eustace, a doctor and contemporary of Sir Arthur Conan Doyle, to write the mystery *The Documents in the Case,* which has a unique format. The story is told as a series of letters and is about a poisoning in which a synthetic substance that has the same characteristics as a deadly mushroom is used. At first, the death of the character George Harrison, which occurs after he has eaten a stew of wild mushrooms, is ruled accidental. The coroner thinks that Harrison picked the poisonous fungus, *Amanita muscaria* by mistake and put it in his stew. Harrison's son does not believe this scenario because his father was very knowledgeable about edible and nonedible mushrooms. Chemical tests show that the poison in the stew was a synthetic form of the poison found in *Amanita muscaria*, not the natural form of the poison. The clever murderer is caught as a result of the chemical testing.

Sayers spent the last years of her life doing what she enjoyed most—giving lectures on religion, philosophy, and medieval literature and translating Dante's works. *See also* Wimsey, Peter Death Bredon.

References

Hitchman, Janet. *Such a Strange Lady.* New York: Harper & Row, 1975.

Papinchak, Robert Alan. "Dorothy L. Sayers." *Mystery and Suspense Writers,* ed. Robin Winks. New York: Charles Scribner's Sons, 1998, vol. 2, pp. 805–826.

Penzler, Otto, Chris Steinbrunner, and Marvin Lachman, eds. *Detectionary.* Woodstock, NY: Overlook Press, 1977.

Steinbrunner, Chris, and Otto Penzler, eds. *Encyclopedia of Mystery and Detection.* New York: McGraw-Hill, 1976.

Scanning Electron Microscope. *See* Electron Microscope.

Scarpetta, Dr. Kay

Dr. Kay Scarpetta, fictional, fortyish chief medical examiner of the state of Virginia, is the feature character in the novels of Patricia Cornwell. Cornwell's best-selling, award-winning first novel *Postmortem* (1990) established the Scarpetta character, and each successive book has revealed more about her. The forensic specialist was born and reared in Miami. Her father died of leukemia when she was 12, but a small grocery store provided for the family. Scarpetta attended Catholic schools in the city before moving on to Cornell for her undergraduate degree. This was followed by a medical degree from Johns Hopkins and law school at Georgetown. She was appointed chief medical examiner about 1983. During the course of the novels, she becomes a consulting pathologist for the FBI's Behavioral Science Unit, allowing her involvement in more diverse cases and situations.

Scarpetta is assisted in her medical and scientific investigation by some recurring characters. Pete Marino, a transplanted New Yorker pushing 50 in *Postmortem*, is the medical examiner's link to the Richmond, Virginia, police department. Romantic tension is supplied by FBI agent/profiler Benton Wesley. Though he is married with three children, there is a strong emotional relationship between the divorced Scarpetta and the agent. Lucy, Scarpetta's adoring niece in the novels, begins the series as a high-school math/science whiz, attends the University of Virginia, and eventually secures a position as a computer expert with the FBI. Trevor Gault, a wealthy serial killer, is Scarpetta's nemesis.

Scarpetta uses scientific and medical knowledge in all her investigations. She is particularly adept at using autopsies to solve crimes but is often hampered by the mistakes of an inexperienced medical examiner. She recognizes the need to degrease and deflesh bones to better study how a body was dismembered. Aware of the hazards involved in doing autopsies, she is careful to don protective clothing to avoid serious infections.

Each case that she investigates involves a specific aspect of forensic science such as arson, arson-murder, infectious disease, poison, and so on. The investigations are realistic, and emphasis is placed on the thrill and pleasure derived from the use of scientific techniques and thinking in solving the difficult puzzles created by many crimes. *See also* Cornwell, Patricia.

References

Cantwell, Mary. "How to Make a Corpse Talk." *New York Times Magazine,* July 14, 1996.

Passero, Kathy. "Stranger Than Fiction: The True-Life Drama of Novelist Patricia Cornwell." *Biography,* May 1998.

Schick, Elizabeth A., ed. "Patricia Cornwell." *Current Biography.* New York: H.H. Wilson Co., 1997.

Scotland Yard

Scotland Yard was the name given to the first headquarters of London's Metropolitan Police Service when the force was established in 1829 by Home Secretary Sir Robert Peel. The name derives from the fact that the headquarters was built on an old palace in which Scottish royalty had been housed during visits to London before the unification of the two countries. The headquarters moved twice, first in 1890 and again in 1967, so that it is now more properly called the New Scotland Yard. Today, "Scotland Yard" often refers to the Criminal Investigation Department (CID) of the London force.

Scotland Yard has a jurisdiction of about 780 square miles of London and its environs. It is not a general police authority, has no national status, and has never exercised any governance over other police departments in Great Britain. In fact, Scotland Yard cannot operate outside the London area unless it is specifically requested to do so by local police. There are two exceptions to the jurisdictional limits. First, Scotland Yard does keep a record of all known criminals in the country and acts as an information clearinghouse for other law-enforcement agencies. It also serves as a link to Interpol, the international police organization. Second, through its Special Branch, Scotland Yard is charged with providing protection for the royal family, cabinet ministers, and visiting dignitaries, as well as monitoring the activities of dangerous foreign nationals.

Prior to the advent of Scotland Yard, the amateur and informally trained Bow Street Runners represented law enforcement in London. Never more than 15 men, the Bow Street Runners was the brainchild of magistrate and novelist Henry Fielding. By 1820, the semiprofessionals were overwhelmed not only by an increasing volume of crime but by corruption in their own ranks. During his first term as home secretary, Sir Robert Peel realized the need for a professional force, but his vision was not shared by Parliament. Fearing that the police would interfere with individual liberty and become an army for a tyrannical government, opponents defeated Peel's plan. During his second term as home secretary, Peel steered the Metropolitan Police Act of June 1829 through Parliament.

The police nickname Bobbies or Peelers was in tribute to the pivotal role Peel played in the passage of the bill.

In 1842, following a highly publicized murder and the questionable 10-day search for the killer, Scotland Yard established its first plainclothes detective bureau of eight men, who were resented by their fellow officers because of their higher pay and viewed by the people as spies. The purpose of the new branch was to infiltrate the London underworld. In 1877, the Yard was rocked by a police corruption scandal in which three senior detectives were sent to prison. The following year, the Criminal Investigation Department (CID) was established. At first, men were especially recruited from outside the force for the new department, but that was abandoned for appointment from the ranks in 1884.

Within the CID are a number of subagencies that have created Scotland Yard's international reputation. One of the best-staffed and equipped forensic labs in the world is under the CID umbrella. The personnel of the lab are not professional policemen but rather civil servants. A second agency is the fingerprint and photo division, which maintains files on criminal offenders throughout the nation. The CID also includes fraud and drug squads, a criminal intelligence branch, a training school for detectives, the flying squad, which has a multipurpose group of select officers ready to respond to any crisis, and the Special Branch. The Special Branch provides protection for people at the direction of Parliament and is concerned with terrorist activity and national security.

References

"About Us." Metropolitan Police (London, UK). http://www.met.police.uk. (the official Web site of Scotland Yard.)

Browne, Douglas G. *The Rise of Scotland Yard.* Toronto: George G. Harrap & Co., 1956.

Jeffers, H. Paul. *Bloody Business: An Anecdotal History of Scotland Yard.* New York: Pharos Books, 1992.

Parker, Kenneth. "Scotland Yard." *Encyclopedia of Crime and Justice*, ed. Sanford H. Kadish. New York: Free Press, 1983.

Secret Service

Though the Secret Service is primarily recognized for its role in protecting the president of the United States, it also has major responsibilities for the discovery and apprehension of forgers and counterfeiters. A uniformed branch of the agency provides security at the White House, the Executive Office Building, the vice president's residence, foreign embassies, and the vaults and buildings of the Treasury Department.

The Secret Service was established by Abraham Lincoln at his last cabinet meeting in 1865. It was conceived as a general law-enforcement agency designed to combat forgery and counterfeiting of government checks and bonds as well as currency. Because the Secret Service was the investigative branch of the Treasury Department, its functions were expanded at the discretion of the president or by default when there was a question of departmental jurisdiction. As a result, the responsibilities of the Secret Service came to include investigations of land fraud, interdiction of opium smuggling, and grappling with the Ku Klux Klan. It is not unusual even today to find Secret Service agents assigned to an organized-crime task force because those groups often finance counterfeiting operations.

The assassination of William McKinley in 1901, the third president to die at the hands of a killer in 36 years, permanently changed the public perception of the Secret Service. In that year, the Secret Service was designated to protect the nation's president, a responsibility that was made permanent in 1951. Over the years, the protective function of the agency has been expanded to include the president-elect (in 1913), the first family (in 1917), and the vice president–elect (in 1962). In 1968, President Lyndon Johnson signed a bill that extended Secret Service protection to major presidential candidates. At the same time, protection was extended to widows of presidents until they remarried or died and to the president's children until the age of 16. *See also* Counterfeiting; Disputed Documents.

Reference

Johnson, David. *Illegal Tender.* Washington, DC: Smithsonian Institution Press, 1995.

Serial Killers

Forensic scientists distinguish among serial killers, mass murderers, and spree killers. A serial killer kills one person at a time in different places and with a significant time interval between murders. This distinguishes him from a mass murderer, who kills many people simultaneously in one location, and from a spree killer, who kills people one at a time in different places with relatively little time between murders.

Forensic psychologists have established a typical psychological profile of a mass murderer. He is a 30- to 50-year-old whose life is filled with failures that he blames on many people other than himself. His desire for revenge can be set into action by loss of a job or a girlfriend, foreclosure on a loan, or some similar event. People described as "going postal" are sometimes typical mass killers. The term arises from the number of incidents of postal employees who, after being fired, returned to the office where they worked to kill many of their former fellow employees.

A spree killer, unlike a serial killer, has a very short recovery time. He kills individuals in different locations, but he kills them in rapid succession.

Some 3,000 to 5,000 murders in the United States each year are the work of serial killers, who are almost always men. Their victims are usually women or children. Organized serial killers are generally intelligent. They often seem to live normal lives. For example, Albert DeSalvo, believed to be the Boston Strangler, was married and had two children. During the sentencing of Theodore Bundy, the judge commented on what a tragedy it was to see such a bright young man waste his life. These killers know ahead of time what weapon they will use, where they will use it, and on whom they will use it. Often the victims of a serial killer are each found in a characteristic position and have been murdered in a very similar manner. Albert DeSalvo apparently killed his victims with articles of their own clothing, such as their stockings. He would then leave the victim's body with the article of clothing tied in a bow around her neck. Others will cut off and take away an ear, a finger, a nipple, or some other body part.

Serial killers are sociopaths who have no remorse for their deeds even though they recognize that their actions are evil and depraved. Because they are organized, there is usually a "cooling-off" period as they prepare for their next killing. On the average, a serial killer attacks about once a month.

Disorganized serial killers are generally of average or below average intelligence who do not plan their murders carefully. They are generally unskilled loners who have failed in life and are immature both socially and sexually. The bodies of their victims are usually found where the murder took place and reflect a sloppy, stressful killing that is often associated with overkill such as multiple stab or gunshot wounds. Their victims are often randomly chosen and from a limited geographical area because these killers tend to operate from a home base and seldom travel to distant places. Some, such as David Berkowitz, "Son of Sam," kill because they hear, or claim to hear, voices. Sometimes they claim that the voice of God tells them to kill. Others believe that they are on a mission. They take it upon themselves to rid the world of some evil.

The five known victims of Jack the Ripper, London's nineteenth-century serial killer, were all prostitutes. He might be classified as disorganized because his killings were very much centered in one section of London and he eviscerated the women he killed. However, the eviscerations were done so skillfully and precisely that many believe that he was a surgeon who had planned the "operation" he performed before he slit the victim's throat. Some kill for very selfish or sensual reasons. They may be excited by or receive pleasure from their killings in a manner similar to that of the pleasure obtained by a drug addict; they may receive money from the serial kill-

ing of spouses; or they may be driven by lust to rape and then murder. Finally, some serial killers seem to be people who need to be in tight control and flourish by being in control of another's life or death. These killers frequently torture their victims before murdering them. *See also* Berkowitz, David; Bundy, Theodore; DeSalvo, Albert; Jack the Ripper; Psychological Profiling; Williams, Wayne.

References

Cyriax, Oliver. *Crime: An Encyclopedia.* North Pomfret, VT: Trafalgar Square, 1996.

Newton, Michael. *Hunting Humans: An Encyclopedia of Modern Serial Killers.* Port Townsend, WA: Loompanics, 1990.

Zonderman, Jon. *Beyond the Crime Lab: The New Science of Investigation.* New York: Wiley, 1990.

Serial-Number Restoration

Serial-number restoration is a technique used by forensic scientists when criminals have removed the serial numbers from a machine that may offer evidence related to a crime. Many machines, such as automobile engines, firearms, chain saws, and other machinery, have a serial number stamped on them. The number is usually stamped on a metal plate by hard steel dies that strike the metal with such force that each number leaves an impression. In an attempt to prevent the machine being traced when it is stolen or used in a crime, serial numbers are often destroyed by filing or grinding them away.

Forensic scientists are able to restore serial numbers because when metallic crystals are subjected to intense pressure, they are permanently strained. If the area where the serial numbers have been obliterated is cleaned and polished and an appropriate etching agent is added to the metal, the regions where the crystals have been strained will dissolve more readily. The original serial numbers can then be discerned.

Helle Crafts, wife of Richard Crafts of Newtown, Connecticut, disappeared in November 1986. One piece of evidence that led a jury to convict her husband Richard Crafts (the woodchipper murderer) of her murder was the serial numbers on a chain saw that police believed Crafts had used to cut up the frozen body of his wife before passing those pieces through a woodchipper. A witness recalled seeing a woodchipper on a bridge over the Housatonic River in mid-November. Divers searching for clues in the freezing water beneath the bridge recovered a chain saw, which was immediately sent to the forensic lab. There, between the teeth of the chain saw, scientists found fragments of bleached, dyed human hair that was consistent with hair taken from the crime site and strands from Helle Crafts's own brush. Inside the chain saw's housing were bits of blue-green fabric as well as traces of flesh and blood. Tests proved that the blood was human and type O. The chain saw, which had been submerged for weeks, was corroded, and the model's serial numbers had been filed off. However, scientists were able to restore the numbers, enabling police to trace the machine. It belonged to Richard Crafts. *See also* Trace Evidence; Woodchipper Murder.

Reference

Saferstein, Richard. *Criminalistics: An Introduction to Forensic Science.* Upper Saddle River, NJ: Prentice Hall, 1998.

Serology. *See* Blood.

Sheppard, Sam (1924–1970)

Sam Sheppard was accused of brutally beating his wife to death in 1954 and served 10 years of a life sentence before being acquitted in a second trial in 1966. Sheppard's acquittal was partially based on a forensic technique, then new, that enabled investigators to analyze the shape, locations, and distribution of the blood at a violent-murder crime scene in order to determine the number of blows, the force of those blows, and the dominant hand of the killer. As recently as April 2000, the case was still in the news as Sheppard's son, Sam Reese Sheppard, tried unsuccessfully to completely clear

his father's name based on new DNA techniques.

Dr. Sam Sheppard and his wife Marilyn had been high-school sweethearts. They lived in Bay Village, an upper-class suburb of Cleveland. Sheppard, his two brothers, and their father ran the Bay View Hospital, an osteopathic facility.

In the early morning of July 4, 1954, Marilyn Sheppard was bludgeoned to death in her bedroom. The Sheppards had entertained another couple the previous evening. After midnight, when the guests had left, Marilyn Sheppard went up to bed while Sam Sheppard stayed downstairs to watch a movie. At some point during the night, Sam Sheppard said that he had awakened to hear his wife calling him. He went upstairs to the bedroom and was struck from behind. When he came to, he saw his wife covered in blood and checked her pulse. There was none. He then checked his son in the next bedroom and found that he was sleeping soundly. He responded to a noise downstairs and saw a form progressing rapidly toward the lake. He later described the form as about 6'3" tall, middle-aged with bushy hair, and wearing a white shirt. He chased the form, and a struggle ensued until he lost consciousness. He did not know how long he was unconscious, but when he came to, he went back to the house, checked his wife again, and called a neighbor, Bay Village mayor Spencer Houk, to report the murder.

During the struggle, Sheppard had damaged his spinal cord and had received bruises and lacerations to his face. His brother, Steve, wanted him to recuperate from these injuries before being questioned by the police. The investigators saw this as uncooperative stalling, which was how it was reported in unsympathetic newspapers. The media became extremely influential as the proceedings developed. During a six-month period, there were close to 400 articles about the Sheppard case in the *Cleveland Press*, one of the three Cleveland newspapers. The inquest was broadcast on television, and during the trial the media were allowed to sit next to the jurors. Most of the publicity was prejudiced against Sheppard. At the trial, which began on November 4, 1954, Sheppard's lawyer, William Corrigan, made a motion for postponement and a change of venue because of prejudice against the defendant created by the biased news coverage. It was denied.

Some of the facts that had come out during the investigation and trial were that Marilyn Sheppard had been pregnant when she was murdered, that there had been some talk of divorce, and that Sam Sheppard had been involved in an affair. Sheppard denied the affair during the inquest but later acknowledged it during the trial. However, he resolutely denied killing his wife and injuring himself to establish an alibi. It was later said, "Sheppard was tried for murder and convicted of adultery." He was found guilty of second-degree murder and received a life sentence.

During Sheppard's 10 years in prison, Corrigan made several unsuccessful attempts to get a new trial. After Corrigan died in 1961, F. Lee Bailey took over the fight. (Erle Stanley Gardner, detective-story writer, attorney, and organizer of the Court of Last Resort, had convinced Bailey that Sheppard was innocent.) In 1963, Bailey filed a habeas corpus motion in federal court charging that Sheppard had been denied his constitutional rights. He said that in the first trial the prejudiced publicity and the denial of a change of venue had prevented Sheppard from receiving a fair trial. The U.S. Supreme Court in an 8–1 ruling agreed that excessive prejudicial publicity had prevented Sheppard from receiving a fair trial consonant with the due-process clause of the 14th Amendment.

At the second trial, Bailey had Paul Kirk, a criminologist, testify that some blood found in the bedroom did not belong to either Sheppard or his wife. Kirk also said that the shape, locations, and distribution of the bloodstains indicated that a left-handed assailant with female strength had beaten Marilyn Sheppard. Sam Sheppard was right-handed. At this retrial, Sheppard was acquitted because the jury could not establish beyond a shadow of doubt that he had killed his wife.

On April 6, 1970, Samuel Holmes Sheppard died. The cause was ruled liver failure. Sheppard had become a heavy drinker after his release from prison. The murder trial and publicity made it impossible for Sheppard to resume his medical practice, and he became a professional wrestler.

After 45 years, the Sheppard case is still in the courts of law. The murder of Marilyn Sheppard has never been solved. The Houks, neighbors of the Sheppards, were F. Lee Bailey's suspects at the second trial. They are both dead. Sam Sheppard's son believes that DNA tests on the blood found in the bedroom will show that Richard Eberling, a former window washer at the family's home who had been convicted of another murder in 1984, was guilty of the crime. Richard Eberling is also dead. In a civil case concluded in April 2000, Sheppard's son claimed that his father was wrongfully imprisoned and was innocent of the crime, which is a stronger legal standard than a not-guilty verdict. However, a jury concluded that the evidence presented did not prove that Sheppard was innocent.

The Sheppard murder case is believed to have been the inspiration for the 1963–1967 television series *The Fugitive*, although Roy Huggins, the creator of this series, denies it. In the series, Dr. Richard Kimble, played by David Janssen, had been wrongfully convicted of killing his wife. He escaped and tried to find the real killer. There was also a 1996 movie with the same title starring Harrison Ford, and in 2000 a new television series called *The Fugitive* was introduced that bears a striking resemblance to the 1960s series. *See also* Blood; Gardner, Erle Stanley; Simpson, Orenthal James.

References

"Attorney Says DNA Test Supports Sam Sheppard's Innocence." *U.S. News*, March 5, 1998. http://www.cnn.com/US/9803/05/dna.sheppard/index.html.

Barrett, Sylvia. *The Arsenic Milkshake and other Mysteries Solved by Forensic Science.* Toronto: Doubleday, Canada, 1994.

Crime Library Home Page. *Sam Sheppard.* Dark Horse Multimedia, 1998. http://

www.crimelibrary.com/sheppard/htm. Accessed Aug. 21, 2001.

McGunnagle, Fred. *Sam Sheppard*. Dark Horse Multimedia, 1998. http://www.crimelibrary.com/com/sheppard/sheppard.htm.

McKnight, Keith. *Haunting Questions—The Sam Sheppard Case*. BJE Special Project: Beacon Journal Publishing Co., 1996. http://www.ohio.com/bj/projects/sam/sam1–8.html.

Shoe Prints. *See* Footprints.

Simpson, Keith. *See* Camps, Francis.

Simpson, O.J. (1947–   )

After June 12, 1994, football great Orenthal James (O.J.) Simpson became best known to the public for his role in a courtroom trial that was nationally televised for nine months before he was acquitted. His defense lawyers were successful in bringing into question the handling of forensic evidence by the Los Angeles Police Department. Simpson was on trial for the brutal murders of his ex-wife Nicole Brown Simpson and her friend Ron Goldman. Prior to that day in June, O.J. Simpson had been known as an outstanding football player who had won the Heisman Trophy in 1968 and had gone on to stardom as a professional National Football League (NFL) player in Buffalo. After his playing days were over, he became a sports commentator and appeared in a number of television advertisements and movies.

The trial had America discussing race and police misconduct. These factors were critical to the fate of O.J. Simpson. As in the trials of Bruns Richard Hauptmann and Sam Sheppard, the media coverage played a major role in Simpson's trial. The trial was televised, and both the prosecution and defense teams tried to use the coverage to their advantage. The trial was a soap opera unfolding. The tragedy was that it was real life, and two people had been viciously stabbed to death.

On the basis of evidence found at the murder scene and at his home, Simpson was accused of the crime and promised to surrender, but he disappeared. He was later spotted in his white Ford Bronco driven by his friend and former teammate Al Cowling. A low-speed chase by the police ensued and was televised across the nation. Finally, after negotiations on the car phone, Simpson surrendered at his home and was taken into custody.

There had been no eyewitnesses to the murders, and the police never found the murder weapon, but the prosecution entered the courtroom with hundreds of pieces of evidence aimed at proving that Simpson had killed his ex-wife and her friend. Blood evidence was the most damning. Chief prosecutor Marcia Clark stated, "The mere fact that we find blood where there should be no blood in the defendant's car, in his house, in the driveway, and even on the socks in his very bedroom at the foot of his bed, that trail of blood from (the crime scene) through his own Ford Bronco into his house at Rockingham is devastating proof of his guilt" (CNN.com, *The Prosecution: Its Case against Simpson*).

The prosecution had DNA experts testify that the blood found at the crime scene, other than the blood of the victims, had to be Simpson's. The experts stated that the blood could have come from only 1 person, black or white, out of 170 million people. One DNA expert, Robin Cotton, told the courtroom that the blood on the sock found at the foot of Simpson's bed matched Nicole Brown Simpson's blood.

The prosecution argued that Simpson had a motive to kill. It introduced evidence of domestic abuse, citing frantic phone calls to the police from Nicole Brown Simpson during the marriage and after the divorce. It had Nicole Brown Simpson's sister, Denise Brown, testify about physical abuse she had witnessed Simpson perpetrate on her sister. It painted a picture of O.J. Simpson as a man who displayed jealous and obsessive behavior toward his ex-wife.

When the defense began its case, it dropped a bombshell on the prosecution. Defense lawyers accused one of the prose-

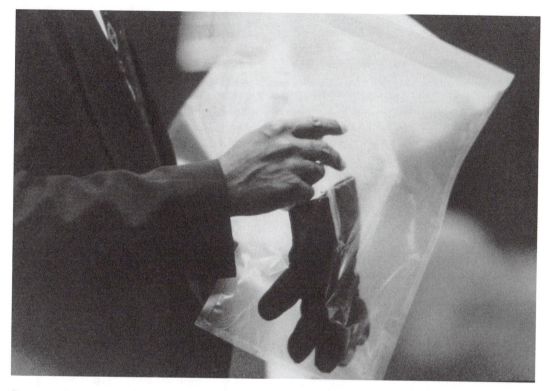

Simpson crime scene—glove in bag. © Corbis/Bill Nation.

cution's witnesses of being a racist. Detective Mark Fuhrman had testified that he found a bloody glove at Simpson's home. A matching glove had been found at the murder scene. But the defense alleged that Fuhrman had planted the glove at Simpson's home in order to frame Simpson because Fuhrman was a racist. Fuhrman denied that he was a racist, but the jury listened to two tapes in which Fuhrman was heard using many racial epithets. Despite Fuhrman's racist words, prosecutors argued that there was no proof that he had planted evidence.

The defense strategy was also to question the credibility of the Los Angeles Police Department's investigation of the murders, the evidence, and the witnesses. It wanted to show that the evidence gathered had been contaminated, planted, or tampered with. It argued that the damning blood evidence had been planted. It had its own blood expert, Herbert MacDonnell, state that the blood found on the sock in Simpson's bedroom was not actually splattered. He said that the

blood had been transferred in the form of a compression stain. In other words, someone had pressed Nicole Brown Simpson's blood into the sock.

The defense tried to show that the DNA results were wrong because the blood had been mishandled or contaminated. One of the prosecution's DNA experts, Michael Baird, countered that if the blood samples had been contaminated they would not have given the wrong result, they would have given no result. One cannot change one person's DNA into another person's DNA. But the defense had made its point. People inside and outside the courtroom were questioning the handling of the evidence, and suspicions were roused concerning the possibility that evidence had been planted.

Race relations with the Los Angeles Police Department were tense, and police were on alert when the 12-person jury consisting of 10 women and 2 men, 9 African Americans, 2 whites, and 1 Hispanic, adjourned to contemplate the fate of O.J. Simpson. On October 3, 1995, O.J. Simpson was acquitted of the murders of Nicole Brown Simpson and Ronald Goldman. However, a year and a half later, on February 4, 1997, Simpson was found liable for the deaths in a civil case that was not televised and in which photographers were banned from the courtroom. The jurors in the civil trial had to agree only that there was a preponderance of evidence linking O.J. Simpson to the murders. They did not have to prove that he was guilty beyond a reasonable doubt, as in the criminal trial. Furthermore, the jury's decision did not have to be unanimous; only 9 of the 12 jurors needed to agree on the verdict. The verdict, however, was unanimous. It found Simpson liable for the death of Ronald Goldman and for committing battery against his ex-wife Nicole Brown Simpson. (The Brown family had not sought damages.) The jury awarded compensation of $8.5 million in damages to the Goldman family and Ron Goldman's biological mother. *See also* DNA Evidence; Lindbergh Kidnapping and Trial; Sheppard, Sam.

References

CNN.com. *The O.J. Simpson Main Page.* http://www.cnn.com/US/OJ/.

CNN.com. *The Prosecution: Its Case Against Simpson.* http://www.cnn.com/US/OJ/verdictr/prosecution/index.html.

Bugliosi, Vincent. *Outrage.* New York: W.W. Norton, 1996.

Petrocelli, Daniel. *Triumph of Justice.* New York: Crown Publishers, 1998.

Shapiro, Robert. *The Search for Justice.* New York: Warner Books, 1996.

Skeletal Evidence. *See* Forensic Anthropology.

Smith, Sir Sydney (1883–1969)

Sir Sydney Smith believed in a multifaceted approach to crime-scene analysis and was involved in the development of many new forensic-science techniques. His career in forensic medicine extended over 50 years.

Smith, born in New Zealand, went to Scotland in 1908 and graduated from the University of Edinburgh with first-class honors. Thanks to a research scholarship, he became an assistant to Harvey Littlejohn, considered the leading specialist on forensic medicine in Scotland. In 1914, he received his master's degree with honors as well as a diploma in public health.

With the outbreak of World War I, Smith returned to New Zealand to serve in the army and take on the role of civilian medical officer of health. In 1917, he took a position with the Egyptian government as the principal medicolegal expert and was also a professor of forensic medicine at the University of Egypt, where he helped Egypt establish one of the best laboratories of forensic medicine in the world. There was much unrest in Egypt because Egyptian nationalists were upset with English rule. There were many murders, and witnesses were hesitant to come forward. Consequently, his department was doing hundreds of autopsies.

In 1925, he published his first book, *Textbook of Forensic Medicine*. In this book, Smith included ballistics as a brand-new branch of forensic medicine. He had started studying firearm identification in Egypt because of witnesses' unwillingness to talk, so he decided to let the bullets and shells do the talking. His book documented his studies in firearm identification, in which he learned that each weapon produced its own characteristic pattern. He had heard of two Americans, Charles E. Waite and Philip O. Gravelle from New York, putting together two microscopes in order to compare bullets in 1924. Smith began to do the same and succeeded in making his own comparison microscope. He used the comparison microscope to match hairs and fibers as well as bullets. In 1934, he collaborated with another Scot, John Glaister, on a murder case. Using microchemical and spectroscopic analysis, they matched hairs and fibers found in a sack that held a victim, eight-year-old Helen Priestley, with those found in the apartment of the accused, Mrs. Jeannie Donald.

When his former professor, Harvey Littlejohn, died in 1927, Smith came back to Edinburgh as professor of forensic medicine. He held the position for 25 years and was knighted for his good works in 1949. During his long career, he saw the need for specialization. With all the scientific discoveries and innovations, forensic medicine was becoming too complicated for one person to know it all. He knew that it would take a multifaceted analysis to find the answer to most crime scenes as weapons and criminals became more sophisticated. His autobiography, *Mostly Murder*, was published in 1959, ten years before his death. *See also* Comparison Microscope.

References

Evans, Colin. *The Casebook of Forensic Detection: How Science Solved 100 of the World's Most Baffling Crimes.* New York: Wiley, 1996.

Lane, Brian. *The Encyclopedia of Forensic Science.* London: Headline, 1992.

Thorwald, Jürgen. *The Century of the Detective.* New York: Harcourt, Brace & World, 1965.

Soil Evidence

Soil evidence can often be found on the clothes and/or shoes of a suspect in a criminal case. Forensic geologists can often trace that soil to a crime scene. Soil that can serve as evidence from a crime scene, from a suspect, or from locations claimed as alibis by a suspect can be carefully collected, placed in marked vials, and brought back to a forensic laboratory for analysis. The same is true of the tires or underside of a suspect's car. A forensic geologist may be able to identify the location from which the soils came. When the soil is viewed under a microscope, it may reveal plant and animal matter or mineral particles that will enable a geologist or biologist to identify a unique site. Tiny bits of brick, plaster, or concrete may link a crime to a particular building site.

In studying soil samples, examiners compare the specimens for color, texture, and composition. They can also compare soil samples, such as those from a suspect and a crime scene, by measuring the density gradients of the soils. This is done by filling two long narrow tubes with layers of liquids that have different densities. The liquid with the greatest density will be on the bottom, and the least dense liquid will be at the top. The two soil samples are then added to the tubes, and the particles in the soil samples will settle until they reach a liquid equal to or greater than their individual densities. With the tubes side by side, it is easy to see whether or not the soils contain particles of comparable densities. If they do not, then the sample taken from the suspect does not match that from the crime scene. If they do match, then the samples could have come from the same place, but unless there are some particles unique to the crime scene, the evidence cannot indisputably identify the crime scene.

The first use of neutron-activation analysis that was accepted as evidence in court involved soil. Federal agents had seized 2,400 gallons of whiskey from a truck in New York State that had traveled from Georgia. Soil taken from the bottom of the tractor-trailer carrying the whiskey was irradiated at a nuclear reactor in Washington, D.C., at the request of the IRS's Alcohol and Tobacco Tax Laboratory. A similar sample from an illegal still in Georgia was irradiated in the same way. When both were tested using neutron-activation analysis, the two samples were found to be virtually identical.

During a two-year investigation (1990–1992) of an organized-crime family in New York City, police collected tools and soils associated with the shallow graves of five homicide victims. The evidence was sent to the FBI Crime Laboratory for analysis. Two shovels and a pick taken as evidence from suspects were packed separately in brown paper bags. Soil samples from each grave were placed in 35-mm film canisters. The tools and soil samples were accompanied by carefully drawn map sketches of the graves and nearby landmarks. Using the map, laboratory personnel collected additional soil samples from sites at distances of 100 yards to one-half mile from the grave sites. They also collected "alibi" samples from the grounds of the residences from which the tools were confiscated. The purpose of these samples was to compare them with soil samples taken from the tools where they might have been used for gardening, transplanting trees, and so forth. Soil from one of the shovels filled about half a film canister. Soil specimens from the other shovel and the pick were contaminated by oil and rust, making them unsuitable for comparison. In the crime lab, forensic scientists demonstrated that the soil on the shovel was similar to the soil at one of the burial sites. On the other hand, the soil on the shovel was very different from the soil (the alibi samples) found at the residence where the shovel was kept. This evidence was vital to the prosecution in seeking a conviction. *See also* Crime Scene; Neutron-Activation Analysis; Trace Evidence.

References

American Jurisprudence Proof of Facts Annotated. Vol. 15. San Francisco: Bancroft-Whitney Co., 1964, 115–148.

Hall, Bruce Wayne. "The Forensic Utility of Soil." *FBI Law Enforcement Bulletin,* Sept. 1993, pp. 16–18.

Saferstein, Richard. *Criminalistics: An Introduction to Forensic Science.* Upper Saddle River, NJ: Prentice Hall, 1998.

Son of Sam. *See* Berkowitz, David.

Spectrophotometry

Spectrophotometry is a technique that involves the selective absorption or emission of visible, ultraviolet, or infrared light to identify substances. The wavelengths of light absorbed or emitted, which actually appear as bright or dark lines, may be recorded on photographic film or converted to electrical signals and recorded graphically in a spectrophotometer. Evidence collected at a crime scene is often analyzed by forensic scientists using spectrophotometry.

Light consists of electromagnetic waves that are produced when charged particles accelerate. Visible light is made up of wavelengths that extend only from about 400 to 750 nanometers (4.0×10^{-7} to 7.5×10^{-7} m). The shortest visible wavelengths are seen as violet and the longest as red. Even shorter

wavelengths, invisible to our eyes, extend from ultraviolet to X rays and gamma rays, whose wavelengths can be smaller than trillionths of a meter. Wavelengths longer than 750 nanometers, also invisible to our eyes, extend from infrared (7.5×10^{-7} to 1.0×10^{-3} m) to radio waves, which can have wavelengths thousands of meters long.

The speed of light in air is a constant 300,000 km/s. Since the speed of any wave is equal to its wavelength times the frequency at which the waves are produced, the frequency decreases as the wavelength increases. Thus light at the violet end of the visible light spectrum has a frequency of $7.5 \times 10^{14} s^{-1}$, while the frequency of light at the red end of the spectrum is $4.0 \times 10^{14} s^{-1}$.

When light energy passes through matter, the atoms and molecules in that matter absorb discrete amounts of energy from the light. The particular packets of energy they absorb are equal to the differences in the energy levels of the electrons within the atoms and molecules and the discrete vibrational and rotational energy levels of the molecules. For example, a continuous spectrum of visible light passing through sodium vapor will have dark lines that correspond to wavelengths of 589.0 nm and 589.6 nm because light with energies corresponding to these two wavelengths are absorbed by sodium atoms.

When matter is vaporized and raised to high temperatures, each element in the matter will produce an emission spectrum as it emits light. The wavelengths emitted are characteristic of the elements in the sample of matter. The energy of the light radiated corresponds to the differences in the possible energy levels of the electrons in the atoms of the elements in the sample, so the energies of the light in the emission spectrum of an element correspond to the energies absorbed by the atoms of the same element in an absorption spectrum. Sodium atoms at high temperatures will emit visible light with wavelengths of 589.0 nm and 589.6 nm, the same two energies that sodium absorbs when light passes through its vapor.

When light traverses a transparent colored substance such as stained glass, certain wavelengths of light are absorbed by the glass. For example, blue glass absorbs most of the red and green wavelengths and transmits those that we see as blue and perhaps violet. Light reflected from an opaque red shirt consists primarily of wavelengths from the red region of the spectrum. Shorter wavelengths in the blue and green regions of the spectrum are absorbed by pigments in the cloth.

In absorption spectrophotometry, a light source that emits a wide range of wavelengths is used to project a beam. For visible light, a bulb with a tungsten filament is usually satisfactory. Other sources are required for ultraviolet and infrared light. After traversing a vaporized sample of matter, the beam passes through a diffraction grating or prism that spreads the light into a spectrum such as one a person sees when he or she looks through a prism. A narrow slit can be moved across the spectrum to select a particular wavelength. That light enters a detector that converts the radiant energy to an electrical signal that can be measured on a recorder and plotted as light intensity or its inverse, the amount of light absorbed (absorbency), on the vertical axis as a function of wavelength or frequency on the horizontal axis.

In the case of an emission spectrum, the matter being analyzed is raised to a high temperature, usually by means of a high voltage applied to the vaporized substance. The light emitted by the matter then passes through a diffraction grating or prism, and the visible lines are photographed or recorded.

Microspectrophotometry

The combination of microscope, computer, and spectrophotometer has provided a new tool for the forensic scientists, the microspectrophotometer. Although spectrophotometry is unsatisfactory for analyzing trace evidence, microspectrophotometry is designed to do just that. Visible, infrared, or ultraviolet light is passed through a small speck of paint, fiber, or ink before entering the microspectrophotometer, where the absorption of different wavelengths is measured by a computer and the absorbency is displayed graphically as a function of the wavelength or frequency of the light. Using microspectrophotometry, forensic scientists, for example, can analyze the absorbency patterns of the inked lines on currency and, thereby, differentiate between counterfeit and real bills.

Ultraviolet and Infrared Spectrophotometry

Glass vessels can be used to hold samples being examined with visible light, but for ultraviolet light the sample is placed in a quartz container because glass is opaque to ultraviolet light. For infrared spectroscopy, the sample is generally placed in a vessel made of sodium chloride because both glass and quartz are opaque to infrared wavelengths. In some cases, the sample is compressed into a thin wafer and placed in a sodium chloride holder before being examined with infrared light.

The infrared absorption spectra of different substances have numerous peaks and are distinctly different, so that positive identification of substances is possible with infrared spectrophotometry. On the other hand, there is much less absorption in the ultraviolet region, so that spectrophotometry using these shorter, higher-energy wavelengths reveals fewer peaks. Nevertheless, ultraviolet spectroscopy can provide useful but tentative identification. For example, if a sample of dissolved white powder shows an absorption peak at 278 nanometers, it is quite likely that the substance is heroin. Neither sugar nor starch, which are commonly used to thin heroin, absorb ultraviolet light. A forensic scientist seeing such a spectrum would immediately seek a confirming test for heroin. An infrared absorption spectrum of the substance would provide a very distinct pattern with many peaks that could be compared with the known spectrum for heroin. *See also* Chromatography; Mass Spectrometry.

References

Dickerson, Richard E., Harry B. Gray, Marcetta Y. Darensbourg, and Donald J. Darensbourg.

Chemical Principles. 4th ed. Menlo Park, CA.: Benjamin/Cummings, 1984.

Saferstein, Richard. *Criminalistics: An Introduction to Forensic Science.* Upper Saddle River, NJ: Prentice Hall, 1998.

Spilsbury, Sir Bernard H. (1877–1947)

Sir Bernard Spilsbury was an English pathologist who has been called the first great medical detective. The physically imposing, naturally charming, and well-dressed Spilsbury was so well respected that the authority and prestige attached to his name alone influenced juries. One disgruntled attorney referred to him as "St. Bernard." For 35 years, he gave forensic evidence in nearly every major murder case in southern England. He was dubbed the "perfect witness" due to his unflappable courtroom demeanor and skill in explaining scientific data and evidence so simply that any layman could understand it. It is estimated that he performed 25,000 autopsies during his tenure with the Home Office, prompting one lawyer to ask when the last time was that Spilsbury had examined a live patient. He was responsible for 110 criminal convictions. Spilsbury was a fellow of the Royal Society of Medicine, a lecturer at London's University College Hospital and St. Bartholomew's Hospital, and the author of *The Medical Investigation of Crimes of Violence.*

Spilsbury was born on May 16, 1877, the oldest son of a chemist. He was an adequate, but unspectacular, student at Oxford and showed no real scientific inclination. With the idea of establishing a general practice, Spilsbury enrolled at St. Mary's Hospital Medical School and had the good fortune to work with three pioneers of forensic medicine: A.P. Luff, Sir William Willcox, and, most important, his mentor, Augustus J. Pepper. Spilsbury's teachers were impressed with his attention to detail, patience, and phenomenal memory. His interest in what was termed the "beastly science" of forensics delayed his graduation, but in 1905, at age 28, he completed his studies and was appointed resident assistant pathologist at St.

Mary's under Pepper. When Pepper retired in 1908, Spilsbury succeeded him as the chief pathologist at the Home Office.

Spilsbury was recognized as a thorough and competent pathologist when the Crippen case (1910) thrust him into the national limelight and established his reputation across England. The abrupt disappearance of Cora Crippen, wife of Dr. H.H. Crippen, caused concern among her friends. Their anxiety was heightened by the appearance of Dr. Crippen at several social functions with his secretary and bookkeeper, Ethel Le Neve, who was observed wearing Cora Crippen's clothes and jewelry. In fact, Le Neve was living with Crippen. To police inquiries, the doctor answered that his wife had returned to Los Angeles to attend to a sick relative and had taken ill herself. According to her husband, Cora Crippen died and was cremated in the United States. After the questioning, Crippen and his new love fled England. A police search of the fugitive's home unearthed human remains buried in lime in the cellar. The body had been dismembered, the head was missing, and there was nothing anatomical to indicate the sex of the victim. Spilsbury was part of the St. Mary's team called in to examine the mutilated evidence. From a piece of abdominal skin measuring an inch and a half by one inch with an old scar, and the fact that Cora Crippen had had an appendectomy prior to her marriage, Spilsbury deduced that the remains were those of Cora Crippen. Dr. Crippen was apprehended and eventually executed for murder based on the forensic evidence.

Yet another example of Spilsbury's expertise was the famed "Wives of Bath" trial (1915). George Smith had three wives in rapid succession. All three died suspiciously by drowning in a bathtub. At the trial, Smith protested his innocence by saying that there were no signs of violence on the bodies. Spilsbury was called in and arranged a spectacular courtroom demonstration. He set up a full tub in the courtroom and called in a nurse clad in a bathing suit. She sat in the tub, and Spilsbury quickly flipped her legs up, submerging her head. The nurse passed out from shock and the rush of water up her nose. Smith received the death penalty.

In 1925, Spilsbury faced the crisis of his professional life. For the first time, he came into major conflict with his forensic colleagues. Norman Thorne was accused of murdering his fiancée in a love triangle. In his defense, Thorne stated that his despondent girlfriend had hung herself in his hutlike home. When he discovered the body, Thorne panicked and was convinced that he would be accused. He tried to cover up the death by sawing the body into four parts and burying it. Spilsbury studied the body, noticing particularly bruises to the head, and concluded that the woman had been murdered. He found no forensic clues that indicated hanging; only shock and blows to the head could account for the death. The defense supplied its own experts, eight of whom refuted Spilsbury. His colleagues criticized Spilsbury for not doing a complete microscopic study of the neck bruises, which Spilsbury said were natural skin creases. Despite the Crown pathologist's questionable study, the jury took only 25 minutes to convict Thorne. The press raised the issue of Spilsbury's apparent infallibility and the fact that his reputation might unduly influence a jury.

The 1940s were devastating for the renowned scientist. He developed a number of physical maladies, including crippling arthritis, which limited his work, and coronary disease, and in 1940 he suffered a mild stroke. In addition, Spilsbury's two sons died. Increasingly depressed by his inability to work as he wanted and grieving the loss of his sons, Spilsbury took his own life in his London lab on December 19, 1947. *See also* Crippen Murder Case.

References

Browne, Douglas and E.V. Tullett. *Bernard Spilsbury: His Life and Cases.* London: Penguin Books, 1955.

Humphreys, Christmas. "Sir Bernard Henry Spilsbury." *Dictionary of National Biography.* New York: Oxford University Press, 1959.

Shew, E. Spencer. *A Companion to Murder.* New York: Knopf, 1961.

Spree Killers. *See* Serial Killers.

Stabbings

Stabbing is the second most common form of homicide in the United States, exceeded only by shootings. In Great Britain, where gun laws are more strict, the majority of homicides are due to stab wounds.

With guns, there is usually a distance between killer and victim, but stabbing requires an attacker to touch his or her subject. As a result, the murderer is almost always stained with the victim's blood.

Wounds resulting from knives are of two types, stab or incision. Stab wounds are those caused by the point of a blade preceding its length. Incision wounds are the result of a blade's edge cutting through flesh. Incised wounds require a sharp blade and are usually accompanied by extensive bleeding. Stab wounds, which may result from a more blunt instrument such as scissors or an awl, generally produce less external blood. They may, however, cause extensive internal bleeding as well as damage to organs such as the heart, lungs, liver, and stomach.

Homicide, Suicide, or Accident?

Stab wounds are seldom accidental, although they may happen if someone should stumble and fall on a pair of scissors or a knife that he or she is carrying. Suicides by stabbing are becoming increasingly rare now that less violent means, such as overdosing on drugs, are available. Nevertheless, some suicides are caused by knives. They are most commonly due to a chest wound at a point the victim believes to be in front of the heart. Usually, there is a single wound, and often clothing covering the site has been removed. Of course, the location and angle of the wound must be consistent with the victim's ability to self-inflict the wound.

Homicidal stabbings usually result in multiple wounds inflicted through clothing. In addition, the victims will commonly have multiple cuts on their hands and arms as a result of trying to protect themselves from the attacker's weapon.

Evidence and Cases

Sometimes the shape of a stab wound is useful in determining the nature of the weapon. For example, a knife with two sharp edges will produce a wound with acute angles at each end. A blade with one sharp edge will leave a wound with one acute and one blunt side. Such wound shapes are clearest when a sharp weapon was used. Bruises around the wound suggest that a blunt instrument made the opening.

When actor Sal Mineo was stabbed to death on February 12, 1976, Dr. Manuel R. Breton, Los Angeles County's deputy medical examiner, ordered that the chest section where the wound had been inflicted be preserved and stored at the county's forensic center. Two years later, police learned that Lionel Williams, a prisoner in a Michigan prison, had boasted to a guard that he had killed Mineo. When police talked to Williams's wife, she reported that her husband had returned that February night covered with blood and had told her that he had killed Mineo with a hunting knife. On the basis of testimony from his wife, police were able to obtain a knife identical to the one owned by Williams. By systematically dissecting the preserved section of Mineo's chest, forensic scientists were able to produce a cast of the wound. The cast provided a perfect match for the knife blade that was identical to the weapon owned by Williams. This evidence was crucial in a trial that led to Williams's conviction and life sentence.

In a case involving the August 1, 1999, stabbing death of Joseph Bates at the Stardust Ranch brothel in Ely, Nevada, an autopsy had an effect opposite to that of the Williams case. The autopsy revealed that the fatal wound could not have been inflicted by a knife owned by Patricio Osvaldo, who had been arrested. As a result, charges against Osvaldo were dropped and he was released.

References

Cyriax, Oliver. *Crime: An Encyclopedia*. North Pomfret, VT: Trafalgar Square, 1996.

Evans, Colin. *The Casebook of Forensic Detection: How Science Solved 100 of the World's Most Baffling Crimes*. New York: Wiley, 1996.

Lane, Brian. *The Encyclopedia of Forensic Science*. London: Headline, 1992.

"Ranch Hand to Stand Trial for Stabbing Death at Ely Brothel." Associated Press, Nov. 3, 1999.

Statutory Rape. *See* Rape.

Strangulation

Strangulation, which leads to asphyxiation, can result from murder, suicide (usually by hanging), or accident. It is the third most common form of homicide in the United States, exceeded only by shootings and stabbings. Approximately 1,400 deaths each year are caused by strangulation. Where it is the suspected cause of death, investigators will look for a furrow around the neck and signs of asphyxia such as a bluish coloring of the face due to lack of blood flow to the head, hemorrhages in the membranes of the eye, and loss of bowel control. If the ligature was a wire or something else narrow in width such as a thin rope, the furrow will be deep and readily noticed. A wider ligature, such as a scarf, will leave a less noticeable mark. If the strangulation was manual, there will be bruises left by thumbs and fingers, and if the murderer had long fingernails, semicircular indentations may be apparent in the skin about the neck. The bruising will be evident in the muscles of the neck and possibly in the tongue, lower mouth, epiglottis, and lining of the larynx. In most cases where strangulation can be attributed to homicide, the victim's hyoid bone will be broken. This is seldom the case where strangulation was due to suicide or accident.

In December 1988, a woman's body was found in a swamp behind an inn in Surrey, British Columbia. Her body was covered with cuts and bruises, but the abrasions and bruises along her neck showed a linear pattern indicating strangulation with a thin rope, cord, or some similar ligature. During the autopsy that followed, it was evident that the bruises in the neck region extended deep into the flesh above the hyoid bone at the base of the tongue. In addition, the right ventricle of the victim's heart was dilated, a common result of airway obstruction. A pathologist used a microscope to examine the bruised flesh and could tell that the woman had been severely beaten about 20 minutes prior to her death by strangulation. However, it was the bite mark above the victim's right breast that led police to her killer. The teeth of a suspect matched within a high degree of certainty the teeth marks in the victim's bruised flesh.

The Christmas 1996 slaying of JonBenet Ramsey was the result of strangulation, according to the coroner. There were deep cuts and bruises around her neck. Several pathologists, however, believe that a blow to the head that left a skull fracture 8½ inches long and 4 inches wide was the cause of death. It is likely that either wound alone would have resulted in death. There were also indications of sexual abuse that included blood, bruising, and abrasions around the vagina; however, according to Dr. Richard Krugman, a child-abuse specialist at the University of Colorado School of Medicine, general physical abuse could produce such injuries. Regardless of the cause of death, the murder was brutal. At the time this is written, the person who murdered six-year-old JonBenet Ramsey remains at large. It is perhaps the most famous of America's unsolved crimes.

In the spring of 1997, Melissa Drexler, an 18-year-old New Jersey high-school student, entered the ladies' room at the beginning of her senior prom. There, within 20 minutes, she gave birth to a six-pound, six-ounce baby boy. According to John Kaye, Monmouth County prosecutor, she then strangled the infant and placed him in a plastic bag, which she deposited in a trash can before returning to the dance. Dr. Cyril Wecht, a forensic pathologist, told then coanchor for *Good Morning America* Joan Lunden that it would be difficult to prove that the baby was born alive because an attempt was made to resuscitate the infant. As a result, any air found in the lungs or gastrointestinal tract could have been introduced artificially and not as a result of natural breathing. Furthermore, any markings on the baby's throat could be attributed to the mother's grasping the child's head and neck in an effort to pull it through the birth canal. Rather than face a murder trial that might result in a prison sentence of 30 years or more if she were found guilty, Drexler pleaded guilty to aggravated manslaughter in the summer of 1998 and received a three-year sentence. *See also* Cause of Death; Crime Scene.

References

Barrett, Sylvia. *The Arsenic Milkshake and Other Mysteries Solved by Forensic Science.* Toronto: Doubleday Canada, 1994.
Cyriax, Oliver. *Crime: An Encyclopedia.* North Pomfret, VT: Trafalgar Square, 1996.
Gibson, Charles. "Violent Death of JonBenet Ramsey." *ABC Good Morning America,* July 15, 1997.
Lane, Brian. *The Encyclopedia of Forensic Science.* London: Headline, 1992.
Lunden, Joan. "Drexler Baby Autopsy." *ABC Good Morning,* June 25, 1997.

Stratton Brothers. *See* Deptford Murders.

Strychnine

Strychnine is an alkaloid poison obtained from the seeds of the *Strychnos nux-vomica* plant. It is a bitter, colorless, crystalline powder that, in small amounts, has been used in medicines that serve as "tonics." In larger quantities (100 mg or more), it produces effects similar to tetanus (lockjaw). The extensor muscles contract violently, producing a rigid body that rests only on head and heels. Convulsions and death by asphyxiation or exhaustion from the muscle spasms may follow. Rigor mortis is immediate because the body remains rigid with the eyes open.

Treatment consists of pumping the stomach to remove the poison followed by the administration of activated charcoal to adsorb any strychnine remaining in the stomach. Because it is important to maintain respiration, a ventilator may be needed. Valium or succinylcholine may be administered intravenously. Stimuli such as loud noises or bright or flashing lights should be avoided because they can initiate a spasm.

Although strychnine is widely distributed as a rat poison, it is seldom used as a means of murder. Its bitter taste would cause most people to spit it out before swallowing. However, in one famous murder case involving forensic scientist Edward O. Heinrich, strychnine was the poison used.

On April 29, 1929, Carroll Rablen, a World War I veteran who had been deafened by a war wound, waited in his car outside a school drinking coffee and eating sandwiches. His wife, Eva, had brought him the coffee and food a few minutes earlier and had then returned to the schoolhouse where she was enjoying square dancing, a recreation in which her deaf husband could not participate. Suddenly, Rablen was screaming between convulsions on the floor of his car. He died before a doctor or ambulance could reach the scene.

Perhaps Rablen, who was subject to mood swings, had committed suicide. An autopsy revealed nothing, and the medical examiner concluded that he had died of natural causes. Steve Rablen, Carroll Rablen's father, was convinced that his son had been murdered. His primary suspect was his daughter-in-law, Eva Rablen, who was the beneficiary of Carroll Rablen's life-insurance policy. He adamantly demanded that the sheriff carry out an investigation. In an effort to placate the elder Rablen, the sheriff conducted a search of the schoolhouse and car. Under the schoolhouse steps, he found a bottle with a strychnine label. It had been purchased at a nearby drugstore by a "Mrs. Joe Williams" who had claimed that she was going to use it to kill groundhogs. When the druggist was able to identify Eva Rablen as the so-called Mrs. Williams, the sheriff arrested her for murder.

Prosecutors asked renowned forensic scientist Edward Heinrich to investigate. Heinrich found traces of strychnine in Carroll Rablen's stomach, on his car's upholstery, and on the cup from which he had drunk. A handwriting expert was ready to testify that the "Mrs. Williams" who had signed the poison register at the drug store was Eva Rablen. However, defense lawyers had found people, including Carroll Rablen's first wife, who were willing to testify that Rablen was suicidal. They also had a witness who claimed to have seen Eva Rablen at another place at the time the poison was purchased. Carroll Rablen could have added the poison to his coffee. No one had seen Eva Rablen add the poison to the cup. But Heinrich was a brilliant investigator. If Eva Rablen had walked across a crowded dance floor carrying coffee and sandwiches, it was reasonable to suspect that she might have spilled coffee on someone.

The sheriff began the tedious task of interviewing everyone who had attended the dance. He soon struck pay dirt. One young woman remembered that Eva Rablen had bumped into her and some coffee had spilled onto her dress. She willingly gave the dress, which had not been washed, to the sheriff. Heinrich found traces of the poison on the dress. It was clear now that the coffee Eva Rablen had carried to her husband had contained strychnine. Shortly before the trial was to begin, news of Heinrich's investigation reached the defense lawyers. To avoid a likely conviction, Eva Rablen changed her plea to guilty and was sentenced to life in prison rather than the death sentence she would probably have received had her case gone to trial. *See also* Atropine; Heinrich, Edward Oscar; Opium; Poisons.

References

Cyriax, Oliver. *Crime: An Encyclopedia.* North Pomfret, VT: Trafalgar Square, 1996.

Evans, Colin. *The Casebook of Forensic Detection: How Science Solved 100 of the World's Most Baffling Crimes.* New York: Wiley, 1996.

Magill's Medical Guide: Health and Illness. Pasadena, CA: Salem Press, 1995.

The Merck Manual of Medical Information. Home Ed. Robert Berkow, MD. Whitehouse Station, NJ: Merck Research Laboratories, 1997.

The Oxford Companion to Medicine. Vol. II. Ed. John Walton, Paul B. Beeson, and Ronald Bodley Scott. Oxford and New York: Oxford University Press, 1986.

Sudden Infant Death Syndrome (SIDS)

Sudden infant death syndrome (SIDS) is the sudden death of an infant who had no serious illness and no apparent cause of death. Most states now require an autopsy, an examination of the scene of death, and a review of the child's medical history before deciding on SIDS as a cause of death. This is done to rule out the possibility of infanticide. The usual method of infanticide is suffocation, which, from an autopsy, is indistinguishable from SIDS. An estimated 2 to 10 percent of reported SIDS deaths are believed to actually be due to infanticide.

SIDS is also called crib death since it happens while the infant is sleeping. The typical age of infants who die from SIDS is between one and four months. About 8,000 infants die this way in the United States each year. SIDS strikes more male than female infants. It is more common among babies born prematurely or with a low birth weight. The incidence is also greater among Native Americans and blacks, and it occurs most frequently during the colder months of the year.

Some doctors believe that there is not a single factor accounting for SIDS, but that it is caused by several different factors. These may include problems with sleep arousal and an inability of the infant's body to sense or respond to a buildup of carbon dioxide in the blood. Among the conditions thought responsible but never proven to be causes are virus inflammations of the heart; smoking by parents; metabolic or electrolyte disturbances; endocardial fibroelastosis (a condition in which the heart muscle is replaced by fibrous tissue); a deficiency in the infant's immune mechanism; hypersensitivity to cow's milk; neurological disorders causing the heart to slow or stop or breathing to cease; sleep apnea, in which babies stop breathing; toxic gases from mattresses containing fire retardants or mold, which inhibit breathing and heart function; or some mechanical suffocation due to an enlarged thymus or to bedclothes.

In the late nineteenth century, toxic gases were implicated in the deaths of hundreds, perhaps thousands, of children. In 1892, an Italian chemist, Gosio, discovered that children of Western Europe were dying because of a toxic gas, arsine or alkyl homologues. The gas was being produced by a fungus interacting with copper arsenate, a chemical found in green pigments in wallpaper, and arsenious oxide, used as a preservative in wallpaper glue.

A famous 1972 SIDS study by Dr. Alfred Steinschneider in Syracuse, New York, was largely discounted when, in 1994, Waneta Hoyt confessed to killing her five children when they were babies. Steinschneider had based much of his study on Hoyt's children, who he had thought had all died of SIDS caused by sleep apnea. When his study suggested keeping baby monitors near babies' cribs, the baby-monitor industry greatly increased. Dr. Linda Norton, a Dallas forensic pathologist specializing in violent death who had led the team that examined Lee Harvey Oswald's exhumed body, was one of the experts questioning Steinschneider's study. Reading the SIDS study in 1976, Norton believed that the children had in fact been murdered, and her work, and that of others, eventually led to the arrest in 1994 of Waneta Hoyt.

The incidence of SIDS has decreased over 40 percent since 1992, when the American Academy of Pediatrics recommended that babies be put on their side or back instead of their stomach when sleeping. Another recommendation came in 1999 from the U.S. Consumer Product Safety Commission. It recommended that babies be placed on a firm, tight-fitting mattress since some infant deaths have been associated with suffocation in soft bedding. *See also* Infanticide.

References

"False Alarm: The Failed Promise of Apnea Monitors." Syracuse.com. May 5–7, 1996. http://www.syracuse.com/features/apnea.

"Scientists Link Smoking with Crib Death." http://www.junkscience.com/news2/sidstemp.htm.

"Sudden Death: The Search for the Truth about Cot Death." BBC Online. Feb. 25, 1999. http://www.bbc.co.uk/science/horizon/sudden death.shtml.

"Sudden Infant Death Syndrome." http://health.yahoo.com/health/Diseases_and_Conditions/Dise . . . /Sudden_infant_death_syndrome.

T

Taylor, Alfred Swaine (1806–1880)

English physician Alfred Swaine Taylor was the author of *The Principles and Practice of Medical Jurisprudence* (1873), the first comprehensive document on forensic medicine in the English language. He also wrote *Elements of Medical Jurisprudence* (1836) and *A Manual of Medical Jurisprudence* (1844). Taylor became professor of forensic medicine at Guy's Hospital (London) Medical School in 1834 after receiving his training in Paris, France.

Despite his extensive writing in the field, Taylor is most often remembered for a forensic mistake. In 1859, he made an error in the murder trial of a Dr. Smethurst. Smethurst's second wife had died under suspicious circumstances, and Taylor suspected poisoning. Having analyzed the dead woman's vomit, he testified at the trial that he had found arsenic in his chemical analysis of the fluid. However, the defense pointed out that he had used copper mesh during his analysis and that the copper contained arsenic. Taylor had to admit that the arsenic he found could have been introduced into the test extract by the copper mesh. His mistake was widely publicized in England and led the British people to be skeptical of forensic medical findings for a long time afterward. *See also* Arsenic Poisoning.

References

Morland, Nigel. *An Outline of Scientific Criminology.* New York: St. Martin's Press, 1971.

Thorwald, Jürgen. *Crime and Science.* New York: Harcourt, Brace & World, 1967.

Thorndyke, Dr. John Evelyn

Dr. John Evelyn Thorndyke was the fictional scientific detective created by R. Austin Freeman for a series of novels and short stories beginning with *The Red Thumb Mark* (1907). Nicknamed "The Great Fathomer," Thorndyke epitomized the use of scientific methods in solving crime.

Thorndyke was in his late 30s in 1907 and was about 50 when he made his final appearance 38 years later. The renowned doctor studied at St. Margaret's Hospital Medical School in London and lived at 5a Kings Bench Walk in London, which also served as his lab, workshop, and museum. A lawyer and doctor, Thorndyke eventually became a widely respected lecturer in medical jurisprudence. His most important crime-solving tool, besides his considerable intellect, was his invaluable little green case. This portable lab, for use at a crime scene, contained smaller versions of items like a microscope and test tubes. Supposedly the French Sûreté, a special police department, adopted the traveling research case and established its labs based on the Thorndyke model.

Thorndyke is assisted in his medical and scientific endeavors by several recurring characters. The most regular is Dr. Christopher Jervis. Jervis, a classmate of Thorndyke's at St. Margaret's, observes and records the facts of his friend's sleuthing. Nathaniel Polton is not only Thorndyke's servant and lab assistant but also a capable photographer. Other characters that appear frequently include Dr. Jardine, a former Thorndyke student, who often refers cases to his mentor; Inspector Badger of the Criminal Investigation Department at Scotland Yard; and Mr. Singleton, a fingerprint expert.

Thorndyke was considered by many critics and historians to be the first "scientific" fictional detective. His investigations reflected realism. He acquired the evidence and details required to track down a criminal slowly and methodically and focused on scientific methods of investigation as a means of establishing guilt. His work often revealed the difficulties associated with criminal investigations. For example, in *The Red Thumb Mark* Thorndyke explored fingerprint forgery and the notion that fingerprinting techniques were fallible. *See also* Fingerprints; Freeman, R. Austin; Time of Death.

Time of Death

The time of death of a victim or suspected victim is one of the tasks that confronts a medical examiner or coroner. The indicators most commonly used to determine time of death are rigor mortis, lividity, and body temperature. Rigor mortis (stiffening of the muscles) usually occurs about three hours after death. It starts in the facial and eye muscles and spreads slowly to the limbs. After about 12 hours, the entire body is rigid. Physical activity just prior to death will reduce the time before the onset of rigor mortis. In most instances, the rigidity begins to fade after another 48–60 hours.

Lividity, the discoloration of the skin resulting from gravity's action on blood, begins shortly after the heart stops beating. A person who dies lying on his back will show bruise-like discolorations in the small of the back and on the back of the neck and thighs. A prone corpse will show similarly colored skin on the anterior parts of the body.

Failure to oxidize food following death means that the body is no longer generating the heat needed to maintain body temperature. Consequently, the temperature of the corpse begins to fall toward the temperature of its surroundings. Since the rate of heat loss is proportional to the temperature difference between the body and its surroundings, the temperature drops more slowly as time passes. A rough approximation of time since death can be obtained from the formula

$$\frac{98.6°F - \text{rectal temperature}}{1.5} =$$

number of hours since death.

Of course, a corpse will cool much faster in a snowbank than in a warm room, and a corpse insulated by fatty tissue or heavy clothing will cool more slowly than a thin or naked body.

All three of these methods of determining time of death are approximate and depend on a number of variables that cannot be controlled. However, if a corpse is found outside where insects have been able to reach it, forensic entomologists can often determine time of death to within an hour. Food found in the stomach can also help to establish time of death, as in the Hendricks case discussed later.

A more recent test developed by Dr. John Coe involves changes in the potassium-ion content of the eyeball's vitreous humor. Coe observed that the breakdown of red blood cells after death leads to a predictable increase in the level of potassium in the vitreous fluid that is not affected by temperature. Coe's discovery may provide a more valid test for time of death than the three traditional methods.

Cases

The case of Petrus Hauptfleisch in 1925 in South Africa was decided in part on the basis of misplaced lividity. Hauptfleisch had killed his mother. After leaving her in a prone position for several hours, he turned her on her back to fit an accident he staged. The coroner became suspicious when he found evidence of lividity on the front of her body. Hauptfleisch was convicted and hanged.

In early November 1983, businessman David Hendricks was on a business trip in Wisconsin. Because he had been unable to reach his family in Bloomington, Illinois, by phone for several days, Hendricks called the local police, who agreed to check the home. They entered the house to find that Susan Hendricks and her three children had been cut to pieces by someone wielding an ax and a butcher knife. Both weapons, which had been carefully cleaned, were found at the foot of a bed.

When Hendricks returned, he was too upset to enter the house, but police questioned him and examined his car and clothes for evidence of bloodstains. They could find no evidence of blood. According to Hendricks, he had taken his children out for a pizza dinner the evening he left for Wisconsin. They finished dinner about 7:30. After playing games in the pizza parlor's game room, they returned home and were in bed by 9:30. Susan Hendricks reached home about an hour later following a baby shower. At midnight, Hendricks kissed his wife goodbye and drove to Wisconsin, where he would meet customers the next morning.

Police had no reason to suspect Hendricks, but the next day, still having not returned to his home, he spoke to reporters and mentioned details of the crime scene that he could not have known without having been there. Police became increasingly suspicious when autopsies revealed that the children's stomachs contained pizza that had not begun to be digested. Since food normally leaves the stomach and enters the small intestine within two hours of being consumed, police concluded that the children had been killed soon after they ate. Hendricks's alibi did not fit the evidence. Following a lengthy trial, he was convicted and sentenced to life for each of the four murders.

The Body Farm

In order to develop a database that would enable investigators to better determine time of death of corpses found outside under different conditions, forensic anthropologist William Bass established the Tennessee Anthropological Research Facility (TARF) at the University of Tennessee in 1971. In a fenced-off four acres of wooded hillside, human corpses are left to decompose under various conditions—buried at different depths in the ground, on top of the ground, in plastic bags, on concrete, in car trunks or seats, and so on. Each corpse is carefully watched and photographed over a period of months or years depending on the rate of decay. The daily records include temperature, weather and soil conditions, changes in the corpse, effects on DNA, insects and other fauna associated with the body, and other useful data. In addition, TARF serves as a training facility for dogs that police departments use to find buried bodies. The animals' keen sense of smell enables them to find corpses that would otherwise remain undetected. The facility was nicknamed "the Body Farm" by mystery novelist Patricia Cornwell.

Bass has had no difficulty obtaining bodies. Corpses of derelicts that are unclaimed and people who donated their bodies to science but are too old or too diseased at death to be of use can be of value to the body farm. Their decomposing corpses provide useful information for forensic scientists who are frequently asked, "How long ago did this person die?" *See also* Forensic Anthropology; Forensic Entomology.

References

The American Medical Association Encyclopedia of Medicine. Charles B. Clayman, ed. Heidi Hough, editor-in-chief. New York: Random House, 1989.

Cyriax, Oliver. *Crime: An Encyclopedia.* North Pomfret, VT: Trafalgar Square, 1996.

Evans, Colin. *The Casebook of Forensic Detection: How Science Solved 100 of the World's Most Baffling Crimes.* New York: Wiley, 1996.

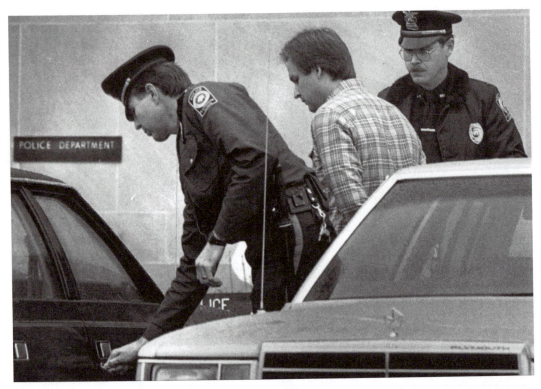

David Hendricks, center, with handcuffs, was taken by Bloomington, Illinois, police to McLean County Jail after being charged with four counts of murder. © *The Pantagraph*, Bloomington-Normal, Illinois.

Tire Prints.

See Footprints.

Tool Marks

Tool marks are impressions, cuts, or marks left on a surface by a tool. Such marks are commonly found in burglary cases where there was forcible entry. Police may find indented wood or marks on a wooden door or window frame if the closure was pried open with a screwdriver or crowbar. Such marks are usually of little value unless there are imperfections in the tool that produce unique impressions. When that is the case, investigators often photograph the scene or make casts of the impression using plaster of paris or dental stone that can be used for comparison with evidence that may be uncovered later.

Occasionally, a tool used in a crime leaves scrape marks. If these marks are clear, an ex-

aminer will be able to see a pattern of irregular ridges and valleys through a microscope, and it may be possible to match the pattern with those on a tool suspected of being used at the crime scene. Tools may also provide trace evidence in the form of paint chips that adhered to the metal during the crime. In analysis by pyrolysis gas chromatography, paint or metal particles on a suspect tool may be compared with those at the crime scene where the tool was used.

In 1981, both Milton and Leah Rosenthal disappeared. Leah Rosenthal had last been seen at the home of her 27-year-old son, Danny Rosenthal, who lived in London. Danny Rosenthal's father, Milton Rosenthal, who lived in Paris, disappeared in the same year shortly after Danny visited him. In Paris, police found the dismembered remains of someone whose characteristics resembled those of Milton Rosenthal. Unfortunately, the hands and head, which could have provided more definite identification, were missing. In Milton Rosenthal's apartment, there were traces of blood and a hacksaw blade that held tiny traces of human bone and tissue. In Danny Rosenthal's London apartment, police had also discovered bloodstains and a hacksaw blade bearing bone fragments and tiny bits of flesh. Dr. Mike Sayce, a French forensic scientist, together with pathologist Michel Durignon, used the hacksaw blade found in the apartment to saw through a piece of paraffin. They then brushed the tool marks left on the paraffin by the teeth of the hacksaw blade with carbon-black powder to improve their visibility. When they compared these tool marks with those left on the dismembered bones, the similarity was striking. Police were able to trace the hacksaw blades to a store where they had been purchased by Danny Rosenthal. The tool-mark evidence was vital to the conviction of the schizophrenic son of the Rosenthals.

At the trial of Terry Nichols, accused of helping Timothy McVeigh construct the bomb that destroyed the Alfred P. Murrah Federal Building in Oklahoma City on April 19, 1995, firearms and tool-mark expert James Cadigan testified that there were striking similarities between the marks on a drilled-out padlock at a burglarized Kansas explosives storage area and those left by a drill bit found in Nichols's home. Authorities believed that Nichols and McVeigh had obtained the explosives needed to detonate the ammonium nitrate and nitromethane used in their bomb from the storage area. Cadigan used Nichols's drill bit to make tool marks on a lead sample. He then used a comparison microscope to compare the marks on the sample with those on the padlock. He testified that the marks were virtually identical.

The case of Unabomber Theodore Kaczynski, who was responsible for a 17-year reign of terror involving 16 bombings, 3 deaths, and 23 major injuries, never reached trial. Kaczynski accepted a plea bargain just before opening arguments were to begin. He admitted his complicity and was charged with 13 counts of transporting explosive devices with intent to kill and maim. On May 5, 1998, he was sentenced to four consecutive life terms in prison without parole. One piece of evidence the FBI had obtained in the Unabomber case was the tool marks left by Kaczynski's stapler on the staples used to fasten together some of the messages he had sent to people and newspapers. Forensic experts were able to show that the marks left on those staples were very similar to the marks left by sample staples obtained by using the stapler found in Kaczynski's cabin. *See also* Ballistics; Chromatography; Comparison Microscope; Crime Scene; Oklahoma City Bombing; Unabomber.

References

Chronis, Peter G. "Tool Expert Testifies." *Denver Post*, Nov. 8, 1997.

Saferstein, Richard. *Criminalistics: An Introduction to Forensic Science.* Upper Saddle River, NJ: Prentice Hall, 1998.

Schehl, Sally A. "Firearms and Toolmarks in the FBI Laboratory." *Forensic Science Communications* 2 (2), April 2000. http://www.fbi.gov/programs/lab/fsc/backissu/april2000/schehl1.htm.

Zonderman, Jon. *Beyond the Crime Lab: The New Science of Investigation.* New York: Wiley, 1990.

Toppan, Jane (1857–1938)

Boston-born Honora Kelley was an American mass poisoner who confessed to 31 murders, but a more realistic total seemed to be between 70 and 100. Kelley spent her early years in an orphanage after her mother's death and her tailor father's commitment to a mental institution for trying to sew his eyes shut. At the age of 5, she was adopted by the Abner Toppans of Lowell, Massachusetts, who renamed her Jane. Her life was uneventful until, at age 24, her fiancé left her at the altar on their wedding day. A nervous breakdown and suicide attempt followed the embarrassment. She entered nursing school at a Cambridge, Massachusetts, hospital, but was dismissed when patients in her care began dying. Hired by another hospital, she was again discharged when hospital authorities discovered that Toppan had not completed her nursing studies. The murderess then forged her own nursing certificate and began private-duty work with some of New England's finest families.

Nurse Toppan was considered kind, sympathetic, sensitive, and very competent. Her poor patient record went unnoticed for 15 years. In 1901, the particularly tragic circumstances of the Davis family of Cape Cod, who were wiped out in a matter of six weeks, aroused suspicion. The husband of the last Davis daughter, Mary D. Gibbs, demanded an autopsy. The procedure confirmed that Mary Gibbs had been poisoned. A full-scale investigation, including exhumations, by the state police led to an indictment on three charges of murder. Toppan, the "Angel of Death," was arrested in Amherst, New Hampshire, in October 1901.

At the trial in June 1902, Toppan confessed to slowly and methodically murdering 31 people with her own concoction of morphine and atropine that she passed off as medicine. The atropine was necessary to counteract the symptoms associated with morphine poisoning. She began with small dosages, which she increased until the victim died. Her only apparent motive was that she enjoyed watching people die. Toppan was sentenced to life imprisonment in Massachusetts at the Taunton State Asylum for the Criminally Insane. In 1938, 36 years later, the serial killer died a natural death at the mental institution in Taunton. *See also* Atropine; Opium; Poisons; Serial Killers.

References

"Jane Toppan, Poisoner of 31, Dies in Hospital at Age 81." *Boston Globe*, Aug. 18, 1938.

Kohn, George C. *Dictionary of Culprits and Criminals*. Mehuchen, NJ: Scarecrow Press, 1986.

Toxicologists

Toxicologists are involved in forensic science because they are skilled in detecting and quantifying the presence of poisons and drugs in the bodies of victims of serious crimes. They are often guided in their search by evidence found at the crime scene, such as chemicals, pill bottles, syringes, and the like. Brief accounts of several of the pioneers in forensic toxicology are given later in this entry.

Lacking any evidence to guide them, forensic toxicologists begin by conducting screening tests to search for what may be minuscule amounts of substances in body tissues. The analytical scheme is designed to detect, isolate, and identify any toxic drugs in blood, urine, or tissues. Screening tests consist of thin-layer chromatography, gas chromatography, and immunoassay. Confirmation of suspected substances makes use of a combination of gas chromatography and mass spectrometery. The mass spectrum for most compounds is unique and provides positive identification.

On rare occasions, toxicologists will screen for heavy metals such as arsenic, bismuth, antimony, mercury, and thallium. Ions of these metals in a solution of hydrochloric acid will produce a silvery or dark coating on a strip of copper inserted into the solution (the Reinsch test). Evidence of such metals can be confirmed by emission spectroscopy, X-ray diffraction, and atomic-absorption spectrophotometry.

If carbon monoxide poisoning is suspected, the concentration of the gas in blood can be determined by spectrophotometry or by using a reagent that releases the gas from a known volume of blood. The carbon monoxide released can then be measured using gas chromatography. The blood of victims found at a fire is often tested for blood carbon monoxide. If the level of the gas in the blood is low, the victim was probably dead when he or she was placed there because he or she did not breathe the toxic fumes.

Orfila, Mathieu Joseph Bonaventure (1787–1853)

See separate entry on Orfila.

Hamilton, Alice (1869–1970)

Alice Hamilton, a physician who lived to be a centenarian, was the first woman to study the effects of industry on the health of its employees and the first woman appointed to the Harvard Medical School faculty. Her

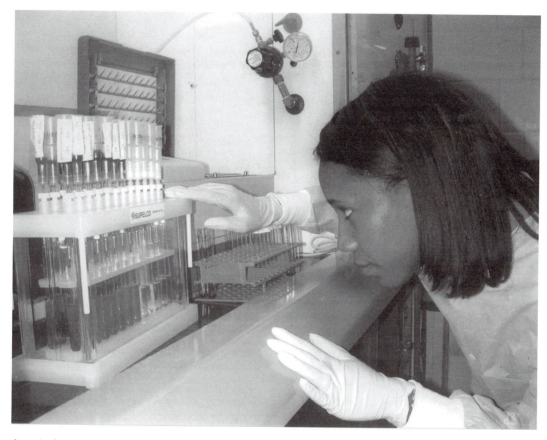

A toxicology scientist uses a solid phase extraction apparatus to isolate drugs from blood or urine in preparation for confirmatory testing. Courtesy of the Georgia Bureau of Investigation, Division of Forensic Sciences.

research on how the manufacture of paint, dyes, explosives, and other materials impacts the well-being of those involved in the process led to legislation designed to protect workers from toxic effects.

Gettler, Alexander O.

In 1918, Alexander Gettler was the first toxicologist at New York City's newly established Medical Examiner's Office, the place where American forensic toxicology was born. Gettler's laboratory was the first to determine blood-alcohol levels and to recognize alcohol's effect on the reflexes of automobile drivers. It was Gettler, too, who, at New York University, established the first graduate program in forensic toxicology. *See also* Arsenic Poisoning; Atropine; Carbon Monoxide; Chromatography; Cyanide; Immunoassay; Poisons; Strychnine; Toxicology.

References

Cyriax, Oliver. *Crime: An Encyclopedia*. North Pomfret, VT: Trafalgar Square, 1996.

Saferstein, Richard. *Criminalistics: An Introduction to Forensic Science*. Upper Saddle River, NJ: Prentice Hall, 1998.

Wilson, Colin. *Clues! A History of Forensic Detection*. New York: Warner Books, 1991.

Toxicology

Toxicology is the study of substances such as drugs and poisons—their effects, detection, and antidotes. Toxicologists rank substances according to the information available and their danger to the public. The ratings include unknown, nontoxic, slightly toxic, moderately toxic, and severely toxic. Of course, even nontoxic substances such as table salt (sodium chloride) can be harmful if large quantities are ingested. The toxicity of a substance depends on a number of factors, including how the substance entered the body, the state of the toxicant, and the condition and size of the individual who received the drug or poison. Forensic toxicologists are primarily concerned with deaths, injuries, or sicknesses that appear to be related to drugs and poisonous substances that may have been deliberately taken or administered to victims.

Because the use of illegal drugs has become widespread, toxicologists are often faced with the task of trying to determine whether a suspect has taken a drug, and, if so, what drug was consumed. To aid toxicologists, some police departments have a trained drug-recognition expert (DRE). By carrying out a number of tests, the expert can establish whether the suspect has taken a drug, and if he or she has, further testing can help determine which drug or drugs have most likely been used. While the DRE is helpful in guiding the toxicologist, the ultimate analysis rests with the toxicologist and the results of the chemical tests he or she conducts.

In 1996, Hillory Farias of La Porte, Texas, was nauseated and suffering from a severe headache following her return from a dance club near her home. By the next day the high school athlete was dead. No evidence of al- cohol or other common drugs were detected during an autopsy, but further testing by suspicious toxicologists led them to conclude that Farias had died from an excessive dose of hydroxybutyrate (GHB). *See also* Arsenic Poisoning; Atropine; Carbon Monoxide; Cyanide; Poisons; Strychnine; Toxicologists.

Reference

Saferstein, Richard. *Criminalistics: An Introduction to Forensic Science.* Upper Saddle River, NJ: Prentice Hall, 1998.

Trace Evidence

Trace evidence consists of small pieces of material left by a criminal at the scene of his or her crime. The basic principle on which forensic science is founded is that a criminal always takes something to the scene of a crime and always leaves something there, or, more simply, in the words of Edmond Locard, "Every contact leaves a trace."

Sometimes the "trace" is quite visible. It might be a bullet casing, a knife, bloodstains, or some similar piece of evidence. But often it is not something that would be found without careful examination. A valuable instrument for finding such trace evidence is the magnifying glass commonly seen in the hand of Sherlock Holmes. With such a simple instrument, detectives may find hair, fibers, tiny pieces of glass, bits of skin, flakes of paint or metal, pieces of wood fiber, and

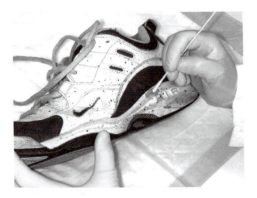

Blood is examined on a shoe. Courtesy of the Biology Section, Centre of Forensic Sciences, Toronto, Canada.

Part of the investigative and prosecutorial team for the Helle Crafts homicide. Standing (left to right): Dr. Allan Reiskin, Det. Shawn Byrne, Atty. Bryan Cotter, Dr. H. Wayne Carver, Dr. John Reffner, Dr. Albert Harper, Atty. Walter Flannagan, Dr. Bruce Hoadley, Atty. Robert Satti, Mr. Robert Horn, Lt. James Hiltz. Sitting: Inspector Anthony Dalessio, Assistant Director Elaine Agliaro, Dr. C.P. Karazulas, and Sgt. Martin Oradan. Courtesy of Dr. Henry Lee, Connecticut State Forensic Science Laboratory.

many other bits and pieces of matter that would not be seen in cursory examination.

In many cases, the crime scene is vacuumed. The material collected is then carefully examined under a microscope for possible evidence that may be sent to the crime laboratory, where greater magnification is possible. An ordinary light microscope can provide images 1,000 times larger than the objects, but electron microscopes can provide magnifications on the order of 150,000. Identification of very small amounts of material is possible through the use of spectrography and mass spectrometry.

Trace evidence was vital to prosecutors in the "woodchipper trial" of Richard Crafts, who was accused of killing his wife, Helle Crafts, and then disposing of her body by running it through a woodchipper on the banks of Connecticut's Housatonic River. Dr. Henry Lee, director of Connecticut's forensic laboratory, found traces of human flesh, hair, and bone chips among wood chips in the trunk of a car Crafts had used to transport his chain saw. One of Lee's assistants, Elaine Pagliaro, found human flesh, hair, and blue fibers matching those in Crafts's home on the teeth of a chain saw recovered by police divers in the icy waters of the Housatonic River. It was on the shore of this river that Crafts was believed to have shredded his wife's frozen body. Harold Deadman, an expert in hair and fiber analysis from the FBI Crime Laboratory, testified that strands of hair found in Helle Crafts's hairbrush matched those found among wood chips and tiny human body parts along the riverbank.

John Reffner, another forensic scientist, testified that blue-green cotton fibers found mixed with human remains on the riverbank matched fibers found on the teeth of the chain saw that Elaine Pagliaro had analyzed. The serial numbers on the saw had been filed off, but forensic scientists were able to restore them. They matched those on the sales receipt of a saw that had been purchased by Richard Crafts several years before. Other receipts revealed that Crafts had made a down payment on a chest freezer five days before his wife, a flight attendant, returned from a trip and that he had rented a woodchipper. He returned the freezer three days after his wife disappeared.

A powder made from some of the recovered bone fragments was used by Albert Harper, a University of Connecticut anthropologist, to produce antibodies and antigens. The results showed that the bone was from a person who, like Helle Craft, had type-O-positive blood. Forensic dentist Dr. C.P. Karazulas, after taking hundreds of X rays at various angles, was able to match a tooth and a gold crown with those from the dental records of Helle Crafts. His work was confirmed by another forensic dentist, Dr. Lowell Levine.

Police theorized that Crafts had murdered his wife and had placed her body in a freezer to make it easier to cut it into sections with a chain saw. He had then run the body segments through the woodchipper and into the river. The trace evidence convinced the jury,

and Richard Crafts was convicted of murder and sentenced to 50 years in prison.

In the case of African American Dale Devon Scheanette, trace evidence led police astray. In September 1996, a third-grade teacher, Christine Vu, was murdered in her East Arlington, Texas, apartment. In December 1996, another elementary-school teacher, Wendie Prescott, who lived in the same apartment complex, was murdered in similar fashion. Both were found bound with duct tape lying face down in several inches of water in their bathtubs. Hairs found in the fingers of Vu and in her pubic hair were clearly those of a person of Caucasian descent. Consequently, local police were searching for a white man when the FBI informed them that a fingerprint taken from Scheanette following his arrest for burglary matched prints found in the apartments of Vu and Prescott. The Caucasian hairs, police surmised, could have been there prior to the murder, or they could have been picked up when the naked body lay on a carpeted floor.
See also Crime Scene; Electron Microscope; FBI Crime Laboratory; Fiber Evidence; Glass Evidence; Hair Evidence; Locard, Edmond; Paint Evidence; Serial Number Restoration; Soil Evidence; Williams, Wayne; Woodchipper Murder.

References

Evans, Colin. *The Casebook of Forensic Detection: How Science Solved 100 of the World's Most Baffling Crimes.* New York: Wiley, 1996.

Trahan, Jason. "Suspect Didn't Fit Police Profile: Inquiry Focused on White Man in Arlington Slayings." *Dallas Morning News*, Sept. 8, 2000.

Zonderman, Jon. *Beyond the Crime Lab: The New Science of Investigation.* New York: Wiley, 1990.

U

Ultraviolet Radiation

Ultraviolet light, known as the blue light special, consists of electromagnetic waves that are shorter (10–400 nm) than those the human eye can detect (400–700 nm). Forensic scientists use ultraviolet light to detect the natural fluorescence found in body fluids, such as blood or the oils left by fingerprints. If ultraviolet light is shined on a wall, any fingerprints left on the wall will fluoresce. When the use of ultraviolet light is combined with heterodyning, the principle used in radio receivers by which two frequencies are combined and made more detectable, and a special camera, the wavelengths from body fluids can be distinguished from the ultraviolet wavelengths in ordinary daylight. The technique needs refining to see if it can be made to differentiate old fluorescence from new.

The Jesse Watkins Case (1927)

On the night of August 21, 1927, Jesse Watkins entered the bedroom of Henry Chambers at the Presidio, a military post in San Francisco. Chambers had recently fired stablehand Watkins, but Watkins wanted the pension money due him. When he awakened Chambers, Chambers grabbed his gun and fired a shot that struck Watkins in the cheek.

Watkins then seized the gun and used it to beat Chambers to death.

Bloodied by the encounter and his gunshot wound, Watkins washed his blood-stained shirt and sent it to the laundry. Aware that Chambers had discharged Watkins, police arrested Watkins and turned to forensic scientist Edward Heinrich at the University of California to analyze the evidence. When Heinrich placed Watkins's laundered shirt under ultraviolet light, bluish green spots, indicative of blood, were clearly evident. Furthermore, Heinrich found that a heel print left in the dust on Chambers's floor matched the heel of one of Watkins's shoes. He also used a comparison microscope to discover that the bullet a roommate had extracted from Watkins's cheek came from Chambers's gun. The evidence was sufficient to convict Watkins, who later confessed to the crime. He was sentenced to life in prison.

The Roger Payne Case (1968)

The use of ultraviolet light played a significant role in solving the murder of Claire Josephs in Bromley, England, on February 7, 1968. The throat-slashed body of Mrs. Josephs, who was wearing a red woolen dress, was discovered by her husband when he returned from work. Police authorities were quite certain the victim knew her killer because there was no evidence of forcible entry, and the presence of coffee cups and cookies on the kitchen table suggested she had entertained her killer.

The investigation that followed led police to interview Roger Payne, a bank clerk who had known the Josephs. Scratch marks on his hands led them to investigate Payne more closely. Using ultraviolet light they discovered 61 red wool fibers on one of his suits that were consistent with those on Josephs's dress. Additional fiber and blood evidence led to Payne's conviction and he was sentenced to life in prison. *See also* Spectrophotometry.

References

Barrett, Sylvia. *The Arsenic Milkshake and Other Mysteries Solved by Forensic Science.* Toronto: Doubleday Canada, 1994.

Evans, Colin. *The Casebook of Forensic Detection: How Science Solved 100 of the World's Most Baffling Crimes.* New York: Wiley, 1996.

Gardner, Robert. *Crime Lab 101: Experimenting with Crime Detection.* New York: Walker, 1992.

Goodman, Brenda Dekoker. "Scene of the Crime: High-Tech Ways to See and Collect Evidence." *Scientific American*, Mar. 1998, pp. 35–36.

Wilson, Colin. *Clues! A History of Forensic Detection.* New York: Warner Books, 1991.

Unabomber (1978–1995)

Unabomber was the FBI code name for the serial bomber Theodore Kaczynski, who was responsible for a 17-year reign of terror involving 16 bombings, 3 deaths, and 23 major injuries. The Unabomber case was the longest and most costly law-enforcement investigation in the nation's history. Its conclusion revealed that even criminals with genius IQs can be apprehended by the persistent effort of forensic investigators. In Kaczynski's case, his writing style was his downfall.

Theodore Kaczynski was born in Chicago in 1942 and lived a comfortable middle-class life. Testing at a genius level, Kaczynski skipped two grades in school and enrolled at Harvard when he was 16. Despite his academic success, the math major was a bookish loner and appeared to have no real interest in social intercourse of any sort. The same pattern of academic success but lack of social life or contacts continued at the University of Michigan, where Kaczynski completed his doctorate. The 25-year-old accepted a position as an assistant professor of mathematics at the University of California at Berkeley. After two years at Berkeley, Kaczynski abruptly resigned and, in 1971, retreated to a primitive 10' × 12' cabin in Montana. The "Harvard Hermit" lived in Big Sky country doing odd jobs and reading at the library, all the while avoiding social contact with others. No one had any reason to believe that Kaczynski was other than one of many Montana eccentrics. Authorities concluded that the anonymity and isolation Kaczynski cultivated in Montana allowed him to go about his bomb making and testing undetected.

The Unabomber's coming out occurred on March 26, 1978, when a security guard at Northwestern University found a wooden box containing a bomb in a parking lot. A second bomb injured a graduate student at the same school the following year. In 1979, there was a cargo-hold explosion aboard an American Airlines flight out of Chicago. Other devices exploded at the computer departments of Vanderbilt University and the University of California at Berkeley, and a bomb was defused in a University of Utah classroom in 1981. The home of an airline's president, the offices of Boeing Company, and the University of Michigan provided still more targets. The Unabomber claimed his first victim in December 1985 when an explosion outside a computer store in Sacramento, California, killed the owner. The FBI created the acronym Unabom by using letters based on what seemed to be the killer's preferred targets, *Un* (universities), *a* (airlines), and *bom* (bomber).

There was alarmingly little to help investigators in their search for the bomber. Agents could not figure out any pattern to the attacks or motive. They initially described the suspect as a "junkyard bomber" because he patched things together to create his primitive, but effective, devices. Authorities looked to the bomb remnants for direction but could find no clues. Experts were impressed by the bomber's meticulous precautions in building his mechanisms. He made his own explosives out of readily available chemicals. There were no fingerprints, despite the fact that most of the components were handmade. The parts were all numbered, and authorities believed that the bombs were frequently taken apart and rebuilt before they were mailed or delivered. The bombs were also packed in handcrafted wooden boxes that became a signature of the bomber. While this information was interesting, none of it led to any identification.

In February 1987, investigators finally got a break in the case. A witness saw a man place a package in front of a Salt Lake City computer store shortly before it was destroyed by an explosion. The witness's description of a man in his 40s, about six feet tall, with light hair and a sandy moustache resulted in a widely circulated composite drawing, but not much else. A promising lead became one of many dead ends. Suddenly, the bombings stopped. From February 1987 to June 1993, there were no events, and the inquiry slowed. Speculation suggested that the suspect might have died or possibly have been put in prison on another charge. Despite a massive amount of evidence in the Unabom Room at FBI headquarters, the agents were no closer to identifying the killer.

June 1993 shattered the nation's calm and signaled the beginning of the bomber's deadliest period. On the 22nd of that month, a geneticist at the University of California at San Francisco was seriously injured by a package bomb, and two days later a Yale professor was severely maimed as he opened a parcel at his New Haven office. In September 1993, two FBI behaviorists, William Tafoya and Mary Ellen O'Toole, provided a third profile of the bomber. The handmade mechanism of the bomb and the attention to detail signaled an obsessive-compulsive personality who was most likely a loner, given the labor-intensive work the devices required. To the FBI, the more sophisticated works of the bombs meant that the builder was self-taught and getting better at his work. O'Toole and Tafoya indicated that the bomber was most likely hostile to the new technology, probably due to job difficulties. Their profile also suggested someone with a degree in "hard" science and a penchant for wood construction that might signify an environmentalist with a back-to-the-earth philosophy. The investigation ground on under a newly formed Unabomber Task Force of 30 persons from the FBI, the Bureau of Alcohol, Tobacco, and Firearms, and postal inspectors. On December 10, Thomas Mosser, an advertising executive, was killed by an explosion at his New Jersey home.

Beginning in July 1995, the bomber went public for the first time with a series of letters to the *San Francisco Chronicle* and the *New York Times*. The author lamented that technology was destroying society and stated that his purpose was to encourage others who disliked the current trend to join him in rebellion. As if to prove a point and taunt investigators, on April 24, 1995, one of his mail bombs killed the president of the California Forestry Association at the group's headquarters.

What turned out to be the final act of the Unabomber affair was foiled by the most revealing of his letters. The Unabomber proposed a truce of sorts. He wanted a major media organization to publish his manifesto. In return, the bomber promised to end his terror against people while reserving the right to continue the fight against the property of the technological elite. The 56-page, single-spaced, typewritten volume was entitled "Industrial Society and Its Future." The manifesto was an indictment of modern technological society, which the author claimed crushed the common man for the benefit of a corporate and governmental elite. He went on to blame international problems on the Industrial Revolution and forecast a dire future world dominated and run by computers. Only a violent revolution would stem the onslaught.

Publishers were faced with a dilemma. Should they print the manifesto and save lives or risk public indignation at furthering the designs of a terrorist? Law-enforcement officials urged that the letter be printed in the hopes that the style, language, or sentiments would trigger some sort of recognition among the public and jump start the beleaguered probe. Ultimately, the *New York Times* and the *Washington Post* published 3,000 words of excerpts, and the task force waited.

David Kaczynski, a youth worker in upstate New York, uneasily recalled helping his mother move out of her Chicago home. In the course of the move, he had uncovered some of his brother Theodore's journals and letters. The papers sounded very similar to the diatribe in the Unabomber's manifesto. Through a friend, Kaczynski made contact with the FBI and explained his concerns. Theodore Kaczynski fit the Unabomber profile, and the agents immediately moved to Lincoln, Montana, to subtly begin an investigation and stakeout of the 54-year-old mathematician's small cabin. Other agents, with the Kaczynski family's permission, began an analysis of the suspect's papers. They found that many of the topics were similar to those in the manifesto. Even the language was similar.

A leak to CBS News about the Kaczynski brothers forced the government into more aggressive action. On April 3, 1996, after two weeks of surveillance, authorities moved on the cabin. They discovered a veritable bomb factory. There were jars of chemicals used in the preparation of explosives, copper piping similar to that used by the Unabomber, several homemade wooden boxes, books on bomb making, and an unfinished bomb. A closer inspection of the cabin uncovered 10 three-ring binders that became the backbone of the government's case. In the journals, Kaczynski admitted his role in the bombings with notes such as "I mailed the bomb." More devastating was the discovery of a draft of the manifesto in the cabin's loft and three typewriters. FBI specialists ascertained that one of the typewriters had produced the manifesto as well as the letter to the *New York Times* demanding publication of the manuscript. With the mountain of evidence secured by government agents, there was little doubt that Kaczynski was the Unabomber.

The Unabomber case never reached trial. In a controversial decision, Kaczynski accepted a plea bargain just before opening arguments were to begin. Rather than allow his lawyers to use the insanity plea, Kaczynski admitted his complicity and was charged with 13 counts of transporting explosive devices with intent to kill and maim. On May 5, 1998, pursuant to the agreement, he was sentenced to four consecutive life terms in prison without the possibility of parole. *See also* Disputed Documents; Psychological Profiling.

References

Douglas, John E. *Unabomber: On the Trail of America's Most-Wanted Serial Killer*. New York: Pocket Books, 1996.

Gibbs, Nancy, and the editorial staff of *Time*. *Mad Genius: The Odyssey, Pursuit, and Capture of the Unabomber Suspect*. New York: Warner Books, 1996.

Graysmith, Robert. *Unabomber: A Desire to Kill*. Washington, DC: Regnery, 1997.

Kelly, John F., and Phillip K. Wearne. *Tainting Evidence*. New York: Free Press, 1998.

V

Voiceprints

Voiceprints, or spectrograms, were first investigated at Bell Telephone Laboratories in 1941. During World War II, U.S. military intelligence used voiceprints to try to identify voices heard on German military broadcasts.

Voiceprints are made using a sound spectrograph that converts the sounds produced by speech into a visual display called a voiceprint. The spectrograph is used by forensic scientists to compare tape recordings of a suspect with one used as evidence. An electronic filter selects certain frequencies that are converted to electrical signals and recorded on special paper using an inked stylus. The resulting print, a series of closely spaced lines, reveals the frequencies recorded over a 2.5-second period. The horizontal axis is a measure of time. The vertical axis reveals the frequencies and loudness of the voiced sounds.

After the war, Lawrence Kersta, who had worked at Bell Labs, became convinced that voice spectrograms could be used to identify individual voices. They could become, he believed, as useful as fingerprints. Kersta maintained that the likelihood that the tongue, teeth, larynx, lungs, mouth, and pharynx—the parts of the body used in making sounds—of any two people would be identical was so small that two people with

matching voiceprints would rarely, if ever, be seen. Many would agree that voiceprints are unique, but, unlike fingerprints, some believe they may change during a person's lifetime. However, the case that follows suggests that voiceprints may indeed be unchanged by time.

In 1971, Clifford Irving convinced McGraw-Hill publishers that recluse Howard Hughes had authorized him (Irving) to write Hughes's biography. After publicity reached Hughes at his Bahama hideaway, he announced that the biography was a fraud and that he had never met Irving. NBC arranged a teleconference between Hughes and a select group of reporters who questioned him to be certain that they were talking to Hughes and not an impostor. To add certitude to the conference, NBC hired Lawrence Kersta to make a voiceprint of Hughes and compare it with a recording of Hughes's voice made nearly 30 years earlier. Kersta's analysis convinced him that the voice on the phone was that of Howard Hughes, and his analysis was confirmed by a critic of Kersta who had to admit that the voice was undoubtedly that of Howard Hughes.

During the 1996 Summer Olympics in Atlanta, Georgia, a bomb exploded in Centennial Olympic Park. About 27 minutes before the explosion, someone, quite likely the bomber, telephoned the park and said,

"There is a bomb in Centennial Park. You have 30 minutes." Using the tape of the call, the FBI made a voiceprint and determined that the call had been made by a white male with an indistinguishable accent. However, the caller used only 11 words; at least 20 words are needed to establish a match between a voiceprint used as evidence and another of a suspect. Furthermore, convincing evidence would require that the two voiceprints be made using the same words. Another problem is the quality of the recording tape. Police often use the same tape to record 911 calls too many times, which results in sounds that are of poor quality. In addition, calls are often made from pay phones or cell phones where there is a lot of background noise that makes it difficult to separate different voices or the voice from environmental noise. In this case, the voiceprint and other evidence led the FBI to arrest Richard Jewell, but it became clear, as more evidence was discovered and more witnesses were questioned, that Jewell was not the bomber, and he was later released.

In the spring of 2000, Microsoft acquired I/O Software's Biometric API technology and SecureSuite authentication technology. It is expected that Microsoft will use the technology to replace passwords. In place of passwords, computer users will employ voiceprints, eyeprints, or fingerprints to access programs. *See also* Fingerprints.

References

Evans, Colin. *The Casebook of Forensic Detection: How Science Solved 100 of the World's Most Baffling Crimes*. New York: Wiley, 1996.

Saferstein, Richard. *Criminalistics: An Introduction to Forensic Science*. Upper Saddle River, NJ: Prentice Hall, 1998.

"Voiceprint Analysis Puts a Face to Words." *All Things Considered*, National Public Radio, July 31, 1996.

Zonderman, Jon. *Beyond the Crime Lab: The New Science of Investigation*. New York: Wiley, 1990.

Vollmer, August (1876–1955)

August Vollmer, the chief of police of Berkeley, California, from 1909 to 1932, is considered by many to be the father of the modern police force. In 1907, while Berkeley's marshal, he established the first formal training school for policemen in Berkeley so that his police officers could gain expertise in evidence procedures and the law. In an effort to establish the best law-enforcement department possible, he recruited men with college experience and a background in science. Although he was not a scientist, he was well read on the latest scientific developments and is credited with being one of the first police chiefs in the United States to apply scientific methods to criminal investigations. In 1923, he set up a small crime laboratory behind his office. Despite the limited size of his lab, it was efficient and was consulted by many other departments on matters of classification and development of evidence.

Vollmer had been a Philippine scout in the U.S. Army and a mailman before becoming a police officer. In 1905, he was elected Berkeley's town marshal, and in 1909 he was appointed Berkeley's first chief of police. His innovative ideas and demands for a high level of honesty and integrity from his men strengthened the force. He established a basic records system in his department and developed the first modus operandi (MO) system. Recognizing the need for good communication among police officers on duty, he devised a method whereby patrol cars could answer emergency calls quickly. He had special signal lights installed at primary intersections that were connected to police headquarters. When a crime was telephoned in to headquarters, the signal lights in the area of the crime were blinked on and off. Officers seeing the flashing light would telephone headquarters to obtain detailed information about the exact location and nature of the emergency.

Vollmer was very interested in lie detection and in 1921 asked a colleague, John Larson, to develop a lie detector (polygraph) that the department could use. Later, Vollmer had another student from the University of California, Leonard Keeler, perfect the polygraph, which became known as the Keeler Polygraph. Some of his other innovations included the establishment of the first police motorcycle patrol in 1911 and one of the first fingerprinting systems in the nation in 1924. In 1925, he hired the first policewoman. Vollmer was also involved in the development of the first school of criminology at the University of California at Berkeley in 1916. During his retirement, Vollmer wrote about and lectured on police work. By the 1940s, his professional influence had been felt across the nation because 25 police chiefs at that time had served under Vollmer in Berkeley. *See also* Larson, John; Polygraph Test.

References

Berkeley, CA. *Police Department History Pages.* http://209.232.44.30/bpd_history.htm.

Evans, Colin. *The Casebook of Forensic Detection: How Science Solved 100 of the World's Most Baffling Crimes.* New York: Wiley, 1996.

Sifakis, Carl. *The Encyclopedia of American Crime.* New York: Smithmark, 1992.

W

Watermarks. *See* Disputed Documents.

Webster, John: Murder (1849)

John Webster was a professor of chemistry at Harvard Medical College who became involved in the first case in which dental evidence played a crucial role. It was late in the autumn of 1849 when Dr. George Parkman visited Webster's laboratory at Harvard. Parkman was a respected wealthy physician nicknamed "Chin" because of his prominent jaw. Webster, a gambler and high liver, had borrowed money from Parkman, who dropped by to collect. When Parkman failed to return home that evening, his family began to search for him. Webster said that the doctor had visited him, had received the money owed him, and had then left. Police talked to Webster, but had no reason to suspect him until they talked to Ephraim Littlefield, a janitor at the college. Littlefield told them that on the day Parkman visited the college, he had noticed that the wall next to the oven in the laboratory was very warm. When he told Webster about his observation, the chemist became pale and said that he was conducting experiments in the oven. The next day, Littlefield's suspicions were further aroused when Webster gave him a Thanksgiving turkey and told him to take the rest of the week off. (Thanksgiving did not become an official national holiday until 1863.)

Acting on a hunch, Littlefield went into the pit below Webster's lab. The pit opened to the river that carried away chemicals and other waste from the laboratory. There he found a human pelvis. When the janitor told the police what he had found, they conducted their own search of the lab. They discovered a torso in a wooden box, various bone fragments scattered about the lab, and in the oven they found a set of false teeth.

A team of doctors from the college was asked to examine the remains. The group included anatomist Oliver Wendell Holmes, whose son became a Supreme Court justice, and Nathan Keep, who was Parkman's dentist. They concluded that the remains were those of a 50–60-year-old male who was about 70 inches in height. Parkman was 60 years old and 71 inches tall. They did not conclude that the remains were definitely those of Parkman. However, at the trial that followed, Nathan Keep explained to the jury that Parkman's protruding jaw had forced him to prepare a special cast when he had fitted his dentures four years earlier. Still, Parkman had complained of irritation, and Keep had had to file the dentures to relieve the pain. Keep then demonstrated how the dentures found at the crime scene fit perfectly the mold he had made for Parkman's dentures. He also showed the jury the file marks he had made to adjust the fit and relieve the discomfort Parkman had experienced.

Webster was convicted and hanged on August 30, 1850. Despite a reported confession by Webster before the hanging, some historians question the validity of the evidence. They believe that Littlefield may have been the actual culprit who, by framing Webster, collected the $3,000 reward offered to anyone who could provide evidence about the missing Parkman. *See also* Forensic Anthropology; Luetgert, Adolphe L. Case; Odontology.

References

Evans, Colin. *The Casebook of Forensic Detection: How Science Solved 100 of the World's Most Baffling Crimes.* New York: Wiley, 1996.

Thomas, Peggy. *Talking Bones: The Science of Forensic Anthropology.* New York: Facts on File, 1995.

Weiner, Alexander (1907–1976)

Alexander Weiner was an American forensic expert who specialized in hematology and serology. He made many discoveries in these areas that benefited forensic investigation and scientific study. Together with Dr. Karl Landsteiner, he completed a comparative study of antigens in the blood of men and monkeys. Their study led to the discovery of the Rh, or rhesus, factor in blood in 1940.

Weiner became head of the Blood Transfusion Division of the Jewish Hospital of Brooklyn in 1935. He and his lawyer father helped write bills for the New York legislature that would help courts order blood-group tests in cases of questioned paternity. In 1938, Weiner was appointed head of the first serological laboratory in New York by the chief medical examiner, Thomas Gonzales. Weiner believed strongly in the value of serological investigation to help solve crimes.

In 1943, Weiner was instrumental in solving a difficult murder case. A woman, Alice Persico, had been found strangled in an al-

leyway. Police determined that she had not been murdered there. They suspected that a man named John Manos had murdered her in his apartment and had then moved her to the alleyway, but they had no proof. Some articles of clothing had been found near the murder victim. Weiner took blood samples from two knotted handkerchiefs and an undershirt. He also took a soiled shirt and handkerchief from Manos's apartment. Using the absorption method on samples from Manos's items, he found that the anti-B serum was so heavily absorbed that Weiner was certain that Manos had type-B blood. He confirmed this with a blood sample. The items found near the victim were also tested and were determined to come from a person with type-B blood, but this was not absolute proof since 11 percent of the population has type-B blood.

Manos had told police that no one else had been in his apartment. The apartment was very neat and clean, but Weiner noticed some spots on the linoleum floor. He made a smear from the spots and found a large number of mucous cells. These cells could have come from edemic fluid that seeps out of the mouth after strangulation. Testing showed that the secretion had come from a person with type-A blood, the same type as Alice Persico's. Again, this did not prove that Manos had murdered her, but it did prove that he had lied when he said that no one else had been in his apartment. Confronted with the evidence, Manos confessed that he had committed the crime. *See also* Blood; Gonzales, Thomas A.

References

RH FACTOR. http://www.fwkc.com/encyclopedia/low/articles/r/r022000675f.html.

Söderman, Harry, and John J. O'Connell, revised by Charles E. O'Hara. *Modern Criminal Investigation.* 5th ed. New York: Funk & Wagnalls, 1962.

Thorwald, Jürgen. *Crime and Science.* New York: Harcourt, Brace & World, 1967.

Wilson, Colin. *Clues! A History of Forensic Detection.* New York: Warner Books, 1991.

Williams, Wayne (1958–)

Wayne Williams is an American serial killer arrested and convicted largely on the basis of fiber evidence established by FBI forensic scientists. Beginning in July 1979, he terrorized Atlanta, Georgia, for nearly two years. The bodies of 11 young black males of slight build had been discovered. Most of them had been strangled. Another 18 young men were reported missing. Following reports in Atlanta newspapers in February 1981 stating that police had found fibers on the victims' bodies, the killer had been stripping his targets and throwing their bodies in local rivers. It was clear that he was reading accounts of his deeds and trying to reduce his chances of being caught. However, it was his attempt to better conceal evidence that led to his arrest.

Because of the killer's recent pattern of dropping bodies into rivers, police had established surveillance crews at bridges around Atlanta. On May 22, 1981, police near the Chattahoochee River heard a splash. They stopped a car driven by 23-year-old photographer Wayne Williams as it was leaving the bridge but could find no reason for detaining him. Two days later, the 12th known victim, Nathaniel Cater, was found on the river bank a mile downstream. It was then that police obtained a warrant to search Williams's car and apartment.

Forensic scientists at the FBI, together with Du Pont chemists, had established that the fibers found on the victims were made by the Wellman Corporation and were used in carpeting manufactured for one year by the West Point Pepperell Corporation. The carpets had been sold for two years in 10 southeastern states. Total sales of the carpeting amounted to 16,397 square yards. Assuming identical sales in all 10 states and that the average quantity sold to a buyer was 20 square yards (12 ft × 15 ft), it was determined that only 82 (16,397/20/10) homes in Georgia would have such a carpet. The probability of finding such a carpet in an Atlanta apartment was 1 in 7,792.

Fibers found on victim Jimmy Ray Payne matched those from Williams's 1970 Chevy station wagon. From General Motors manufacturing figures and the number of vehicles of that model in the Atlanta area, police established that the probability of finding matching fibers on a victim was 1 in 3,828.

There was other evidence as well—fibers from other carpets, hairs from Williams's head and from his dog, and fibers matching victims' clothing on a glove—but the crucial evidence was the probability of finding fibers from both the carpet in the accused murderer's bedroom and the carpeting in his car. The prosecution, using 40 charts and 350 photographs, carefully reviewed the evidence before the jurors and explained to them that the probability of finding victims with both the bedroom and car fibers was 1 in 7,792 × 3,828 or 1 in 29,827,776. With such odds against him, the jury took less than 12 hours to find Wayne Williams guilty of murder. He was sentenced to life in prison. *See also* Serial Killers; Trace Evidence.

References

Deadman, H.A. "Fiber Evidence and the Wayne Williams Trial: Conclusion." *FBI Law Enforcement Bulletin*, 53(5), 10–19, 1984.

Evans, Colin. *The Casebook of Forensic Detection: How Science Solved 100 of the World's Most Baffling Crimes.* New York: Wiley, 1996.

Saferstein, Richard. *Criminalistics: An Introduction to Forensic Science.* Upper Saddle River, NJ: Prentice Hall, 1998.

Zonderman, Jon. *Beyond the Crime Lab: The New Science of Investigation.* New York: Wiley, 1990.

Wimsey, Peter Death Bredon

Lord Peter Death Bredon Wimsey, the self-proclaimed aristocrat of detectives, was the creation of mystery writer Dorothy L. Sayers for a series of novels and short stories beginning with *Whose Body?* (1923). Wimsey, who once described himself as a "Sherlock Holmes disguised as a walking gentleman," was 32 in the first novel and 48 when the saga ended in *Busman's Honeymoon* (1937). Sayers also wrote a short story, "Talboys," which was not published until the early 1970s. The story showed Wimsey at 52, married for seven years with three sons. In 1998, Jill Paton Walsh completed an unfinished Sayers mystery "Thrones, Dominations" involving Wimsey's early married life.

Assisting Wimsey in his adventures were a number of colleagues. Mervyn Bunter, Wimsey's wartime companion, was a "gentleman's gentleman" and his chief partner early in the series. Aside from his normal butler's duties, Bunter also had some forensic skills such as photography and taking fingerprints that were especially useful to the detective. The butler could also do things that Wimsey's class pretensions simply would not allow. It was Bunter who fraternized with the servants or who frequented the pubs in search of information. Wimsey's connection to Scotland Yard was Charles Parker, a workmanlike officer given to reading New Testament commentaries when he was not on duty. Eventually he married Wimsey's sister, Lady Mary.

Perhaps the most important associate of Wimsey did not appear until the sixth novel of the series, *Strong Poison* (1930). In this story, Harriet Vane is accused of murdering her lover by poison, and Wimsey works successfully, using forensic evidence, to free her. It is love at first sight, and the heartsick peer spends the next three novels in which Vane appears trying to convince her to marry him. He is finally successful in the last Wimsey novel, *Busman's Honeymoon* (1937). Harriet Vane was Dorothy Sayers's alter ego. There was a remarkable physical resemblance between the two women, and both were mystery writers. The 29-year-old Vane had lived with a man out of wedlock and did not marry, him, as had Sayers. Both were fiercely independent, highly educated, and very bright.

Wimsey made the big screen twice. In 1935, *The Silent Passenger*, not based upon any specific Sayers book, was filmed. In 1940, *Busman's Honeymoon*, starring the American actor Robert Montgomery, played in theaters with mixed reviews. By far the greatest success for the British detective was on television. Between 1973 and 1977, PBS Television's *Masterpiece Theatre* aired several Wimsey mysteries. *See also* Sayers, Dorothy L.

References

Hitchman, Janet. *Such a Strange Lady*. New York: Harper & Row, 1975.

Papinchak, Robert Alan. "Dorothy L. Sayers." *Mystery and Suspense Writers*, ed. Robin Winks. New York: Charles Scribner's Sons, 1998, vol. 2, pp. 805–826.

Woodchipper Murder (1986)

The woodchipper murder is the story of the murder of Helle Crafts. It was one of the most gruesome, grisly, and forensically demanding cases in Connecticut legal history. The tragedy began on November 18, 1986, when Pan Am stewardess Rita Buonanno dropped off coworker Helle Crafts at her home in Newtown, Connecticut. Buonanno never saw Crafts again.

During the next few days, the dutiful Crafts missed work and could not be located. Richard Crafts, Helle Crafts's husband of 11 years, appeared unmoved about the absence and did not bother to report it to the authorities. For several days, Buonanno and others tried to reach Helle Crafts by phone. Richard Crafts claimed that he did not know where she was. Later, he told callers that she had gone to visit her mother in Denmark, but when her mother was contacted, she said that she had not seen Helle Crafts.

As word of the disappearance became public, Helle Crafts's lawyer reported that she was considering a divorce because of her husband's repeated extramarital affairs. She had cautioned her attorney not to take it as a coincidence if she disappeared. She had also told a number of friends that if anything happened to her, they should not consider it an accident. Local police, however, remained unconcerned despite the reports. They were aware of the marital difficulties, but Richard Crafts, an Eastern Airlines pilot and, more significantly, an auxiliary policeman who was serving on a part-time basis in neighboring

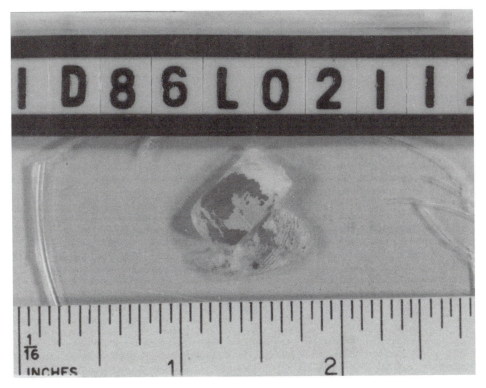

Fingernail with polish and portion of skin. Courtesy of the Connecticut Department of Public Safety, Division of Scientific Services, Forensic Science Laboratory.

Southbury at the time of the disappearance, was an unlikely suspect.

A number of circumstances finally led police to conclude that something was wrong. Crafts's explanations about his wife's disappearance kept changing—she was visiting her mother, then her sister, she had run off with another man, she had deserted the family— he did not know. In addition, his credit-card receipts revealed a suspicious series of purchases. Five days before Helle Crafts disappeared, Richard Crafts made a down payment on a large freezer, which was delivered on the day she returned to Newtown. On that same day, he rented a woodchipper in Darien, a good hour's drive from his home. Why would he go so far away to rent a woodchipper?

The Crafts's nanny/housekeeper added another suspicious clue. She had discovered a brownish, grapefruit-sized stain on the rug in the couple's bedroom. Richard Crafts dismissed it as a kerosene spill from a portable heater used during a recent storm. He pulled the carpet up and took it to the dump. It was never recovered.

In the face of mounting questions and concerns, the case was turned over to the state police, who issued a search warrant. Soon after, forensic experts discovered bloodstains on one side of the mattress. Henry Lee, chief of the state police forensic lab and a blood-spatter expert, was called in to interpret the scene. From the size and distance of the stains, Lee determined that a blow from a blunt object had produced the evidence. Tests showed that the blood came from someone with type-O blood—the same type as Helle Crafts's. In the trunk of Richard Crafts's car, in which he had carried a chain saw, Lee found human hairs and tiny pieces of flesh and bones. Police were now convinced that Helle Crafts was dead and that she had been murdered by her husband. Despite the theory, there was no hard proof, and the overwhelming question remained,

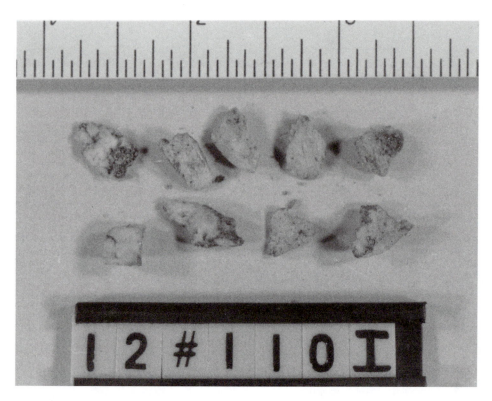

Bone chips recovered in the Helle Crafts murder investigation. Courtesy of the Connecticut Department of Public Safety, Division of Scientific Services, Forensic Science Laboratory.

where was the body? In response to police questioning, Crafts agreed to a polygraph test. The results indicated that he was telling the truth about his innocence.

Despite the polygraph results, detectives following the woodchipper lead began questioning town road workers. One worker recalled seeing someone operating what might have been a woodchipper on River Road beside the Housatonic River in Southbury. He was particularly struck by the incident because it was at night during a storm.

The state police began an intensive search of the area. There they found traces of blonde hair, two caps from teeth, a fingernail, bone fragments, a small amount of blood, and a toenail, none of which could be conclusively identified as the victim's. On the basis of the initial results, the state police established a field lab at the site by the river and called in the state's forensic dental expert in the hope of finding definitive evidence.

In the search for more witnesses, police encountered one citizen who recalled seeing a woodchipper on a bridge over the Housatonic River in mid-November. For four days in freezing temperatures, divers searched for clues under the bridge. They recovered a chain saw, which was immediately sent to the forensic lab. Scientists checking the chain saw found fragments of bleached, dyed human hair that matched hair from the River Road site and strands from the victim's own brush. Inside the chain saw's housing were bits of blue-green fabric as well as traces of flesh and blood. Tests proved that the blood was human and type O. Though the chain saw was corroded and the model's serial numbers had been filed off, the scientists were able to restore the numbers, enabling police to trace the machine. It belonged to Richard Crafts.

The final piece of evidence was the discovery of a tooth at the River Road location. Forensic dental experts were able to prove

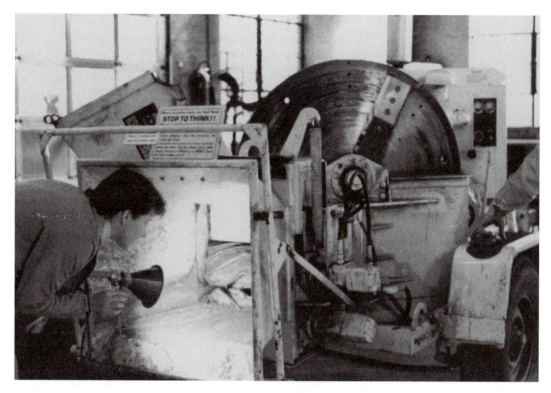

Criminalist examining the woodchipper used by Richard Crafts. Courtesy of the Connecticut Department of Public Safety, Division of Scientific Services, Forensic Science Laboratory.

that it was definitely Helle Crafts's tooth. From the condition of the tooth, they deduced that a tremendous force, such as a saw or woodchipper, had broken the tooth from its seat. The tooth, together with skin and hair evidence, allowed the medical examiner to declare Helle Crafts dead two months after her disappearance.

From all the evidence, police surmised that Crafts had killed his wife with a blunt instrument in the bedroom. He had then frozen her body in the large freezer to facilitate cutting her into smaller parts with a chain saw. Finally, he had run those body pieces through a woodchipper in an effort to destroy any evidence. Without a body, he probably thought, a conviction was impossible. The tiny pieces of flesh, blood, bone, and teeth, however, provided all the evidence needed to bring him to justice. Richard

Crafts was arrested and indicted for his wife's murder. With less than six ounces of bodily remains, prosecutors made a convincing case. After one mistrial, Richard Crafts was convicted and received a 50-year prison sentence. *See also* Hair Evidence; Odontology; Serial-Number Restoration; Trace Evidence.

References

Evans, Colin. *The Casebook of Forensic Detection: How Science Solved 100 of the World's Most Baffling Crimes.* New York: Wiley, 1996.

Gardner, Robert. *Crime Lab 101: Experimenting with Crime Detection.* New York: Walker, 1992.

Grogan, David W. "The Lady Vanishes, and a Woodchipper Leaves Just a Shred of Evidence." *People Weekly*, Aug. 1, 1988.

Herzog, Arthur. *The Woodchipper Murder.* New York: Henry Holt, 1989.

Zonderman, Jon. *Beyond the Crime Lab: The New Science of Investigation.* New York: Wiley, 1990.

X

X-Ray Diffraction

X-ray diffraction occurs when a beam of X rays strikes the orderly arranged layers of atoms in a crystal. The pattern produced when the X rays are diffracted can be used by forensic scientists to identify crystals taken as evidence from a crime scene. They have also been used to establish the identity of substances in disputes regarding patents held by various drug companies.

X rays are generated when fast-moving electrons strike a metal target. The acceleration of the electrons resulting from the sudden change in their velocity as they strike the target produces high-energy electromagnetic waves (X rays) that have a wavelength on the order of 0.001 to 1 nanometer (10^{-12} to 10^{-9}m). When a beam of X rays strikes the orderly arranged layers of atoms in a crystal, the waves reflected from each layer overlap to form a series of high-energy and low-energy bands like the bright and dark bands one sees when one looks at a fluorescent light through a narrow slit. These bands make up a diffraction pattern. Where the crests from different reflected beams overlap, the intensity of the X rays is strong; where crests and troughs overlap, the waves cancel one another and the intensity is low. In visible light, this would correspond to bright and dark bands.

The diffraction pattern can be recorded on film and provides a "fingerprint" because each crystal produces a unique pattern. By comparing a sample of crystals taken as evidence with X ray diffraction patterns of known crystals, the identity of the sample can be readily determined. The technique works very well with pure crystals or mixtures where each component makes up a significant fraction of the mixture. It may, however, fail to detect a substance that constitutes only a small portion of a mixture of crystals.

In the case of *Glaxo Inc. and Glaxo Group Limited v. Novopharm Limited, and Granutec Inc.*, some of the evidence cited involved X-ray diffraction tests used to identify the two forms of ranitidine hydrochloride, the active ingredient in Zantac. The court ruled that Novopharm had developed Form 1 of the compound independently and that the evidence presented by Glaxo, including X-ray diffraction analysis, was insufficient to show that its patent on the manufacture of ranitidine hydrochloride had been infringed.

References

Glaxo Inc. and Glaxo Group Limited v. Novopharm Limited, and Granutec Inc. United States District Court, E.D. North Carolina, Western Division, July 5, 1996.

Saferstein, Richard. *Criminalistics: An Introduction to Forensic Science*. Upper Saddle River, NJ: Prentice Hall, 1998.

Sears, Francis W., Mark W. Zemansky, and Hugh D. Young. *College Physics*. 6th ed. Reading, MA: Addison-Wesley, 1985.

Y

Yule Bomb Killer. *See* Osborn, Albert S.

Appendix: Helpful Web Sites

Academy of Behavioral Profiling
http://www.profiling.org/
This association is dedicated to the application of evidence based on criminal profiling techniques within investigative and legal venues. The ultimate aim of the group is to foster the development of a class of practitioners capable of raising the discipline of evidence-based behavioral profiling to the status of a profession.

American Academy of Forensic Science
http://www.aafs.org/
This is a professional society dedicated to the application of science to the law. Membership includes physicians, criminalists, toxicologists, attorneys, dentists, physical anthropologists, document examiners, engineers, psychiatrists, educators, and others who practice and perform research in the many diverse fields relating to forensic science.

American Academy of Psychiatry and Law
http://www.emory.edu/AAPL/
Forensic psychiatry is a medical subspecialty that includes research and clinical practice in the many areas in which psychiatry is applied to legal issues. The academy provides information and professional support in areas in which psychiatry and the law share a common boundary. Such concerns might include correctional psychiatry, juvenile justice, competence, civil and criminal, and criminal responsibility.

American Board of Forensic Anthropology
http://www.csuchico.edu/anth/ABFA/

Forensic anthropology is the application of the science of physical anthropology to the legal process. The identification of skeletal, badly decomposed, or otherwise unidentified human remains is important for both legal and humanitarian reasons. Forensic anthropologists apply standard scientific techniques developed in physical anthropology to identify human remains and to assist in the detection of crime.

American Board of Forensic Document Examiners
http://www.abfde.org/
The Board provides a program of certification in forensic document examination with the dual purpose of serving the public interest and promoting the advancement of forensic science. The site has interesting links to articles on forensic science such as counterfeiting and on important court cases. The developing skill builder section is interesting as well.

American Board of Forensic Entomology
http://www.missouri.edu/~agwww/entomology/index.html
Forensic entomology, or medicocriminal entomology, is the science of using insect evidence to uncover circumstances of interest to the law, often related to a crime. Entomologists are being called upon with increasing frequency to apply their knowledge and expertise to criminal and civil proceedings. They are also recognized members of forensic laboratories and medical/legal investigation teams. This site has links to

APPENDIX

related topics as well as ten interesting entomological case studies.

American Board of Forensic Odontology
http://www.abfo.org/
The professional certification board in forensic odontology registers experts for court testimonies, provides guidelines for the discipline, and maintains a speakers bureau.

American Board of Forensic Psychology
http://www.abpf.com/board.html
The ABFP was organized for the purpose of contributing to the development and maintenance of forensic psychology as a specialized field of study, research, and practice. The mission of the ABFP includes the operation of a continuing education program in forensic psychology, the provision of a forum for the exchange of scientific information, and other activities that enhance or advance profession of forensic psychology.

American Board of Medicolegal Death Investigators, Inc.
http://www.slu.edu/organizations/abmdi/index.shtml
This is a voluntary certification program of individuals who have the proven knowledge and skills necessary to perform medicolegal death investigations as set forth in the *National Guidelines for Death Investigation* published in 1997 by the National Institutes of Justice.

American College of Forensic Examiners
http://www.acfe.com/
Multi-disciplinary in its scope, the society actively promotes the dissemination of forensic information. The college serves as the national center for this purpose and circulates information and knowledge through the official journal (*The Forensic Examiner*), lectures, seminars, conferences, workshops, the Internet, continuing education courses, and home study courses.

The American Polygraph Association
http://www.polygraph.org/
This site is dedicated to providing a valid and reliable means to verify the truth and establish the highest standards of moral, ethical, and professional conduct in the polygraph field. The page contains a police and crime headline section, a FAQ segment, training and accreditation programs, and a current issues search program.

American Society of Crime Laboratory Directors
http://www.ascld.org/
Initiated as a result of an FBI request in 1973, the ASCLD is a nonprofit professional society devoted to the improvement of crime laboratory operations through sound management practices. Its purpose is to foster the common professional interests of its members; to promote and foster the development of laboratory management principles and techniques; to acquire, preserve, and disseminate information related to the utilization of crime laboratories; and to maintain and improve communications among crime laboratory directors. A particular strong point of the site is its advice to students who would like to pursue the profession.

American Society of Forensic Odontology
http://www.asfo.org/
The ASFO is the largest organization dedicated to the pursuit of forensic dentistry. The ASFO publishes a newsletter three times a year.

American Society of Questioned Document Examiners
http://www.asqde.org/
The purposes of the society are to foster education, sponsor scientific research, establish standards, exchange experience, and provide instruction in the field of questioned document examination and to promote justice in matters that involve questions about documents. The library section of the site provides a strong bibliography of related materials.

Association for Crime Scene Reconstruction
http://www.acsr.com/
This organization emphasizes the use of scientific methods, physical evidence, and deductive and inductive reasoning to gain explicit knowledge of the series of events that surround the commission of a crime. Besides the usual links provided by most professional Web Sites, there are two interesting crime scene quizzes at this location. Be careful of the first quiz as it is not for the faint of heart!

Association of Firearm and Tool Mark Examiners
http://www.afte.org/
AFTE was established in recognition of the need for the exchange of information, methods, development of standards, and the furtherance of research in the field. It provides descriptions of new techniques and procedures, reviews instru-

mentation, and aids in the solution of common problems encountered in the field. The table of contents of the most recent copy of the association journal is available at this site.

Australian and New Zealand Forensic Science Society
http://users.bigpond.net.au/anzfss/
This society was formed with the aim of bringing together scientists, police, criminalists, pathologists, and members of the legal profession actively involved in the forensic sciences. The Society's objectives are to enhance the quality of forensic science by providing both formal and informal lectures, discussions, and demonstrations encompassing the various disciplines within the forensic sciences.

Canadian Society of Forensic Science
http://www.csfs.ca/
A professional organization designed to promote the study and enhance the stature of forensic science. It is organized into sections representing diverse areas of forensic examination: Anthropology, Medical, Odontology, Biology, Chemistry, Documents, Engineering, and Toxicology. The last seven issues of the society's journal are available in abstract form.

Department of Forensic Science at George Washington University
http://www.gwu.edu/~forensic/
This academic department of a major university provides answers to frequently asked questions about forensic science and has a very good links section for references.

Evidence Photographers International Council, Inc.
http://www.epic-photo.org/
Epic's primary purpose is the advancement of forensic photography/videography in civil evidence and law enforcement. The council has a journal but it is not readily available on the Web. The site also offers a chat room, current workshops, and certification guidelines.

The Fingerprint Society
http://www.fingerprint-society.org.uk/
This English society aims to advance the study and application of fingerprints and to facilitate the cooperation among persons interested in the field of personal identification.

Forensic Science Resources [Tennessee Criminal Law Defense Resources]
http://www.tncrimlaw.com/forensic
This location provides a series of links to a number of sites, many of which are mentioned in this appendix.

The Forensic Science Society
http://www.forensic-science-society.org.uk/
This is the homepage for England's leading forensic organization, which acts as a multidisciplinary society dedicated to the application of science to the cause of justice. There is a search page for what the society calls one of the most comprehensive links databases in the world, and the last three issues of the annual journal are available online.

High Technology Crime Investigation Association
http://www.htcia.org/
HTCIA is designed to encourage, promote, aid, and affect the voluntary interchange of data, information, experience, ideas, and knowledge about methods, processes, and techniques relating to investigations and security in advanced technologies. Peace officers, investigators, and prosecuting attorneys make up the membership of the organization.

International Association of Arson Investigators
http://www.fire-investigators.org/
The goal of this organization is to foster, support, and promote fire prevention and arson awareness through education and training. The association exists to improve the education and training of those who examine the origin and cause of explosions and fires and to educate members to better protect life and property from crimes of insurance fraud and arson.

International Association of Computer Investigative Specialists
http://cops.org/index.html
This is the homepage of an organization composed of law enforcement professionals dedicated to education in the field of forensic computer science. IACIS members represent federal, state, local, and international law enforcement professionals. Regular IACIS members have been trained in the forensic science of seizing and processing computer systems.

International Association of Forensic Linguists
http://www.iafl.org/

An organization consisting primarily of linguists whose work involves the law. It includes the study of such topics as courtroom interpreting and translating, the readability/comprehensibility of legal documents, interviews with children in the legal system, and authorship attribution for both written and spoken language.

International Association of Forensic Nurses
http://www.forensicnurse.org/
IAFN is the only international professional organization of registered nurses formed exclusively to develop, promote, and disseminate information about the science of forensic nursing nationally and internationally. Forensic Nursing is the application of nursing science to public or legal proceedings and the application of the forensic aspects of health care combined with the bio-psycho-social education of the registered nurse in the scientific investigation and treatment of trauma and/or death of victims and perpetrators of abuse, violence, criminal activity, and traumatic accidents.

International Association of Forensic Toxicologists
http://www.tiaft.org//
The aims of the association are to provide an organization of professionals engaged or interested in the analysis of potentially harmful substances and in the interpretation of the results from such analyses in a judicial context and to encourage research in and the practice of forensic toxicology and allied areas. There is an online library available at the site as well as links to members' homepages. Portions of the site are password guarded.

International Association for Identification
http://www.theiai.org/
IAI strives to be the primary professional association for those engaged in forensic identification, investigation, and scientific examination of physical evidence. There is some access to the association's publication, *Journal of Forensic Identification*.

International Association of Investigative Locksmiths, Inc.
http://www.iail.org/
A professional group of technicians who offer technical and professional advice, council, and investigative services to the community, insurance companies, and law enforcement officials.

International Crime Scene Investigators Association
http://www.icsia.com/
ICSIA is strictly Internet based. It is the place for the "field" people to express ideas, exchange information, and seek the advice of others. Investigators have rarely had the opportunity to directly discuss techniques and tips or share other pertinent information with crime scene personnel throughout the world. The ICSIA hopes to make that possible.

International Forensic Imaging Enhancement Society
http://ourworld.compuserve.com/homepages/Forensic_Expert/internat.htm
The society was started to facilitate the exchange of information among law enforcement agencies, laboratories, and private consultants who utilize forensic digital image enhancement processes. It is a very small site with limited information.

International Society for Forensic Genetics
http://www.isfg.org/
The group is responsible for the promotion of scientific knowledge in the field of genetic markers analyzed with forensic purposes.

Microscopy Society of America
http://www.msa.microscopy.com/
Though not technically a forensic organization, this society is dedicated to the promotion and advancement of the knowledge of the science and practice of all microscopical imaging, analysis, and diffraction techniques used in diverse areas of biological, materials, medical, and physical sciences.

National Academy of Forensic Engineers
http://spionline.org/
This is a professional organization formed to advance the art and skill of engineers who serve as consultants to members of the legal profession and as expert witnesses in courts of law, arbitration proceedings, and administrative adjudication proceedings.

National Association of Document Examiners
http://expertpages.com/org/nade.htm
NADE conducts educational and research programs, maintains a speakers bureau, and offers education, ethics, and professional certification programs. Their newsletter *Communique* is

published bi-monthly. The NADE journal is published quarterly.

National Association of Medical Examiners
http://www.thename.org/index.html
NAME is the national professional organization of physician medical examiners, medical death investigators, and death investigation system administrators who perform the official duties of the medicolegal investigation of deaths of public interest in the United States. The organization has the dual purposes of fostering the professional growth of physician death investigators and disseminating the professional and technical information vital to the continuing improvement of the medical investigation of violent, suspicious, and unusual deaths.

Society of Forensic Toxicologists
http://www.soft-tox.org/
The "introduction to Forensic Toxicology" section of the site is worth visiting. There is an overview of toxicology, its objectives, and its future.

Zeno's Forensic Site
http://www.forensic.to/forensic.html
One of the best known forensic pages on the Web, this is a gateway location for links to other sites. Topic links include general information, such as jobs or conferences, forensic medicine, and links to eighteen specific forensic disciplines.

Sample State and Local Forensic Organizations

California Association of Crime Laboratory Directors
http://www.geocities.com/cacld/
This is an association designed to promote the exchange of information at forensic science laboratories and to improve management methods in forensic science facilities.

California Association of Criminalists
http://www.cacnews.org/

CAC is a professional membership organization of forensic scientists founded in 1954 by sixteen members from various agencies throughout California. They meet to exchange ideas and new testing methodologies and to share case histories.

Iowa Division of the International Association for Identification
http://www.iowaiai.org/
The Iowa Division hopes that through their Web site people will gain a strong appreciation for such scientific disciplines as fingerprinting, forensic photography, DNA forensics, and more. The site contains articles from numerous experts and associations from around the world. There is a lengthy links page to other forensic locations.

New Jersey Association of Forensic Scientists
http://www.njafs.org/
There is a very strong local flavor to this site, which contains a section on New Jersey forensic history and crime labs as well as a FAQ segment. The major event of the organization is the spring seminar, which has covered a wide variety of forensic topics.

Northeastern Association of Forensic Scientists, Inc.
http://www.geocities.com/CapeCanaveral/Lab/5122/
The association hopes to exchange ideas and information within the field of forensic science, foster friendship and cooperation among the various laboratory personnel, and promote recognition of forensic science as an important component of the criminal justice system. There is a brief links page and information about the annual meeting on the Web page.

Southern Association of Forensic Scientists
http://www.southernforensic.org/
A message board, training information, links, annual meeting information, and employment opportunities make this a standard regional Web site.

Bibliography

Print

"Accord Urged in Strangler Case; U.S. Wants an End to Evidence Fight." *Boston Globe*, March 1, 2001, p. B4.

Adler, Jerry, and John McCormick. "The DNA Detectives." *Newsweek*, Nov. 16, 1998, p. 66.

Ahlgren, Gregory and Stephen R. Monier. *Crime of the Century: The Lindbergh Kidnapping Hoax*. Boston: Brandon Books, 1993.

Aiken, C.G.G., and D. A. Stoney, eds. *The Use of Statistics in Forensic Science*. New York: Ellis Horwood, 1991.

Amalgamated Transit Union, Division 1279 v. Cambria County Transit Authority. Civ. A. No. 88–796, U.S. District Court W.D. Pennsylvania, July 19, 1988.

American Jurisprudence Proof of Facts Annotated. Vol. 15. San Francisco: Bancroft-Whitney Co., 1964, 115–148.

The American Medical Association Encyclopedia of Medicine. Charles B. Clayman, ed. Heidi Hough, editor-in-chief. New York: Random House, 1989.

"Arson Detection Dogs Sniff Out Valuable Evidence." *American City and County*, Dec. 1996, pp. 12–13.

Asbaugh, D.R. *Quantitative-Qualitative Friction Ridge Analysis: An Introduction to Basic and Advanced Ridgeology*. Boca Raton, FL: CRC Press, 1999.

Babitsky, Steven, James J. Mangraviti, Jr., and Christopher J. Todd. *The Comprehensive Forensic Services Manual: The Essential Resources for All Experts*. Falmouth, MA: S.E.A.K., Inc., Legal and Medical Information Systems, 2000.

Baden, Michael M. *Unnatural Death: Confessions of a Medical Examiner*. New York: Ivy Books, 1989.

Barrett, Sylvia. *The Arsenic Milkshake, and Other Mysteries Solved by Forensic Science*. Toronto: Doubleday Canada, 1994.

Behn, Noel. *Lindburgh: The Crime*. New York: NAL-Dutton, 1994.

Bell, Allison. "Date Rape." *Teen Magazine*, July 1997, pp. 68ff.

Bergreen, Laurence. *Capone: The Man and the Era*. New York: Simon & Schuster, 1994.

Biasotti, A. "A Statistical Study of the Individual Characteristics of Fired Bullets." *Journal of Forensic Sciences* 4(1), 133–140, 1959.

Black's Law Dictionary. 7th ed. Bryan A. Garner, editor-in-chief. St. Paul, MN: West Group Publishing, 1999.

Block, Eugene B. *Famous Detectives*. Garden City, NY: Doubleday, 1967.

———. *Science vs. Crime: The Evolution of the Police Lab*. San Francisco: Cragmont, 1979.

———. *The Wizard of Berkeley*. New York: Coward-McCann, 1958.

Blundell, Nigel. *Encyclopedia of Serial Killers*. North Dighton, MA: JG Press, 1996.

Blundell, R.H., and Wilson Haswell. "Blood in the Drains: The Ruxton Killings." *Crimes and Punishment: The Illustrated Crime Encyclopedia*. Westport, CT: H.S. Stuttman, 1994.

———. "Buck Ruxton." *Famous Trials*, ed. James H. Hodge. Harmondsworth, Middlesex, Eng.: Penguin Books, 1950.

Bodayla, Stephen D. "Goddard, Calvin Hooker." *Dictionary of American Biography: Supplement Five*. New York: Charles Scribner's Sons, 1977.

Bodziak, W.J. *Footwear Impression Evidence*. New York: Elsevier, 1990.

Boswell, Charles, and Lewis Thompson. *The Carlyle Harris Case*. New York: Collier Books, 1961.

Brice, Chris. "Expert Witness." *Advertiser* (Adelaide, Australia), Aug. 12, 2000, p. M23.

Brook, Paula. "The Ones Who Get Away: Drunks at the Wheel." *Chatelaine*, Vol. 69 Dec. 1, 1996, p. 72.

Browne, Douglas G. *The Rise of Scotland Yard*. Toronto: George G. Harrap & Co., 1956.

Browne, Douglas, and E.V. Tullett. *Bernard Spilsbury: His Life and Cases*. London: Penguin Books, 1955.

Buchanan, Edna. *The Corpse Had a Familiar Face*. New York: Random House, 1987.

Buckley, William F., Jr. "O.J. on Our Mind." *National Review*, June 26, 1995, p. 71.

Bugliosi, Vincent. *Outrage*. New York: W.W. Norton, 1996.

Burton, William C. *Burton's Legal Thesaurus*. 2nd ed. New York: Macmillan, 1992.

Canon, Scott. "Uncharged Murder Suspect Stays in Jail." *Boston Globe*, Sept. 20, 2000, p. 1.

Cantwell, Mary. "How to Make a Corpse Talk." *New York Times Magazine*, July 14, 1996.

Chang, Maria L. "Fly Witness." *Science World*, Oct. 1997, p. 8.

Chang, Raymond. *Chemistry*. 2nd ed. New York: Random House, 1984.

Chaplin, Charles. *My Autobiography*. New York: Simon & Schuster, 1964.

Chemistry and Crime: From Sherlock Holmes to Today's Courtroom. Ed. Samuel M. Gerber. Washington, DC: American Chemical Society, 1983.

Christopher, Milbourne. *Mediums, Mystics, and the Occult*. New York: Crowell, 1975.

Chronis, Peter G. "Tool Expert Testifies." *Denver Post*, Nov. 8, 1997.

Churchill, Allen. *A Pictorial History of American Crime, 1849–1929*. New York: Holt, Rinehart & Winston, 1964.

Clark, Franklin, and Ken Diliberto. *Investigating Computer Crime*. Boca Raton, FL: CRC Press, 1996.

Commonwealth v. Charles E. Smith IV. Appeals Court of Massachusetts, Suffolk, 1993.

Commonwealth v. Peter G. Saunders. Appeals Court of Massachusetts, Plymouth, 1998.

Conklin, John E. *Criminology*, 3rd ed., New York: Macmillan, 1989.

Conlier, Joseph R. "The Charlie Chaplin and Joan Barry Affair—1943." *The People's Almanac 2*, ed. David Wallechinsky and Irving Wallace. New York: Bantum Books, 1978.

"Connecticut Police Find Body of Missing Mother." Associated Press, *Cape Cod Times*, Sept. 3, 2000.

Cooper, C.L., and S.R. Sheppard. *Mockery of Justice: The True Story of the Sheppard Murder Case*. Boston: Northeastern University Press, 1995.

Corsini, Raymond J., ed. *Encyclopedia of Psychology*. 2nd ed. New York: Wiley, 1994.

Crawford, Ralph. "Forensic Casebook's 'No Conclusion' Results in Agreement." *Direct*, April 1, 1999.

Crimes and Punishment: The Illustrated Crime Encyclopedia. Westport, CT: H.S. Stuttman, 1994.

Cullen, Tom. *Crippen: The Mild Murder*. London: Bodley Head, 1977.

Cyriax, Oliver. *Crime: An Encyclopedia*. North Pomfret, VT: Trafalgar Square, 1996.

Darwin, George H. "Sir Francis Galton." *Dictionary of National Biography: Second Supplement*. New York: Oxford University Press, 1958.

Deadman, H.A. "Fiber Evidence and the Wayne Williams Trial: Conclusion." *FBI Law Enforcement Bulletin* 53 (5), 10–19, 1984.

DeHaan, John D. *Kirk's Fire Investigation*. 4th ed. Upper Saddle River, NJ: Brady, 1997.

Dewey, Thomas E. *Twenty against the Underworld*, ed. Rodney Campbell. New York: Doubleday & Garden City, 1974.

Dickerson, Richard E., Harry B. Gray, Marcetta Y. Darensbourg, and Donald J. Darensbourg. *Chemical Principles*. 4th ed. Menlo Park, CA: Benjamin/Cummings, 1984.

DiMaio, Vincent J.M. *Gunshot Wounds: Practical Aspects of Firearms, Ballistics, and Forensic Techniques*. New York: Elsevier Science, 1985.

"DNA Tests Clear Prisoner of Rape, But Implicate His Younger Brother." *News-Times* (Danbury, CT), November 26, 2000.

Donaldson, Norman. *In Search of Dr. Thorndyke: The Story of R. Austin Freeman's Great Scientific Investigator and His Creator*. Bowling Green, OH: Bowling Green University Popular Press, 1971.

Dorland's Illustrated Medical Dictionary. 27th ed. Philadelphia: W.B. Saunders Co., 1988.

Douglas, John E. *Unabomber: On the Trail of America's Most-Wanted Serial Killer*. New York: Pocket Books, 1996.

Duke, Thomas. *Celebrated Criminal Cases of America*. Reprint. Montclair, NJ: Patterson Smith, 1991.

Ellis, David. "Sin of the Father: A Child Conceived in Rape 51 Years Ago Proves Her Mother Was Telling the Truth." *People Weekly*, July 25, 1994, pp. 38ff.

Evans, Colin. *The Casebook of Forensic Detection: How Science Solved 100 of the World's Most Baffling Crimes*. New York: Wiley, 1996.

———. "The Poisonous Umbrella." *Time*, Oct. 16, 1978.

Evett, I.W., and B.S. Weir. *Interpreting DNA Evidence*. Sunderland, MA: Sinauer Associates, 1998.

Feldman, Bernard J. "Elementary Physics in a Real Automobile Accident." *Physics Teacher*, Sept. 1997, pp. 335–337.

Fisher, Barry A.J. *Techniques of Crime Scene Investigation*. 6th ed. Boca Raton, FL: CRC Press, 2000.

Flatow, Ira. "Analysis: Biometric Identification, Using Physical Characteristics Such as Voice Sample or Handprint to Identify a Person." *Talk of the Nation Science Friday*, Sept. 1, 2000.

Foley, Michael. "Forensic Casebook: Electricity and Water: A Volatile Mix." *Electrical Construction and Maintenance*, May 1, 2000.

Follick, Joe. "In Child Support Case, Ex-Husband Plays DNA Card." *Tampa Tribune*, Aug. 19, 2000.

Forrest, D.W. *Francis Galton: The Life and Work of a Victorian Genius*. New York: Taplinger, 1974.

Frank, Laura and John Hanchette. "Convicted or False Evidence? False Science Often Sways Juries, Judges." *USA Today*, July 19, 1994.

Freedberg, Sydney P. "Good Cop, Bad Cop." *St. Petersberg Times*, March 4, 2001.

Friedland, Martin L. *The Death of Old Man Rice: A True Story of Criminal Justice in America*. New York: NYU Press, 1994.

"Fulton Oursler, Author, Dies at 59." *New York Times*, May 25, 1952.

Gannon, Robert. "The Body Farm." *Popular Science*, Sept. 1997, p. 78ff.

Gardner, Robert. *Crime Lab 101: Experimenting with Crime Detection*. New York: Walker, 1992.

Gardner, Robert, and Edward A. Shore. *Math and Society*. New York: Watts, 1995.

Gardner, Robert, and Dennis Shortelle. *From Talking Drums to the Internet: An Encyclopedia of Communications Technology*. Santa Barbara, CA: ABC-CLIO, 1997.

Gaudette, B.D. "A Supplementary Discussioon of Probablities and Human Hair Comparisons." *Journal of Forensic Sciences* 27(2), pp. 279–289, 1982.

Gaudiano, Nicole. "5 Arrested in Undercover Drug Sting." *Record*, Sept. 26, 2000, p. 101.

Gerberth, V.J., ed. *Practical Homicide Investigation: Tactics, Procedures, and Forensic Techniques*. 3rd ed. Boca Raton, FL: CRC Press, 1996.

Gibbs, Nancy, and the editorial staff of *Time*. *Mad Genius: The Odyssey, Pursuit, and Capture of the Unabomber Suspect*. New York: Warner Books, 1996.

Gibson, Charles. "Violent Death of JonBenet Ramsey," *ABC Good Morning America*, July 15, 1997.

Glaxo Inc. and Glaxo Group Limited v. Novopharm Limited, and Granutec Inc. United States District Court, E.D. North Carolina, Western Division, July 5, 1996.

Goodman, Brenda Dekoker. "Scene of the Crime: High-Tech Ways to See and Collect Evidence." *Scientific American*, Mar. 1998, pp. 35–36.

Goodman, Jonathan. *The Crippen File*. London: Allison and Busby, 1985.

Goodman, Jonathan. "The Jigsaw Murder Case." *The Mammoth Book of Murder Science*, ed. Roger Wilkes. New York: Carroll and Graf, 2000.

Gordon, Richard. *A Question of Guilt: The Curious Case of Dr. Crippen*. New York: Atheneum, 1981.

Gorman, Christine. "Liquid X." *Time*, Sept. 30, 1996, p. 64.

Gose, Ben. "Brown University's Handling of Date-Rape Case Leaves Many Questioning Campus Policies; an Accused Student Faced Suspension Although No One Refuted His Claim That the Woman Initiated Sex." *Chronicle of Higher Education*, Oct. 11, 1996, p. A53ff.

Grace Dolan, Administrix v. Commonwealth. Appeals Court of Massachusetts, Suffolk, 1988.

Grano, Joseph. "Lineups." *The Guide to American Law*. St. Paul, MN: West Publishing, 1984, vol. 7, p. 194.

Grant, Patrick, David Chambers, Louis Grace,

Douglas Phinney, and Ian Hutcheon. "Advanced Techniques in Physical Forensic Science." *Physics Today*, Oct. 1998, pp. 32–38.

Grau, Joseph J., ed. *Criminal and Civil Investigation Handbook*. New York: McGraw-Hill, 1981.

Graysmith, Robert. *Unabomber: A Desire to Kill*. Washington, DC: Regnery, 1997.

Greenya, John. *Blood Relations*. New York: Harcourt Brace Jovanovich, 1987.

Grogan, David W. "The Lady Vanishes, and a Woodchipper Leaves Just a Shred of Evidence." *People Weekly*, Aug. 1, 1988.

Grolier Multimedia Encyclopedia. Danbury, CT: Grolier Electronic Publishing, 1995.

Guthrie, Bruce, Linda Kramer, Ellen O'Conner, and Jane Sims. "Up Front: The Heartbreak Kids: Two Fair, Blue-eyed Girls Find Themselves at the Center of a Tragic Baby-switch Case." *People*, Aug. 17, 1998.

Gwynne, S.C. "Genes and Money." *Time*, Apr. 12, 1999, p. 69.

Haber-Schaim, Uri, John H. Dodge, Robert Gardner, Edward A. Shore. *PSSC Physics*. 7th ed. Dubuque, IA: Kendall/Hunt, 1991.

Haime, Kevin. "The Stockwell Strangler: A Crazed Serial Killer Was Murdering Senior Citizens in Southwest London." *Ottawa Sun*, August 13, 2000, p. 16.

Hall, Bruce Wayne. "The Forensic Utility of Soil." *FBI Law Enforcement Bulletin*, Sept. 1993, pp. 16–18.

Hansen, Peter. "Queensland's Own Quincy." *Sunday Mail* (Brisbane, Australia), Sept. 3, 2000, p. 40.

Hawaleshka, Danylo. "A Zip in Time." *Maclean's*, Mar. 27, 2000, p. 58.

Herzog, Arthur. *The Woodchipper Murder*. New York: Henry Holt, 1989.

Hitchman, Janet. *Such a Strange Lady*. New York: Harper & Row, 1975.

Horan, James D. *The Desperate Years*. New York: Bonanza Books, 1962.

Howe, Sir Ronald. *The Story of Scotland Yard*. New York: Horizon Press, 1966.

Hughes, Dorothy B. *Erle Stanley Gardner: The Case of the Real Perry Mason*. New York: William Morrow, 1978.

Humphreys, Christmas. "Sir Bernard Henry Spilsbury." *Dictionary of National Biography*. New York: Oxford University Press, 1959.

Huntington, Tom. "Sherlock Holmes," *Historic Traveler*, Feb. 1997.

Icove, David J., Karl Seger, and Von Storch, William R. *Computer Crime: A Crimefighter's Handbook*. Sebastopol, CA: O'Reilly & Associates, 1995.

"I See . . . Yes! A Backhoe," *New York Times Magazine*, Aug. 25, 1996, p. 8.

"Is It a Crime to Lie about Rape?" *Cosmopolitan*, Nov. 1999, p. 56.

James, S.H., and W.G. Eckert. *Interpretation of Bloodstain Evidence at Crime Scenes*. Boca Raton, FL: CRC Press, 1999.

"Jane Toppan, Poisoner of 31, Dies in Hospital at Age 81." *Boston Globe*, Aug. 18, 1938.

Jarriel, Tom, Connie Chung, and Jack Ford. "Mark of a Killer." *ABC 20/20*, Sept. 20, 1999.

Jeffers, H. Paul. *Bloody Business: An Anecdotal History of Scotland Yard*. New York: Pharos Books, 1992.

Jeffrey, Karen. "Man Dies in Hyannisport Car Fire." *Cape Cod Times*, Aug. 31, 2000, p. 1.

Johnson, Alexander, and Aaron Brown. "Tracking Guns with IBIS Is Faster and Easier." *ABC World News Saturday*, Aug. 9, 1997.

Johnson, David. *Illegal Tender*. Washington, DC: Smithsonian Institution Press, 1995.

Jones, Stephen and Peter Israel. *Others Unknown: The Oklahoma City Bombing Case and Conspiracy*. New York: Public Affairs, 1998.

Junger, Sebastian. *The Perfect Storm*. New York: W.W. Norton, 1997.

"Jury Inspects Site of Shooting." *Nelson Mail* (Nelson, New Zealand), May 18, 2000, p. 2.

Kadlec, Daniel. "Crimes and Misdeminors: A Teenager Shows How Easily Stocks Can Be Manipulated and How Hard It Is to Get Away with It. So Why Are So Many Hailing Him as a Genius?" *Time*, Oct. 2, 2000, pp. 52–54.

Kelly, John F., and Phillip K. Wearne. *Tainting Evidence*. New York: Free Press, 1998.

———. "Whose Body of Evidence?" *Economist*, July 11, 1998.

Kennedy, Ludovic H. *Crime of the Century*. New York: Viking Penguin, 1996.

Kluger, Jeffrey. "DNA Detectives." *Time*, Jan. 11, 1999, pp. 62–63.

Kohn, George C. *Dictionary of Culprits and Criminals*. Metuchen, NJ: Scarecrow Press, 1986.

Kunstler, William. *The Minister and the Choir Singer: The Hall-Mills Murder Case*. New York: Morrow, 1964.

Lane, Brian. *Chronicle of 20th Century Murder*. Vol. 1, *1900–1938*. New York: Berkley Books, 1995.

————. *The Encyclopedia of Forensic Science.* London: Headline, 1992.

Latour, Francie. "This Is the Gun." *Boston Globe,* Sept. 20, 2000, p. 1.

Lewin, Roger. *Principles of Human Evolution: A Core Textbook.* Malden, MA: Blackwell Science, 1998.

Lindsey, Robert. *A Gathering of Saints.* New York: Simon & Schuster, 1988.

Liston, Robert. *Great Detectives.* New York: Platt and Munk, 1966.

Loftus, E.F. *Eyewitness Testimony.* Cambridge, MA: Harvard University Press, 1996.

Lunden, Joan. "Drexler Baby Autopsy." *ABC Good Morning,* June 25, 1997.

Macaulay, David. *The Way Things Work.* Boston: Houghton Mifflin, 1988.

MacDonell, Herbert L. *Bloodstain Pattern Interpretation.* Corning, NY: Laboratory of Forensic Science, 1983.

Magill's Medical Guide: Health and Illness. Pasadena, CA: Salem Press, 6 vols., 1995.

Marcus, Paul. "Eyewitness Identification: Constitutional Aspects." *Encyclopedia of Crime and Justice,* ed. Sanford H. Kadish. Vol. 2. New York: Free Press, 1983.

Marqu, Julie. "Dentist Charged with Felonies in Use of Sedatives; Courts: Pasadena Woman Is Accused of Committing Child Abuse by Giving Overdoses of Chloral Hydrate." *Los Angeles Times,* Apr. 29, 2000.

Martin, Douglas. "Keith Mant Dies at 81; Pathologist Helped Convict Nazis." *New York Times,* Nov. 26, 2000.

Mathis, Ayana. "Stop, Drop, and Swab." *Village Voice,* May 31–June 6, 2000.

McAleer, John. "R. Austin Freeman." *Dictionary of Literary Biography.* Vol. 70. *British Mystery Writers, 1860–1919.* Detroit: Gale Research, 1988.

McCullen, Kevin. "Mirabal Jurors Visit Mountain Crime Scene Wife Was Strangled and Dumped in Canyon." *Denver Rocky Mountain News,* June 21, 2000, p. 26A.

McNamee, Tom. "Monster of 63rd Street Was Terrifying Serial Killer of the 1890s." *Chicago Sun-Times,* May 7, 1996.

The Merck Manual of Medical Information. Home ed. Ed. Robert Berkow, MD. Whitehouse Station, NJ: Merck Research Laboratories, 1997.

Mirsky, Steve. "Fright of the Bumblebee." *Scientific American,* July 1995, pp. 17–18.

Morland, Nigel. *An Outline of Scientific Criminology.* 2nd ed. New York: St. Martin's Press, 1971.

Muir, Andrew Forest. *William Marsh Rice and His Institute: A Biographical Study.* Houston, TX: Rice University, 1972.

Naifeh, Steven, and Gregory White Smith. *The Mormon Murders.* New York: New American Library, 1989.

Nash, Jay Robert. *Almanac of World Crime.* New York: Bonanza Books, 1986.

————. *Bloodletters and Badmen.* New York: M. Evans & Co., 1973.

————. *Encyclopedia of World Crime.* Wilmette, IL: Crime Books, 1990, vol. 3, p. 2172.

Nettels, Elsa. "The Tragedy of Pudd'nhead Wilson." *Benet's Reader's Encyclopedia of American Literature.* New York: HarperCollins, 1991, 886.

Newman, Andy. "Those Dimples May Be Didgits." *New York Times,* May 3, 2001, p. G1ff.

Newton, Michael. *Hunting Humans: An Encyclopedia of Modern Serial Killers.* Port Townsend, WA: Loompanics, 1990.

Nickel, Steven. *Torso: The Story of Eliot Ness and the Search for a Psychopathic Killer.* Winston-Salem, NC: John F. Blair, 1989.

Noguchi, Thomas T. *Coroner.* New York: Simon & Schuster, 1983.

"Now for Another Opium War." *Economist,* Nov. 19, 1988, p. 40.

"NSW: Obsessive Ex Boyfriend Jailed for Murdering Student." *General News* (Australia), July 20, 2000.

Oakes, Larry. "Blom Defense Rests after Presenting Dental Expert." *Minneapolis Star Tribune,* Aug. 12, 2000.

O'Neill, Helen (Associated Press). "A Rape Victim Regrets Role as Witness." *Boston Globe,* Oct. 1, 2000, pp. 18–19.

Oran, Daniel. *Oran's Dictionary of the Law,* 2nd ed. New York: West Publishing Company, 1991.

"Oursler, Fulton." *Current Biography.* New York: H.W. Wilson Co., 1942.

The Oxford Companion to Medicine. Ed. John Waltons, Paul B. Beeson, and Ronald Bodley Scott. Oxford and New York: Oxford University Press, 1986.

Pan, Esther Begley Sharon. "Jefferson's DNA Trail." *Newsweek,* Nov. 9, 1998. p. 66.

Papinchak, Robert Alan. "Dorothy L. Sayers." *Mystery and Suspense Writers,* ed. Robin Winks. New York: Charles Scribner's Son, 1998, vol. 2, pp. 805–826.

BIBLIOGRAPHY

Parker, Kenneth. "Scotland Yard." *Encyclopedia of Crime and Justice*, ed. Sanford H. Kadish. New York: Free Press, 1983.

Passero, Kathy. "Stranger Than Fiction: The True-Life Drama of Novelist Patricia Cornwell." *Biography*, May 1998.

Patalas v. United States. No. 10687, U.S. Court of Appeals, District of Columbia Circuit, 1950.

Pearson, Edmond. *Masterpieces of Murder*. Ed. Gerald Gross. Boston: Little, Brown, 1963.

Penzler, Otto, Chris Steinbrunner, and Marvin Lachman, eds. *Detectionary*. Woodstock, NY: Overlook Press, 1977.

Perez, Gayle. "Pueblo, Colo., Counterfeiting Scheme Exposed." *Pueblo Chieftain* (Pueblo, Colorado), June 22, 2000.

Peters, Paula. "Drug Bust Nets 17: Arrests Follow Sales to Undercover Officers." *Cape Cod Times*, Oct. 5, 2000, p. A3.

Petrocelli, Daniel. *Triumph of Justice*. New York: Crown Publishers, 1998.

"Pursuing the Paranormal." *CQ Researcher*. Washington, DC: Congressional Quarterly, Mar. 29, 1996 vol. 6, no. 12, p. 280.

Purvis, James. *Great Unsolved Mysteries*. New York: Grosset & Dunlap, 1978.

Ragle, L. *Crime Scene*. New York: Avon Books, 1995.

"Ranch Hand to Stand Trial for Stabbing Death at Ely Brothel." Associated Press, Nov. 3, 1999.

Randi, James. *An Encyclopedia of Claims, Frauds, and Hoaxes of the Occult and Supernatural*. New York: St. Martin's Press, 1995.

Raspberry, William. "Lone-Star Lessons." *Washington Post*, Aug. 4, 2000, p. A29.

Raymond, Joan. "Forget the Pipe, Sherlock: Gear for Tomorrow's Detectives." *Newsweek*, June 22, 1998, p. 12.

Reilly, John M., ed. *Twentieth Century Crime and Mystery Writers*, 2nd ed. New York: St. Martin's Press, 1991.

Renstrom, Peter G. *The American Law Dictionary*. Santa Barbara, CA: ABC-CLIO, 1991.

Robinson, David. *Chaplin: His Life and Art*. New York: McGraw-Hill, 1985.

Rosenberg, Debra. "A Murder Case That Will Not Die." *Newsweek*, March 5, 2001, pp. 50–51.

Royte, Elizabeth. " 'Let the Bones Talk' Is the Watchword for Scientist-Sleuths." *Smithsonian*, May 1996.

Rumbelow, Donald. *Jack the Ripper: The Complete Casebook*. London: Penguin. 1988.

Saferstein, Richard. *Criminalistics: An Introduc-tion to Forensic Science*. 4th ed. Englewood Cliffs, NJ: Prentice Hall, 1990.

———. *Criminalistics: An Introduction to Forensic Science*. 6th ed. Upper Saddle River, NJ: Prentice Hall, 1998.

Saker, Anne. "Terri's Fire." *Raleigh News and Observer*, Nov. 15–22, 1998.

Sanminiatelli, Maria. "Insurer's Payoff on Switched Girls Up in Air." *Washington Times*, Aug. 8, 2000, p. C3.

Schechter, Harold. *Depraved: The Shocking True Story of America's First Serial Killer*. New York: Pocket Books, 1994.

Schick, Elizabeth A., ed. "Patricia Cornwell." *Current Biography*. New York: H.W. Wilson Co., 1997.

Schnuth, Mary Lee. "Lip Prints." *FBI Law Enforcement Bulletin*, Nov. 1992, pp. 18–19.

Science and Technology Illustrated. Chicago: Encyclopaedia Britannica, 1984.

Sears, Francis W., Mark W. Zemansky, and Hugh D. Young. *College Physics*. 6th ed. Reading, MA: Addison-Wesley, 1985.

Sears, William. *SIDS: A Parent's Guide to Understanding and Preventing Sudden Infant Death Syndrome*. Boston: Little, Brown, 1995.

Seligman, Jean, and Patricia King. " 'Roofies': The Date-Rape Drug: This Illegal Sedative Is Plentiful—and Powerful." *Newsweek*, Feb. 26, 1996, p. 54.

Serrano, Richard A. *One of Ours: Timothy McVeigh and the Oklahoma City Bombing*. New York: W.W. Norton, 1998.

Shapiro, Robert. *The Search for Justice*. New York: Warner Books, 1996.

Shepard, Michelle. "A Case of Murder." *Toronto Star*, Sept. 30, 2000.

Shew, E. Spencer. *A Companion to Murder*. New York: Knopf, 1961.

Siegal, Nina. "Ganging Up on Civil Liberties." *Progressive*, Oct. 1997, pp. 28–31.

Sifakes, Carl. *The Encyclopedia of American Crime*. New York: Smithmark, 1992.

"Signing with Your Genes." *Popular Mechanics*, Sept. 1995.

Sillitoe, Linda, and Allen Roberts. *Salamander*. Salt Lake City: Signature Books, 1988.

Smith, Richard N. *Thomas E. Dewey and His Times*. New York: Simon & Schuster, 1982.

Smith, W.C., Kinney, R. and D. Departee. "Latent Fingerprints—a Forensic Approach." *Journal of Forensic Identification* 43(6) 563–570, 1993.

Sniffen, Michael J. "Crime in the U.S." Washington, DC: Associated Press, Oct. 17, 1999.

Söderman, Harry, and John J. O'Connell, revised by Charles E. O'Hara. *Modern Criminal Investigation.* 5th ed. New York: Funk & Wagnalls, 1962.

Statsky, William P., Bruce L. Hussey, Michael R. Damond, Richard H. Nakamura. *West's Legal Desk Reference.* New York: West Publishing, 1991.

Steinbrunner, Chris, and Otto Penzler, ed. *Encyclopedia of Mystery and Detection.* New York: McGraw-Hill, 1976.

Stigler, Stephen M. "Galton and Identification by Fingerprints." *Genetics,* July 1995.

Stone, Richard. "Analysis of a Toxic Death. (Dying Patient in Emergency Room Releases Toxic Fumes)." *Discover Magazine,* April 1, 1995, pp. 66–75.

Suhr, Jim. "Manslaughter Trial Begins over 'Date-Rape' Drug." The Associated Press. *Cape Cod Times,* Jan. 31, 2000.

Stover, Eric. "The Grave at Vukovar." *Smithsonian,* March 1997, pp. 40ff.

Suzuki, K., and Y. Tsuchihashi. "Personal Identification by Means of Lip Prints." *Journal of Forensic Medicine,* 1970, pp. 52–57.

Taylor, Chris. "How They Caught Him." *Time,* April 12, 1999, p. 66.

Taylor, Jill and Pat Moore. "Bruises Led to Charges for Couple." *Palm Beach Post,* July 28, 2000.

"Technology Unlocks the Architecture of a Crime." *American City and County,* Aug. 1996, p. 58.

Theoharis, Athan G., ed. *The FBI: A Comprehensive Reference Guide.* Phoenix, AZ: Oryx Press, 1999.

Thomas, Peggy. *Talking Bones: The Science of Forensic Anthropology.* New York: Facts on File, 1995.

Thorwald, Jürgen. *The Century of the Detective.* New York: Harcourt, Brace & World, 1965.

———. *Crime and Science.* New York: Harcourt, Brace & World, 1967.

———. *Proof of Poison.* London: Thames & Hudson, 1966.

Thurber, James. "A Sort of Genius." *Unsolved: Classic True Murder Cases,* selected by Richard Glyn Jones. New York: Peter Bedrick Books, 1987.

Trahan, Jason. "Suspect Didn't Fit Police Profile: Inquiry Focused on White Man in Arlington Slayings." *Dallas Morning News,* Sept. 8, 2000.

Tsuchihashi, Y. "Studies on Personal Identification by Means of Lip Prints." *Forensic Science,* 1974, pp. 233–248.

Turkington, Carol. *The Home Health Guide to Poisons and Antidotes.* New York: Facts on File, 1994.

Twain, M. (Samuel Clemens). "The tragedy of Puddn'head Wilson." *Century Magazine,* 1894.

———. *The Unabridged Mark Twain. Vol II.* Ed. Lawrence Teacher. Philadelphia: Running Press, 1979.

"Voiceprint Analysis Puts a Face to Words." *All Things Considered* (National Public Radio), July 31, 1996.

Vorpagel, Russell. "Role of the Coroner." *Encyclopedia of Crime and Justice.* New York: Free Press, 1983.

Weiner, Eric, and Renee Montagne. "Japanese Poison Suspects Arrested." *Morning Edition* (National Public Radio), Oct. 5, 1998.

Weiss, P. "Magnifier May Crack Crimes, Crashes." *Science News,* July 8, 2000, p. 20.

Western Alliance Insurance Company v. Jarnail Singh Gill and Others. Supreme Judicial Court of Massachusetts, Middlesex, 1997.

Willey, Fay. "Death by Umbrella." *Newsweek,* Sept. 25, 1978.

Wilson, Colin. *Clues! A History of Forensic Detection.* New York: Warner Books, 1991.

———. *Colin Wilson's World Famous Murders.* London: Robinson Publishing, 1993.

———. *The Mammoth Book of True Crime.* New Ed. New York: Carroll & Graf, 1998.

Wilson, Colin and Patrick Pitnam. *Encyclopedia of Murder.* New York: G.P. Putnam & Sons, 1962.

Wilson, Colin, with Damon Wilson and Rowan Wilson. *World Famous Murders.* London: Robinson, 1993.

Wiseman, Richard, Donald West, and Roy Stemman. "Psychic Crime Detectives: A New Test for Measuring Their Successes and Failures." *Skeptical Inquirer,* Jan./Feb. 1996.

Within, George and Katie Helter. "High-Tech Crime Solving." *U.S. News and World Report,* July 11, 1994, p. 30.

Woodruff, Bob, Mark Mullen, and Blake Asha. "Bill Cosby Testifies in Extortion Trial," *ABC World News This Morning,* Jul. 16, 1997.

Wolf, Marvin, and Katherine Mader. *Rotten Apples: Chronicles of New York Crime and Mystery.* (1689 to the Present.) New York: Ballantine Books, 1991.

"Work and Lies in the Promised Land." *World Press Review*, June 1995.

The World Almanac and Book of Facts, 1999. Mahwah, NJ: World Almanac Books, 1998.

The World Almanac and Book of Facts, 2000. Mahwah, NJ: World Almanac Books, 1999.

Yarmey, Daniel. "Eyewitness Identification: Psychological Aspects." *Encyclopedia of Crime and Justice*, ed. Sanford H. Kadish. Vol. 2. New York: Free Press, 1983.

Zoglin, Richard, reported by Charlotte Faltermayer. "Justice: From Here to Paternity: Bill Cosby Offers a Blood Test, and Autumn Jackson Gets Another Father Figure or Two. Is This a Sitcom?" *Time*, Aug. 11, 1997, p. 32.

Zonderman, Jon. *Beyond the Crime Lab: The New Science of Investigation*. New York: Wiley, 1990.

Electronic (Internet)

"About Us." Metropolitan Police (London, UK). http://www.met.police.uk. (The Official Web site of Scotland Yard.)

Aiuto, Russell. *Lindbergh*. Crime Library. Dark Horse Multimedia, 1998. http://www.crimelibrary.com/lindbergh/main.htm.

American Psychiatric Association. Public Information Insanity Defense Fact Sheet. http://www.psych.org/public_info/insanity.cfm.

American Academy of Psychiatry and the Law. http://www.emory.edu/AAPL/org.htm.

"Arthur Conan Doyle (1859–1930)." http://www.kirjasto.sci.fi/acdoyle.htm.

"Arthur Conan Doyle: Doyle vs. Holmes." http://www.letsfindout.com/subjects/art/arthur-conan-doyle.html.

Arthur Conan Doyle Society. http://www.ashtree.bc.ca/acdsocy.html.

"Attorney Says DNA Test Supports Sam Sheppard's Innocence." *U.S. News*, March 5, 1998. http://www.cnn.com/US/9803/05/dna.sheppard/index.html.

"Author Jeffery Dever: The Empty Chair." http://chat.yahoo.com/c/events/transcript/authors.051000deaver.html.

Bardsley, Marilyn. *The Kingsbury Run Murders*. Great Falls, VA: Dark Horse Multimedia, 1998 (Available on Internet Web site http://crimelibrary.com/kingsbury2/kingplog.htm.)

Berkeley, CA. *Police Department History Pages*. http://209.232.44.30/bpd_history.htm.

Bickel, Bill. "A Chat with Jeffery Deaver." http://crime.about.com/crime/library/weekly/aa090299.htm.

"The Boston Strangler, Albert DeSalvo." http://www.crimelibrary.com/bostonmain.htm.

Brenner, Anita Susan. *Cyber-Rights and Criminal Justice: Legal Resources on the Web*. brenner@cyberspace.org.

A Brief Counterfeiting History in U.S. http://www.servicemart.com/money/histr1.htm.

CDC Medical Examiner and Coroner Information Sharing Program. http://www.cdc.gov/nceh/pubcatns/1994/cdc/brosures/me-cbro.htm.

Channel 3000 News. *Woman Arrested in Cyanide Mail Scare*. http://wisctv.com/news/stories/news-980824-081245.html.

"The Chronicles of Sir Arthur Conan Doyle." http://www.siracd.com.

CNN.com. *The Execution of Timothy McVeigh*. http://www.cnn.com/SPECIALS/2001/okc/.

CNN.com. *The O.J. Simpson Main Page*. http://www.cnn.com/US/OJ/.

CNN.com. *The Prosecution: Its Case Against Simpson*. http://www.cnn.com/US/OJ/verdictr/prosecution/index.html.

"Cops Rule Out Ramsey in Sex Case." *AP Online*, Sept. 13, 2000. Bigchalk.com, inc. http://www.crimelynx.com/ramsey.html.

Cybercrime Web site. *Computer Crime and Intellectual Property Section*. http://www.usdoj.gov/criminal/cybercrime/index.html.

"The Crib Death Cover-Up?" Dr. Jim Sprott. http://www.myfreeoffice.com.

Crime Scene Investigation. http://police2.ucr.edu/csi.html.

Davis, Danton. "A Simple Theory—Crib Death." http://www.criblife2000.com/simpletheory.htm.

Decaire, Michael. *Criminal Profiling: An Informal Introduction and Discussion*. http://www.suite101.com/article.cfm/forensic_psychology/21333.

———. *Criminal Profiling in Fiction*. http://www.suite101.com/article.cfm/forensic_psychology/24469.

Deaver, Jeffery. "Jeffery Deaver—The Official Website." http://www.jefferydeaver.

Ever the Twain Shall Meet. joseph@telerama.lm.com.

Eyedentify, Inc. http://www.eyedentify.com.

Federal Bureau of Investigation Web Site. "The Greenlease Kidnapping." http://www.fbi.gov/fbinbrief/historic/farncases/greenlease/greenleasenew.htm.

Forensic Psychology & Forensic Psychiatry: An Overview. http://flash.lakeheadu.ca/~pals/forensics/forensic.htm.

Forensic Science Resources in a Criminal Fact Investigation Index. http://www.tncrimlaw.com/forensic/.

Fraser, Bruce T. "UN Manual on the Prevention and Control of Computer Related Crime." *Computer Crime Research Resources.* Florida State University, Tallahassee. btf1553@mailer.fsu.edu.

Gallagher, Cathy. "Lincoln Rhyme Returns." http://mysterybooks.com/book/library/weekly/aa000525a.htm.

Gartner, Hana. "Gory and Grisly Life of Crime Writer Kathy Rechs." *CBC Infoculture Interview.* http://www.infoculture.cbc.ca/archives/bookswr/bookswr_07231999_reichs.html.

Growing Families International. Spring 1997. http://redrhino.mas.vou.edu/Ezzo/cosleeping.

Held, Dorothy-Anne E. "Handwriting, Typewriting, Shoeprints, and Tire Treads: FBI Laboratory's Questioned Documents Unit." *Forensic Science Communications,* April 2001. http://www.fbiogov/hq/lab/fsc/backissue/april2001/held.htm.

Introduction to Forensic Entomology. http://www.uio.no/~mostarke/forens_ent/introductshtm.

Introduction to Forensic Firearms Identification. http://www.geocities.com/~jsdoyle/A_Welcome.htm.

"Is SIDS Nothing More Than an Accidental Poisoning?" http://www.criblife2000.com/.

Kidnapping. Funk & Wagnall Corp. http://encarta.msn.com/Microsoft Encarta.

Lindbergh Kidnapping:

    http://www.crimelibrary.com/lindbergh/lindcrime.htm.

    http://www.crimelibrary.com/lindbergh/linmain.htm.

    http://www.crimelibrary.com/lindbergh/linthreads.htm.

    http://www.crimelibrary.com/lindbergh/lindtrial.htm.

    http://www.crimelibrary.com/lindbergh/lindinv.htm.

Massire Dollar Counterfeiting Activities. http://www.koreascope.org/ktext/english/sub/2/1/nk9_4.htm.

"Matthew Mirabal Gets Life in Prison for Killing His Wife." *Daily Camera.* http://www.thedailycamera.com/extra/topten00/mirabal.

McGunnagle, Fred. *Sam Sheppard.* Dark Horse Multimedia, 1998. http://www.crimelibrary.com/com/sheppard/sheppard.htm.

McKnight, Keith. *Haunting Questions—The Sam Sheppard Case.* BJE Special Project: Beacon Journal Publishing Co., 1996. http://www.ohio.com/bj/projects/sam/sam1–8.html.

Memoirs of Popular Delusions, Vol. 2—*The Slow Poisoners.* http://www.student.dtu.dk/~c973572/drugs/chemistry/data/e.

Microsoft Encarta. *Insanity.* Funk & Wagnall Corp. 1993. http://encarta.msn.com/.

MSNBC TV News. *Examining the Examiner.* 1998. http://www.msnbc.com/onair/nbc/dateline/shockcoroner.asp.

Mudge, Alden. "Meet the Season's Best Discovery: Kathy Reichs." *First Person Book Page.* http://www.bookpage.com/9709bo/firstperson1html.

Nawrocki, Stephen P. *Ten Basic Points Concerning Human Remains Scenes.* University of Indianapolis Archeology and Forensics Laboratory. http://www.uindy.edu/~archlab.

"Nelson Man Gets Eight Years after Shooting." June 22, 2000. http://www.planet/fm.co.nz.news/june00/220600.htm.

1998 National Drug Control Strategy. http://www.whitehousedrugpolicy.gov/publications/policy/98ndcs/contents.html.

Nolan, William F. *Erle Stanley Gardner* (1889–1970). http://erlestanleygardner.com/nolan.htm.

Questioned Document Examination. http://www.webmasters.net/qde/.

Ramsland, Katherine. "Bite Marks as Evidence to Convict." Crime Library. 2000. http://www.crimelibrary.com/forensic/bitemarks/4.htm.

Redmond, Chris. Sherlockian.Net. http://www.sherlockian.net.

RH FACTOR. http://www.fwkc.com/encyclopedia/low/articles/r/r022000675f.html.

"The Richardson Theory." http://www.criblife2000.com/richardson.htm.

Riley, Donald E. "DNA Testing: An Introduction for Non-Scientists. An Illustrated Explanation." *Scientific Testimony—An Online Journal.* http://www.scientific.org/tutorials/articles/riley/riley.html.

Schehl, Sally A. "Firearms and Toolmarks in the FBI Laboratory." *Forensic Science Communications* 2(2), April 2000. http://www.fbi.gov/programs/lab/fsc/backissu/april2000/schehl1.htm.

The Science of Crime. http://whyfiles.news.wisc.edu–forensic/index.html.

"Scientists Link Smoking with Crib Death." http://www.junkscience.com/news2/sidstemp.htm.

Senate, Richard. *A Brief Biography of Erle Stanley*

Gardner. http://www.erlestanleygardner.com/.

Sherlock Holmes: History of the Mystery. http://www.mysterynet.com/history/holmes/main.shtml.

Sherlock Holmes Museum. http://www.sherlock-holmes.co.uk.

Sherlockiannet. "Arthur Conan Doyle." http://www.sherlockian.net/acd/.

Silet, Charles. "An Interview with Edna Buchanan." http://www.mysterypages.com/buchananiv1-9.html.

Simulation Crime Scene. http://www.crimescene.com/.

Simulation DQ-Alpha. http://cgi-server.shadow.net/~mchinsee/scripts/simscrpt.pl.

"Sir Arthur Conan Doyle." http://www.spartacus.schoolnet.co.uk/jeonan.htm.

"Sir Arthur Conan Doyle." http://pantheon.yale.edu/-yoder/mystery/doyle.html.

"Sudden Death: The Search for the Truth about Cot Death." BBC Online. Feb. 25, 1999. http://www.bbc.co.uk/science/horizon/suddendeath.shtml.

"Sudden Infant Death Syndrome." http://health.yahoo.com/health/Diseases_and_-Conditions/Dise . . . /Sudden_infant_death_syndrome.

Syracuse.com. "False Alarm: The Failed Promise of Apnea Monitors." May 5–7, 1996. http://www.syracuse.com/features/apnea.

"Truth in Justice." http://www.truthinjustice.org.

Turvey, Brent E. *Criminal Profiling Overview.* http://corpus-delicti.com/profile.html.

"Untold Dangers in the Family Bed?" *The Community Perspective (Newsletter)*, Growing Families International, Spring 1997. http://redrhino.mas.vou.edu/Ezzo/cosleeping.

U.S. Census Bureau. *Statistical Abstract of the United States*, 2000. 120th ed. http://www.census.gov/prod/2001pubs/statab/sec05.polf.

U.S. Department of Justice: Bureau of Justice Statistics. http://www.ojp.usdoj.gov/bjs/.

"What Would Adolf Meyer Have Thought of the Neo-Kraepelinian Approach?" http://www.uea.ac.uk/~wp276/meyer.htm.

White, R. "The Trial of Abner Baker, Jr., MD." http://www.cc.emory.edu/AAPL/bull/J183–223.htm.

Index

About the Authors

BARBARA GARDNER CONKLIN is a special education teacher at Plymouth Elementary School, in Plymouth, New Hampshire. She received a B.A. from Bard College and an M.Ed. from Plymouth State College. She has coauthored two books with Robert Gardner, *Health Science Projects about Psychology* and *Science Projects about Sports Health*.

ROBERT GARDNER, former chair of the science department at the Salisbury School, in Salisbury, Connecticut, is a consultant on science education and a distinguished, prolific, and longtime author of science books for children, young adults, and adults. Among his many books are *From Talking Drums to the Internet: An Encyclopedia of Communications Technology*, *Communication*, *Where On Earth Am I?*; *Human Evolution*; *Kitchen Chemistry*, and *Ideas for Science Projects*.

DENNIS SHORTELLE is a history teacher and former chairman of the history department at Salisbury School in Salisbury, Connecticut. He is the coauthor with Robert Gardner of *The Future and the Past*; *The Forgotten Players: The Story of Black Baseball in America*, and *From Talking Drums to the Internet: An Encyclopedia of Communications Technology*.